A MANUAL OF GREEK MATHEMATICS

BY

SIR THOMAS L. HEATH

K.C.B., K.C.V.O., F.R.S.,

Sc.D. Camb., Hon. D.Sc. Oxford, Hon. Litt.D. Dublin;
Honorary Fellow (sometime Fellow) of
Trinity College, Cambridge

ὁ θεὸς ἀεὶ γεωμετρεῖ.

DOVER PUBLICATIONS, INC.
MINEOLA, NEW YORK

Bibliographical Note

This Dover edition, first published in 1963, and republished in 2003, is an unabridged republication of the work first published by Oxford University Press in 1931.
This edition is published by special arrangement with Oxford University Press.

Library of Congress Cataloging-in-Publication Data

Heath, Thomas Little, Sir, 1861–1940.
   A manual of Greek mathematics / Thomas L. Heath.
     p. cm.
   Originally published: Oxford : Clarendon Press, 1931.
   Includes indexes.
   ISBN-13: 978-0-486-43231-1 (pbk.)
   ISBN-10: 0-486-43231-9 (pbk.)
     1. Mathematics, Greek. 2. Mathematics—History. I. Title.

QA22.H42 2003
510'.938—dc22

2003055789

Manufactured in the United States by LSC Communications
43231905   2017
www.doverpublications.com

PREFACE

THE story told in these pages is in substance the same as that unfolded in somewhat greater detail in my *History of Greek Mathematics*, in two volumes, published nine years ago; the aim, however, is slightly different. In the *History* I had in mind the requirements, on the one hand, of the classical scholar who might look for light on the interpretation of passages of mathematical content in Greek authors which came his way, and, on the other hand, of the expert mathematician who might wish so far to assimilate the whole argument of a particular treatise, say of Archimedes, as to be able, on occasion arising, to apply the same method to a different problem. But besides the categories of classical scholars and professional mathematicians there is the general reader who has not lost interest in the studies of his youth and would wish to know how it came about that a Greek of the name of Euclid wrote a text-book which, in an almost literal translation, was used in schools and in the universities in this country as the one recognized basis of instruction in elementary geometry, and on which generations of Senior Wranglers no less than average mortals were brought up, asking nothing better, until some fifty years ago. Who were Euclid's forerunners, and what were their respective contributions to the mass of heterogeneous material which Euclid moulded into a finished treatise covering elementary geometry, geometrical algebra, the theory of numbers, the subject of irrationals, and solid geometry? For, of course, the Elements did not spring fully formed from the brain of Euclid like Athene armed from the head of Zeus. The story of the creation of mathematics as a science by the Greeks and of the development of the Elements in the period ending with Euclid occupies about one-half of the present volume, so that no educated reader need hesitate to tackle it for fear of its making too great a demand on his mathematical knowledge. And who, having got so far, will not wish to know what heights were scaled by Euclid's successors, in the discovery and complete investigation of the conic sections, in the works of Archimedes—who by means of pure geometry alone (following the lead of Eudoxus) practically anticipated the integral calculus, and by arithmetic combined with geometry 'measured the circle' —who again laid the mathematical foundations of statics and created the whole science of hydrostatics—in the astronomy of Hipparchus and Ptolemy (to say nothing of the anticipation of the Copernican theory by Aristarchus of Samos), and in the beginnings of trigonometry and algebra?

The issue of the book has given me a welcome opportunity of correcting some few mistakes and misprints in the earlier *History*. The publication of the fine edition of the Papyrus Rhind by Professor T. Eric Peet in 1923 and the still more magnificent edition in two volumes (1927 and 1929) by Arnold B. Chace and others (under the auspices of the Mathematical Association of America), as well as the details which have become known of the contents of the as yet unpublished Moscow Papyrus, have enabled me to improve on the account of Egyptian mathematics given in the *History*. I have also taken account of the striking and even mystifying results of the study and interpretation of certain ancient Babylonian mathematical texts which have appeared in the last few years, especially in 1929 and 1930.

I have added, in the chapter on Archimedes, an account of a remarkable theoretical method of inscribing a regular heptagon in a circle (or rather of reducing that problem to another of a type common in Greek geometry). It is contained in an Arabic translation by Thâbit b. Qurra of a treatise on the subject by Archimedes which is apparently lost so far as the Greek text is concerned; a copy of this translation exists in Cairo and was studied in 1925 by C. Schoy, who published his results in *Isis* (vol. 8, 1926).

In order that the book may serve as a real Manual, a book of easy reference, I have tried to make the Indices as full as possible.

T. L. H.

October 1930.

CORRIGENDUM
(Page 467)

In describing the Egyptian problem of dividing a total area of 100 square cubits into two squares with sides in the ratio of 1 to $\frac{3}{4}$, I followed, too incautiously, the view of the solution usually taken hitherto, namely that the method used is that of 'false hypothesis'. The actual text as recently reproduced by O. Neugebauer (*Quellen und Studien zur Geschichte der Mathematik*, Abt. B, Band I, Heft 3, 1930) shows that this is not the case. The method is in fact similar to that used in solving the equivalent of simple equations in one unknown, which consists in finding the arithmetical sum of the coefficients of all the terms containing the unknown and then dividing a number given in the problem by the said sum; the only difference in the present case is that there occur, in addition, the operations of squaring and extracting the square root. The problem should be stated as the equivalent of one equation (in x) and not two equations (in x and y), thus:

$$x^2 + \{(\tfrac{1}{2}+\tfrac{1}{4})x\}^2 = 100.$$

The working is as follows (slightly shortened):

$$1^2 + (\tfrac{1}{2}+\tfrac{1}{4})^2 = 1 + \tfrac{1}{2} + \tfrac{1}{16}.$$
$$\sqrt{(1+\tfrac{1}{2}+\tfrac{1}{16})} = 1+\tfrac{1}{4}; \quad \sqrt{(100)} = 10.$$
$$10/(1+\tfrac{1}{4}) = 8;$$
$$x = 8,\ (\tfrac{1}{2}+\tfrac{1}{4})x = 6.$$

CONTENTS

I. INTRODUCTORY 1
   Classification of mathematical subjects . . 5
   Mathematics in Greek education . . . 7

II. NUMERICAL NOTATION AND PRACTICAL CALCULATION 11
   The decimal system 11
   Egyptian numerical notation 11
   Babylonian systems 12
   Greek numerical notation 14
   (a) The 'Herodianic' or 'Attic' system . . 14
   (b) The ordinary alphabetic numerals . . 15
   (c) Notation for large numbers . . . 18
   (d) Archimedes' system for large numbers ('octads') 19
   Fractions 20
   Sexagesimal fractions 23
   Practical calculation 24
   (a) The abacus 24
   (β) Addition and subtraction . . . 28
   (γ) Multiplication 29
   (δ) Division 31
   (ε) Extraction of square root . . . 32

III. PYTHAGOREAN ARITHMETIC . . . 36
   Definitions of the unit and of number . . . 38
   Classification of numbers 39
   'Perfect' and 'Friendly' numbers . . . 41
   Figured numbers 43
   (a) Triangular numbers 43
   (b) Square numbers and gnomons . . . 44
   (c) Gnomons of the polygonal numbers . . 45
   (d) Right-angled triangles with sides in rational numbers 46
   (e) Oblong numbers 48
   The theory of proportion and means . . . 51
   Geometric Means 53
   The irrational 54

x   CONTENTS

Algebraic equations	55
(a) Indeterminate equations of the second degree $2x^2-y^2 = \pm 1$	55
(β) *Epanthema* of Thymaridas	57
(γ) Equation $xy = 2(x+y)$	60
Manuals of 'Arithmetic'	61
Nicomachus of Gerasa	61
Sum of series of cube numbers	68
Theon of Smyrna	70
Iamblichus	71

IV. THE EARLIEST GREEK GEOMETRY. THALES . 73

The 'Summary' of Proclus	73
Egyptian geometry (mensuration)	75
Thales	81
(a) Measurement of height of pyramid	82
(b) Geometrical theorems	83
(c) Thales as astronomer	89
From Thales to Pythagoras	90
Anaximander	90

V. PYTHAGOREAN GEOMETRY . . . 92

(a) Sum of angles of any triangle equal to two right angles	93
(β) The 'Theorem of Pythagoras'	95
(γ) Application of areas and geometrical algebra	100
(δ) The irrational	105
(ϵ) The five regular solids	106
(ζ) Pythagorean astronomy	109
Summary	110

VI. PROGRESS IN THE ELEMENTS DOWN TO PLATO'S TIME 112

Anaxagoras	113
Oenopides of Chios	114
Democritus	115
Hippias of Elis	120
Hippocrates of Chios	121
(a) Quadratures of lunes	122
(b) Reduction of the problem of doubling the cube	131
(c) The Elements as known to Hippocrates	131

CONTENTS xi

Theodorus of Cyrene 132
Theaetetus 133
Archytas of Taras 134

VII. SPECIAL PROBLEMS 139

THE SQUARING OF THE CIRCLE . . . 139
  (a) The quadratrix of Hippias . . . 143
  (b) The spiral of Archimedes . . . 144
  (c) Solutions by Apollonius and Carpus . . 145
  (d) Ancient approximations to π . . . 146

THE TRISECTION OF ANY ANGLE . . . 147
  (a) Reduction to a νεῦσις solved by conics . . 148
  (b) The conchoids of Nicomedes . . . 150
  (c) Another reduction to a νεῦσις . . . 151
  (d) Solution by means of conics . . . 152

THE DUPLICATION OF THE CUBE, OR THE PROBLEM OF
  THE TWO MEAN PROPORTIONALS . . . 154
  Archytas 155
  Eudoxus 157
  Menaechmus 158
  Solution attributed to Plato . . . 161
  Eratosthenes 162
  Nicomedes 164
  Apollonius, Heron, Philon of Byzantium . . 166
  Diocles and the cissoid 167
  Sporus and Pappus 169

VIII. FROM PLATO TO EUCLID . . . 171

Plato and the philosophy of mathematics . . 172
  The hypotheses of mathematics . . . 173
  Definitions 174
Summary of the mathematics in Plato . . 175
  The five regular solids 175
  Geometric means 178
  The two geometrical passages in the *Meno* . . 178
  Solution in integers of $x^2+y^2=z^2$ . . 180
  Incommensurables 180
Plato's astronomy 182

CONTENTS

Successors of Plato 183
Heraclides of Pontus 186
EUDOXUS of Cnidos 187
   Hypothesis of concentric spheres . . . 188
   Theory of proportion 190
   The Method of Exhaustion 191
Zeno's paradoxes 192
Aristotle 194
Sphaeric. Autolycus of Pitane . . . 200

IX. EUCLID 202
   THE *ELEMENTS* 204
   EUCLID'S OTHER WORKS 255
      The *Data* 255
      On *Divisions (of Figures)* 258
      Pseudaria. Porisms 262
      Conics. Surface-Loci 265
      Phaenomena. Optics 266
      Catoptrica 267
      Musical treatises 268
      Supposed mechanical works 269

X. ARISTARCHUS OF SAMOS . . . 270
   Anticipation of Copernicus 270
   On *the sizes and distances of the Sun and Moon* . 272

XI. ARCHIMEDES 277
   Extant works 283
   Other reputed works 285
   Text and editions 286
      The *Method* 287
      On the Sphere and Cylinder . . . 293
      Measurement of a Circle 305
      On Conoids and Spheroids 310
      On Spirals 317
      Plane Equilibriums 323
      The *Sand-reckoner* 327
      Quadrature of a Parabola 330
      On Floating Bodies 332
      The *Cattle-Problem* 336

CONTENTS

On semi-regular solids	337
'Liber assumptorum'	338
On the regular heptagon in a circle	340
ERATOSTHENES	343
Measurement of the earth	343

XII. CONIC SECTIONS 347

Discovery of conics by Menaechmus	347
Euclid and Aristaeus	349
Archimedes	351
APOLLONIUS of Perga	352
The *Conics*	352
Sectio Rationis	363
Sectio Spatii	365
On Determinate Section	366
Contacts or Tangencies	366
Circle touching three circles	367
Plane Loci	371
Νεύσεις, Inclinationes	372
Other works	375

XIII. THE SUCCESSORS OF THE GREAT GEOMETERS 377

Nicomedes	378
Diocles	378
The *Fragmentum mathematicum Bobiense*	379
Perseus and 'spiric sections'	380
Zenodorus	382
Hypsicles	383
Dionysodorus	385
Posidonius	385
Geminus	387

XIV. TRIGONOMETRY: HIPPARCHUS, MENELAUS, PTOLEMY 393

Theodosius' *Sphaerica*	393
HIPPARCHUS	395
Discovery of precession	396
On the Length of the Year	396
Trigonometry	398

CONTENTS

MENELAUS of Alexandria	399
Sphaerica	400
PTOLEMY	402
The *Syntaxis*	403
Preparation of Table of Chords	405
Other works	412
XV. MENSURATION: HERON OF ALEXANDRIA	**415**
Heron's date	415
List of works	417
Commentary on Euclid	417
Mensuration	418
The *Metrica*	419
Area of triangle in terms of sides	420
Approximations to surds	422
Areas of regular polygons	424
Measurement of solids	426
On divisions of figures	428
Quadratic equations	430
On the Dioptra	431
Mechanics	431
Catoptrica	432
XVI. PAPPUS OF ALEXANDRIA	**434**
Date and works	434
The *Collection*	435
Editions	436
Books I, II	437
Book III	437
On problem of two mean proportionals	437
On Means	438
'Paradoxes' of Erycinus	439
On five regular solids	440
Book IV	440
Extension of Pythagoras' Theorem	441
Problems on the ἄρβηλος	442
On spirals, conchoids, and the *quadratrix*	443
A spiral on a sphere	445
On the trisection of any angle	446

CONTENTS xv

Book V 448
   On isoperimetry: digression on bees and honey-
     combs 448
   On the sphere and cylinder . . . 449
   Comparison of five regular solids . . . 450

Book VI 450
   On astronomical treatises. . . . 450

Book VII 451
   On works forming 'Treasury of Analysis'. . 451
   Extension of notion of locus with respect to three
     or four lines (Pappus' Problem) . . 453
   'Theorem of Guldin' anticipated . . . 454
   Lemmas to treatises of Apollonius and Euclid . 455–9

Book VIII 460
   Mechanics: historical preface . . . 460
   On centre of gravity 461
   Construction of conic through five points . . 462
   Problem of seven equal hexagons in a circle . 463

XVII. ALGEBRA: DIOPHANTUS OF ALEXANDRIA. 466
   Egyptian anticipations of algebra . . . 466
   Problems in Anthology 468
     Indeterminate problems, first degree . . 469
   Indeterminate problems from MS. of Heron's *Metrica*. 470

DIOPHANTUS 472
   Date and works 472
   The *Arithmetica* 473
     Lost Books. 'Porisms' 474
     Commentaries and editions . . . 474
     Notation: sign for unknown and its powers . 476
        sign for *minus* 479
     I. Diophantus' treatment of equations . . 482
       A. Determinate Equations . . . 482
         (1) 'Pure' equations 482
         (2) 'Mixed' quadratics . . . 482
         (3) Simultaneous equations involving quad-
            ratics 483

CONTENTS

B. Indeterminate equations	484
(a) Equations of second degree	484
(1) Single equation	484
(2) Double equation	487
(a) of first degree	487
(b) of second degree	490
(b) Equations of degree higher than second	491
(1) Single equations	491
(i) Expressions to be made squares	491
(ii) Expressions to be made cubes	492
(2) Double equations	493
II. Method of approximation to limits	493
III. Porisms and propositions in the theory of numbers	494
Numbers as the sum of two, three, or four squares	496
Characteristic examples and solutions	499
Rational right-angled triangles	505
Treatise on Polygonal Numbers	507

XVIII. COMMENTATORS AND MINOR WRITERS . 510

Cleomedes	510
Theon of Smyrna	512
Serenus of Antinoeia	515
Theon of Alexandria	516
Hypatia	516
Proclus	516
Domninus of Larissa	517
Simplicius	517
Eutocius	518
Anthemius of Tralles	519

APPENDIX ADDITIONAL NOTES:

1. Egyptian mathematics	520
2. Ancient Babylonian mathematics	522
3. Hipparchus and Chaldaean astronomy	530

INDICES

Greek	532
English	537

I

INTRODUCTORY

WHY should we study Greek mathematics? In the first place, it is true generally that, if we would study any subject properly, we must study it as something that is alive and growing and consider it with reference to its origins and its evolution in the past. In the case of mathematics, it is the Greek contribution which it is most essential to know, for it was the Greeks who first made mathematics a science. As Kant once wrote, 'a light broke upon the first man who demonstrated the property of the isosceles triangle (whether his name was Thales or what you will)'; since which time, thanks to 'that wonderful people the Greeks', mathematics has travelled 'the safe road of a *science*'. The Greeks in fact laid down the first principles in the shape of the indemonstrable axioms or postulates to be assumed, framed the definitions, fixed the terminology and invented the methods *ab initio*; and this they did with such unerring logic that, in the centuries which have since elapsed, there has been no need to reconstruct, still less to reject as unsound, any essential part of their doctrine.

The story of Greek mathematics is fascinating in itself, but we can also point to the practical utility of some knowledge of it both to the teacher and to the student. We need only refer to the ordinary terminology in use to-day. Almost all the technical terms in mathematics are either pure Greek or Latin translations from the Greek. Take a word like 'isosceles'; a schoolboy may be shown what an isosceles triangle is, but he will better appreciate the term if he knows that 'isosceles' means 'with equal legs', being compounded of ἴσος, equal, and σκέλος, a leg. A 'parallelogram' again is a figure contained by *parallel lines* (two pairs of them), 'parallel' itself expressing the fact that two parallel lines go *alongside one another* (παρ' ἀλλήλας) all the way. Once more, who can properly appreciate the term *hypotenuse* if he does not know that it is the feminine participle of ὑποτείνειν (which means 'to stretch under' or *subtend*, in the general sense of 'lying opposite to'), and is short for the side in a right-angled triangle which is opposite to the right angle (ἡ τὴν ὀρθὴν γωνίαν ὑποτείνουσα πλευρά, as Euclid says)? Similarly the *latus rectum* of a conic section is not really intelligible unless one has seen it in Apollonius as the 'erect side' of a certain rectangle in the case of each of the three conics; and the same is true of *ordinate* (which is a straight line drawn in the *prescribed* or *ordained* manner, τεταγμένως) and *asymptote* (meaning 'non-meeting' or 'non-secant', however far produced). From this point of view, I cannot but think that it would be an excellent training for boys learning Greek at school to read a Book or two at least of Euclid in the original Greek.

The Greek genius for mathematics was simply one aspect of their genius for philosophy in general. Their philosophy and their mathematics both arose out of the instincts of the race, their insatiable curiosity, their passion for 'inquiry', and a love of knowledge for its own sake which the Greeks possessed in a greater degree than any other people of antiquity. To quote again from a brilliant review of a well-known work:

To be a Greek was to seek to know, to know the primordial substance of matter, to know the meaning of number, to know the world as a rational whole. In no spirit of paradox one may say that Euclid is the most typical Greek; he would know to the bottom, and know as a rational system, the laws of the measurement of the earth. Plato too loved geometry and the wonders of numbers; he was essentially Greek because he was essentially mathematical. And if one thus finds the Greek genius in Euclid and the *Posterior Analytics*, one will understand the motto written over the Academy, ἀγεωμέτρητος μηδεὶς εἰσίτω. To know what the Greek genius meant you must (if one may speak ἐν αἰνίγματι) begin with geometry.

As regards the actual achievements of the Greeks in mathematics, we may, perhaps, here anticipate so far as to mark certain stages in the development of the main subject, pure geometry. In Thales' time (about 600 B.C.) we find the first glimmerings of a science of geometry in the theorems that a circle is bisected by any diameter, that in an isosceles triangle the angles opposite to the equal sides are equal, and (if Thales really discovered this) that the angle in a semicircle is a right angle. Rather more than half a century later Pythagoras was taking the first steps towards the theory of numbers and continuing the work of making geometry a theoretical science; he it was who first made geometry one of the subjects of a liberal education. By the middle of the fifth century the Pythagoreans had developed methods and covered, roughly, the range of Books I, II, IV, and VI (and perhaps III) of Euclid's *Elements* (subject to the qualification that their theory of proportion was inadequate in that it only applied to commensurable magnitudes). In the same fifth century the difficult problems of doubling the cube and trisecting any angle, which are beyond the geometry of the straight line and circle, were not only mooted but solved theoretically. The former, having been first reduced by Hippocrates of Chios to the problem of finding two mean proportionals in continued proportion between two straight lines, was then solved in that form by Archytas, who used a remarkable theoretical construction in three dimensions, while the latter was solved by means of the curve of Hippias of Elis known later as the *quadratrix*. The problem of squaring the circle was also attempted, and Hippocrates, as a contribution to it, discovered and squared three out of the five *lunes* which can be squared by means of the straight line and circle. In the fourth century Eudoxus discovered the great theory of proportion expounded in Euclid, Book V, and laid down the principles of the *method of exhaustion* for measuring areas and volumes; the conic sections and their fundamental properties were discovered by Menaechmus; the theory of irrationals (probably discovered, so far as $\sqrt{2}$ is concerned, by the Pythagoreans) was generalized by Theaetetus; and the geometry of the sphere was worked out in systematic treatises. About the end of the century Euclid wrote his *Elements* in thirteen Books. The next century, the third, is that of Archimedes, who may be said to have anticipated the integral calculus, since, by performing what are practically *integrations*, he found the area of a parabolic segment and of a spiral, the surface and volume of a sphere and a segment thereof, the volume of any segment of the solids of revolution of the second degree, the centres of gravity of a semicircle, a parabolic segment, any segment of a paraboloid of revolution, and any segment of a sphere or spheroid. Apollonius of Perga, the 'great geometer', about 200 B.C., completed the theory of geometrical conics, with specialized investigations of normals as maxima and minima, leading quite easily to the determination of the circle of curvature at any point of a conic and of the equation of the evolute, which with us is part of analytical conics. With Apollonius the main body of Greek geometry is complete, having practically reached the limits of what was possible with the help of pure geometry alone.

CLASSIFICATION OF MATHEMATICAL SUBJECTS

The word 'mathematics' itself is of course Greek, being derived from μάθημα through the adjective μαθηματικός. The literal meaning of μάθημα is simply a 'subject of instruction', but it became very early the tendency to restrict its application to mathematical subjects. We are told that the Pythagoreans were the first to comprehend geometry and arithmetic under one name μαθηματική, while a fragment attributed to Archytas definitely uses the expression τοὶ περὶ τὰ μαθήματα, 'those concerned with μαθήματα', in the sense of 'mathematicians'. Plato, it is true, uses μάθημα in the more general sense, but with a tendency to appropriate it to mathematical subjects. By Aristotle's time the restriction was completely established.

The Pythagorean *quadrivium* distinguished four mathematical subjects, arithmetic, geometry, *sphaeric*, and music. *Sphaeric* here means the geometry of the sphere as studied for the purposes of astronomy, and is therefore equivalent to astronomy. The term 'arithmetic' calls for explanation. It does not mean, as with us, practical calculation, but the Theory of Numbers, the science of numbers in themselves. For practical calculation the Greeks used another word, λογιστική, the art of calculation (λογίζομαι, I reckon, and λογισμός, reckoning). The distinction between *arithmetic* and *logistic* is clearly stated by Plato and is fundamental in Greek mathematics. Arithmetic was what we call the Theory of Numbers; *logistic* covered, in the first place, the ordinary arithmetical operations, addition, subtraction, multiplication, and division, the handling of fractions, &c.; in the next place *logistic* dealt with such problems as, with us, come under elementary algebra, 'problems about sheep (or apples), bowls, and the like', in which description we recognize such problems as we find in the arithmetical epigrams of the Greek Anthology. Several of these ask us to divide a certain number of apples or nuts among a certain number of persons; others deal with the weights of bowls, or of statues and their pedestals, and the like; others again with cisterns filled or emptied by pipes of different calibre, and so on; these problems involve, generally, the solution of simple equations with one unknown, or easy simultaneous equations with two or more unknowns, and occasionally indeterminate equations of the first degree with one unknown.

Geometry was apparently regarded as including solid geometry, though Plato, in his curriculum for the education of statesmen, made a point of interpolating between geometry and astronomy in the Pythagorean programme the separate subject of *stereometry*; this was because, in Plato's opinion, solid geometry had not been sufficiently studied, and he wished to stimulate his pupils to further energetic research in it.

By the time of Aristotle there was also separated out from geometry a distinct subject γεωδαισία, *geodesy*. This was not, as with us, the study of the figure of the earth, but what we call *mensuration*; meaning literally 'land-dividing', the word covered, in addition, the practical measurement of the surfaces and volumes of other objects.

The Greeks distinguished between the subjects belonging to what we call 'pure' and 'applied' mathematics. Aristotle calls optics, harmonics, and astronomy the *more physical* branches of mathematics, observing that these subjects and mechanics depend for the proofs of their propositions upon the pure mathematical subjects, e.g.

optics on geometry, mechanics on geometry or stereometry, harmonics on arithmetic, while *Phaenomena* (observational astronomy) similarly depends on astronomy (theoretical). Under optics (so says Geminus) three branches were distinguished: (1) optics proper, (2) *catoptric*, the theory of mirrors, (3) σκηνογραφική, scene-painting, i.e. applied perspective. There were, according to the same authority, various subdivisions of mechanics and astronomy, while *Canonic*, the Division of the Canon, was the part of music dealing with the musical intervals.

MATHEMATICS IN GREEK EDUCATION

We have few details of the part played by mathematics in Greek education. During the elementary or primary stage lasting till the age of fourteen we gather that practical arithmetic (in our sense), including weights and measures, was taught along with the main subjects of letters (reading and writing, dictation, and the study of literature), music, and gymnastics. Then there were games played with cubic dice or knucklebones to which boys were addicted, and which involved some practice in calculation. Plato (who, however, was an enthusiast) would have had freeborn boys learn the sort of thing which 'vast multitudes of boys in Egypt learn along with their letters', calculations put in a suitable form for boys, and combining amusement with instruction, e.g. distributions of apples or garlands among more or fewer boys, games with bowls containing gold, silver, and bronze (coins?), and the like, besides calculations of the measurements of things which have length, breadth, and depth.

Geometry and astronomy belonged to secondary education, which occupied the years between the ages of fourteen and eighteen. These subjects seem to have been newly introduced into the curriculum in the time of Isocrates (436–338 B.C.), who approved of them, within limits. If they do no other good, he says, they keep the young out of mischief, and no other subjects could be devised more useful or more fitting; but they should be abandoned by the time that the pupil has reached man's estate. Isocrates does not agree with the view of most ordinary people, who would condemn them as being of no use in private or public life, and certain to be forgotten directly because they do not go with us in our daily activities. True, those who specialize in such subjects as astronomy and geometry are not benefited unless they choose to teach them for a livelihood; nay, if they get too deeply absorbed, they become unpractical and incapable of doing ordinary business; but the study of these subjects up to a certain point enables a boy to concentrate, trains his mind, and sharpens his wits, so that he will the more quickly and easily learn other things of greater value.

According to a legend, it was one of the early Pythagoreans unnamed who first taught geometry for money; this was allowed because he had lost his property. Hippocrates of Chios, the first writer of Elements, probably did the same for a like reason. According to one story, he was a merchant, but lost all his possessions through being captured by a pirate vessel; Aristotle has the different version that he allowed himself to be defrauded of a large sum by custom-house officers at Byzantium, thereby proving, in Aristotle's opinion, that, though a good geometer, he was stupid and incompetent in the business of ordinary life.

Mathematics was taught, not only by regular masters in schools, but also by the sophists who travelled from place to place giving lectures, and included arithmetic, geometry,

MATHEMATICS IN GREEK EDUCATION

and astronomy in their wide range of subjects. Hippias of Elis, the inventor of the curve afterwards known as the *quadratrix*, was one of them. There is a sly hit at him put down to Protagoras in Plato's dialogue of that name, where Protagoras is made to say that 'the other sophists maltreat the young, for, at an age when they have escaped the arts, they take them against their will and plunge them once more into the arts, teaching them calculations, astronomy, geometry, and music—and here he cast a glance at Hippias—whereas, if any one comes to me, he will not learn anything but what he comes for'. Of Hippias we are further told that he got no fees for his lectures in Sparta, and that the Spartans could not endure lectures on astronomy or geometry or logistic; it was only a small minority of them who could even count; what they liked was history and archaeology.

We gather from all this that the general attitude towards mathematics in education was not much different in Greece from what it is now. Two more stories, which one would like to believe true, may be added in illustration.

Pythagoras, we are told, anxious to transplant to his own country the system of education which he had seen in operation in Egypt, and the study of mathematics in particular, could get no hearing in Samos. He adopted therefore this plan of communicating his arithmetic and geometry, so that they might not perish with him. Selecting a young man who from his behaviour in gymnastic exercises seemed adaptable and was withal poor, he promised him that, if he would learn arithmetic and geometry systematically, he would give him sixpence for each 'figure' (proposition) that he mastered. This went on until the youth became interested, when Pythagoras rightly judged that he would gladly go on without the sixpence.

He hinted therefore that he himself was poor and must try to earn his daily bread instead of doing mathematics; whereupon the youth, rather than give up the study, volunteered to pay sixpence himself to Pythagoras for each proposition. We must presumably connect with this story the Pythagorean motto, 'a figure and a platform' (from which to ascend to the next higher step), 'not a figure and sixpence'.

The other story is that of one who began to learn geometry with Euclid and asked, when he had learnt one proposition, 'What shall I get by learning these things?' Whereupon Euclid called the slave and said, 'Give him threepence, since he must needs make gain out of what he learns'.

II

NUMERICAL NOTATION AND PRACTICAL CALCULATION

THE DECIMAL SYSTEM

FROM the earliest historical times we find the Greeks, like civilized peoples all the world over, using the decimal system. There are indeed traces, in very early times, of counting by 5 (thus in Homer πεμπάζειν, to 'five', means to count); but this was probably little more than auxiliary to counting by tens, just as in the earliest form of numerical notation used in Greece fifty is a separate category between ten and a hundred, five hundred between a hundred and a thousand, and so on. The quinary reckoning is thus not a variation like the vigesimal system used by the Celts and Danes, traces of which remain in the French *quatre-vingts*, *quatre-vingt-treize*, &c., and in our score, three-score and ten, &c.

No doubt the origin of the decimal system, as of the quinary and vigesimal variations, was the practice of counting, first with the fingers of one hand, then with the fingers of both hands, and then again with the ten toes in addition. This is confirmed by the use of the Greek and Latin words for 'hand', χείρ and *manus*, to denote a number. The so-called geometry of Boëtius similarly records that the ancients called all the numbers, below ten, *digits* (fingers).

The systems of numerical notation which invite comparison with the Greek are the Egyptian and Babylonian.

EGYPTIAN NUMERICAL NOTATION

The Egyptians had a purely decimal system with the following signs, each of which could be repeated up to nine times: I for the unit, ∩ for 10, ⊚ for 100, ⌇ for 1000, for 10000, ⤳ for 100000; 𒁹 stood for 1000000. If the number of any denomination was more than four or five, they saved lateral space by writing them in two or three rows, one above the other. Numbers could be written from left to right or from right to left; in the latter case the above signs were turned the opposite way. The greater denominations preceded the smaller. The fractions written were all submultiples (fractions with unity as numerator), except $\frac{2}{3}$, which was denoted by ⊕

or ⊓. The submultiples were distinguished by writing ⌒ (reduced in hieratic to a dot) over the corresponding whole number; thus

$$\overset{\frown}{\cap\cap|||} = \tfrac{1}{23}, \quad \overset{\frown}{\odot\odot\odot\cap_{||||}} = \tfrac{1}{324}, \quad \overset{\frown}{\underset{\text{X X}}{||}\odot\overset{\cap\cap\cap}{\cap\cap\cap}} = \tfrac{1}{2190}.$$

BABYLONIAN SYSTEMS

Part of the Babylonian notation proceeded on the decimal plan. The unit was the simple wedge 𒁹, which could be repeated up to nine times; when more than three were wanted, they were arranged in rows, as 𒐼 = 4, 𒐌 = 7. The sign for 10 was 𒌋; 11 would therefore be 𒌋𒁹. 100 had the compound sign 𒑰; then, on a multiplicative principle, 𒌋𒑰 meant 10 hundreds or 1000, and again 𒌋𒌋𒑰 10 times (not 2 times) 1000, i.e. 10000. The independent signs for powers of 10 went no farther than that for 1000, 𒌋𒑰; larger numbers could apparently only be expressed, on this system, as multiples of 1000, so that the notation was quite unsuitable for expressing large numbers.

But the decimal notation was supplemented by the

EGYPTIAN AND BABYLONIAN SYSTEMS 13

sexagesimal system which is found in use on the Tables of Senkereh discovered by W. K. Loftus in 1854, and which may go back as far as (say) 2300 B.C. This was practically nothing less than a 'position-value' system, which, by means of columns side by side, one of which contains the units to any number from 1 to 59, and the rest the numbers of each successive power of 60 included in the total, made it possible to express numbers of any size whatever. Only the units representing the number of each denomination (the particular power of 60) contained in the total, that is, some number of units from 1 to 59 expressed in the ordinary wedge notation, was written in the column; the denomination had to be gathered from the context. The column for the units was followed by similar columns appropriated to each of the successive submultiples $\frac{1}{60}$, $\frac{1}{60^2}$, &c., the number of each of these fractions being represented by ordinary wedge-numbers as before. Thus ⟨⟨𝐘 ⟨⟨𝐘𝐘𝐘 ⟨⟨ represents $44.60^2 + 26.60 + 40 = 160,000$; ⟨⟨ 𝐘𝐘𝐘 ⟨⟨𝐘 ⟨⟨⟨𝐘𝐘𝐘 $= 27.60^2 + 21.60 + 36 = 98,496$. Similarly we find ⟨⟨⟨ ⟨⟨⟨ representing $30 + \frac{30}{60}$ and ⟨⟨⟨ ⟨⟨⟨𝐘𝐘𝐘 representing $30 + \frac{27}{60}$; the 27 is here written as $30 - 3$, 𝚪 denoting *lal* or *minus*.

This system only required a definite sign for zero to make it a complete 'place-value' system, and it would appear from sources other than the Senkereh Tables that a gap often indicated a zero, or there was a sign used for the purpose, namely ⟨, called the 'divider'. H. V. Hilprecht's researches show that $60^4 = 12960000$ played an important part in Babylonian arithmetic, and he found

a table containing certain quotients of the number
)⟨ ------ = $60^8 + 10.60^7$ or 195955200000000.

The inconvenience of the system was that it required a multiplication table extending from 1.1 to 59.59. On the other hand, the scale of 60 had the advantage that 60 has a great number of factors, namely 1, 2, 3, 4, 5, 6, 10, 12, 15, 20, 30.

It should be added that there were also separate names and signs for certain numbers connected with the sexagesimal system: e.g. 60 itself was called *sussu* or *susi* (= *soss*), 600 was *ner*, 60^2 was *sar*, and 60^4 was *sar-gal* ('great *sar*').

GREEK NUMERICAL NOTATION

There were two systems in use in classical times : (*a*) the 'Herodianic' signs, (*b*) the ordinary alphabetic notation.

(*a*) The *Herodianic* system is known alternatively as the 'Attic'. The designation 'Herodianic' is due to the accident that the notation is described in a fragment attributed to Herodian, a grammarian of the latter half of the second century A.D., and printed in the Appendix to Stephanus' *Thesaurus*, vol. vii. The signs are not numerals in the proper sense : except for the stroke I representing the unit, they consist of the first letters of the Greek words for certain numbers; thus 5 is represented by Π (the first letter of πέντε), 10 by Δ (for δέκα), 100 by H (for ἕκατον), 1000 by X (χίλιοι), and 10000 by M (μύριοι). The half-way numbers 50, 500, 5000 were expressed by combining Π (5) with the signs for the other factors in each case; thus ᛒ, ᛕ, ᛆ, made up of Π (5) and Δ (10) meant 50; ᛒ, made up of Π and H, = 500, ᛕ = 5000, and ᛆ = 50000. In this system then ΠI = 6, ΔIIII = 14, HΠ = 105, and XXXXᛒHHHHᛒΔΔΔΠIIII = 4999. The system is closely parallel to the Roman, in which, e.g., the last-mentioned number would be MMMMDCCCLXXXXVIIII.

Instances of the 'Attic' system of notation are found in Attic inscriptions from 454 to about 95 B.C. The same system was in use outside Attica, though the precise form of the signs varied with the form of the letters in the local alphabets. Thus in Boeotian inscriptions ⌐ or ⌐ = 50, ⊢E = 100, ⊓E = 500, Ⅴ = 1000, ⌐ = 5000, and ⌐⊓E⊢E⊢E⊢E▷▷III = 5823.

(b) *The ordinary alphabetic numerals.*

The Greeks took their alphabet from the Phoenicians. The Phoenician alphabet had 22 letters, and, in appropriating the different signs, the Greeks had the happy inspiration to use for the vowels, which were not written in Phoenician, the signs for certain spirants for which the Greeks had no use; thus Aleph became A, He was used for E, Yod for I, and Ayin for O. When, later, the long E was differentiated, Cheth, ⊟ or H, was used. Similarly with the superfluous signs for sibilants : out of Zayin and Samech the Greeks made Z and Ξ. There remained two sibilants, Ssade and Shin. From the latter came the simple Greek Σ, while Ssade, a softer sibilant, was taken over in the place it occupied after Π and was written in the form Μ or Ϻ. The form Ͳ (= σσ) found in inscriptions of Halicarnassus (cf. ῾ΑλικαρναΤ[έων] = ῾Αλικαρναϲϲέων) and Teos ([θ]αλάΤης; cf. θάλαϲϲαν in another place) seems to be derived from some form of Ssade. This Ͳ, after it disappeared from the literary alphabet, remained as a numeral, passing through the forms Λ, ɱ, Γ, ѡ, and ⸕ to the fifteenth-century form ⸑, which in the second half of the seventeenth century came to be called Sampi. The original Greek alphabet retained the Phoenician Vau (Ϝ) in its proper place between E and Z, and the Koppa = Qoph (Ϙ) between Π and P. The Phoenician alphabet ended with T, to which the Greeks first added Y (derived apparently from Vau, notwithstanding the retention of F), then the letters Φ, X, Ψ, and finally Ω. This gave 27 letters including the Ϡ or ↷, which made it possible to divide them into three sets of nine, the first set denoting the units 1, 2, 3 ... 9, the second the tens, 10, 20, 30 ... 90, and the third the hundreds, 100, 200 ... 900, thus:

A	= 1	I	= 10	P	= 100
B	= 2	K	= 20	Σ	= 200
Γ	= 3	Λ	= 30	T	= 300
Δ	= 4	M	= 40	Y	= 400
E	= 5	N	= 50	Φ	= 500
C [ς]	= 6	Ξ	= 60	X	= 600
Z	= 7	O	= 70	Ψ	= 700
H	= 8	Π	= 80	Ω	= 800
Θ	= 9	Ϙ	= 90	Ϡ [↷]	= 900

The sign for 6, C, is a form of the digamma F. In the seventh and eighth centuries it came to be written in the form ς, and then, through its similarity to the cursive ϛ (= στ), it came to be called Stigma.

The earliest attested use of the alphabetic numerals seems to be in a Halicarnassus inscription of date not much later than 450 B.C. Two caskets from the ruins of a famous mausoleum at Halicarnassus which are attributed to the time of Mausolus, about 350 B.C., are inscribed with the letters ΨNΔ = 754 and ΣϘΓ = 293. Again, a stone inscription found at Athens and perhaps belonging to the middle of the fourth century B.C. has, in five fragments of columns, numbers in tens and units expressed on the same system, the tens on the right and the units on the left.

ALPHABETIC NUMERALS

It was a long time before the alphabetic numerals found general acceptance. They were not officially used till the time of the Ptolemies, when it became usual to write, in inscriptions and on coins, the year of the reign of the ruler for the time being. They are found on coins, assigned to 266 B.C., of Ptolemy II Philadelphus. A very old Graeco-Egyptian papyrus assigned to 257 B.C. contains the number $\kappa\theta = 29$. The first official use of the alphabetic system in Greece proper dates from about 200 B.C., as is shown by an inventory from the temple of Amphiaraus at Oropus. From this time Athens stood alone in retaining the old 'Attic' system, the last attested use of which was about 95 B.C.; the alphabetic numerals were used there in the time of Augustus and were officially established by A.D. 50.

When a number containing more than one denomination, e.g. units with tens and hundreds, had to be written, the components might be arranged either in descending or in ascending order; the former was usual in European Greece, while in inscriptions in Asia Minor the latter prevailed. Thus 111 might be either PIA or AIP; and even PAI was permissible. But the descending order finally prevailed, perhaps through Roman influence.

The alphabetic notation enabled all numbers from 1 to 999 to be expressed. For thousands up to 9000 the signs for the units from 1 to 9 were used with a stroke attached, thus: $_{\prime}A$ or $'A = 1000$. The stroke might be combined with the letter, as $\lambda = 1000$; and other forms were possible, e.g. $'A = 1000$, $'\Gamma = 6000$. To express myriads, or tens of thousands, $M (= \mu\acute{\upsilon}\rho\iota\omicron\iota)$ was borrowed from the other system; thus 20000 is BM, MB, or $\overset{B}{M}$.

To distinguish letters denoting numbers from the letters forming words in any passage various devices were used.

The numeral might be put between dots, : or :, or space might be left on each side. In Imperial times it became usual to put a horizontal stroke over the numeral, e.g. ἡ βουλὴ τῶν X̄, variations being ·X·, X̌, and the like. In cursive writing the stroke over the letter became the orthodox mark for the numeral or the collection of numeral signs. The following was therefore the scheme:

units (1 to 9)   ᾱ, β̄, γ̄, δ̄, ε̄, ϛ̄, ζ̄, η̄, θ̄;
tens (10 to 90)   ῑ, κ̄, λ̄, μ̄, ν̄, ξ̄, ō, π̄, ϟ̄;
hundreds (100 to 900) ρ̄, σ̄, τ̄, ῡ, φ̄, χ̄, ψ̄, ω̄, ↗;
thousands   ͵ᾱ, ͵β, ͵γ̄, ͵δ̄, ͵ε̄, ͵ϛ̄, ͵ζ̄, ͵η̄, ͵θ̄.

(c) *Notation for large numbers.*

The orthodox way of writing tens of thousands was by means of the letter M with the number of myriads above it, e.g. $\overset{\beta}{M} = 20000$, $\overset{\text{͵ϛροε}}{M}\text{͵εωοε} = 71755875$ (Aristarchus of Samos); or the number of myriads could be put after M or $\overset{\text{Υ}}{M}$ and separated from the thousands, &c., by a dot: thus $\overset{\text{Υ}}{M}\rho\nu.\text{͵ζ}↗\pi\delta = 1507984$ (Diophantus). Or again myriads were sometimes represented by an ordinary numeral with two dots over it; thus $\ddot{\alpha}\text{͵}\eta\phi\mathrm{Ϙ}\beta = 18592$ (Heron, *Geometrica*). To express still higher numbers powers of myriads were used; a myriad was called a *first myriad* (πρώτη μυριάς) to distinguish it from a *second myriad* (δευτέρα μυριάς) which was 10000^2, and so on; the names could be written in full or represented by $\overset{\text{Υ}}{M}$, $\overset{\text{Υ}}{MM}$, &c. Thus δευτέραι μυριάδες ιϛ πρῶται μυριάδες ͵β↗νη $\overset{\text{o}}{M}$ ͵ϛφξ = 16 2958 6560 (Diophantus), the $\overset{\text{o}}{M}$ being used to distinguish the units from the myriads.

The 'tetrads' used by Apollonius, in a work now lost

SYSTEMS FOR LARGE NUMBERS 19

but partly reproduced by Pappus, were these same powers of 10000, but were called μυριάδες ἁπλαῖ, διπλαῖ, τριπλαῖ... (simple, double, triple ... myriads), meaning 10000, 10000^2, 10000^3, &c.; thus $\mu^\Gamma{,}\epsilon\upsilon\xi\beta$ καὶ $\mu^\beta{,}\gamma\chi$ καὶ $\mu^a{,}\varsigma\upsilon$ = 5462 3600 6400 0000. A less convenient notation was to put successive pairs of dots over the signs for ordinary units; thus $\ddot{\overset{..}{\eta}}$ = 9000000, $\ddot{\beta}$ = 2(10000)², $\ddot{\mu}$ = 40(10000)³, and so on (Nicholas Rhabdas, fourteenth century A.D.).

(d) *Archimedes' system for large numbers* ('octads')

A yet more remarkable system, going by 'octads' (powers of 10000^2 or 100000000 = 10^8), was devised by Archimedes in order to show Gelon, King of Syracuse, that, on any reasonable estimate of the size of the universe, and assuming a sphere of that size to be filled with sand, it is not beyond the power of language to express the number of grains of sand which it would contain. This is the subject of Archimedes' tract *Psammites* or *Sand-reckoner*. The system is developed as follows. The numbers from 1 to 100000000 or 10^8 form the *first order* of numbers; the *second order* contains the numbers from 10^8 to 10^{16}, the *third order* those from 10^{16} to 10^{24}, and so on, the 100000000*th order* ending with $(100000000)^{100000000}$ or $10^{8.10^8}$. The aggregate of these *orders* consisting of the numbers from 1 to $10^{8.10^8}$ (= P, say) form the *first period*. The *second period* begins from P, and the *first order* comprised in it contains the numbers from P to $P \cdot 10^8$, the *second order* the numbers from $P \cdot 10^8$ to $P \cdot 10^{16}$, and so on, up to the $10^8 th$ *order* which ends with $P \cdot 10^{8.10^8}$ or P^2. Further *periods* are developed in this way up to the $10^8 th$ *period*, the $10^8 th$ *order* of which ends with the number P^{10^8} or $(10^{8.10^8})^{10^8}$, which is described by Archimedes as 'a myriad-myriad units of the myriad-myriadth order of the myriad-myriadth period'. The prodigious extent of this system will be appreciated when it is realized that the last number in the *first period* would now be represented by 1 followed by 800 000 000 ciphers, and the last number of the 10^8th *period* by 1 followed by 80000 million millions of ciphers. The number of grains of sand which, on Archimedes' assumptions as to the size of the universe, the universe would hold is proved to be actually less than 10000000 units of the *eighth order* of the *first period*, or $10^{56+7} = 10^{63}$.

FRACTIONS

The Egyptians had no notation for fractions other than $\frac{2}{3}$ and the series of submultiples or aliquot parts. When therefore in their arithmetical work they had to write down the result of dividing a smaller number by a greater, they expressed it as a sum of submultiples with or without $\frac{2}{3}$; thus the result of dividing 3200 by 365 they stated as $8\frac{2}{3}\frac{1}{10}\frac{1}{2190}$, and that of dividing 2 by 13 as $\frac{1}{8}\frac{1}{52}\frac{1}{104}$. The Greeks too had a preference for expressing fractions in this way. The ordinary notation for a submultiple was the letter for the corresponding numeral with an accent instead of a horizontal stroke above it; thus $\frac{1}{3}$ is γ', the full expression being $\gamma'\ \mu\acute{\epsilon}\rho o\varsigma = \tau\rho\acute{\iota}\tau o\nu\ \mu\acute{\epsilon}\rho o\varsigma$, 'third part' ($\gamma'$ was really short for τρίτος and was used for the ordinal number 'third' as well as for $\frac{1}{3}$), and similarly for other submultiples. There were special signs for $\frac{1}{2}$, namely ∠', and for $\frac{2}{3}$, w'. We find in Archimedes ∠'δ' = $\frac{1}{2}\frac{1}{4}$ for $\frac{3}{4}$, and in Heron κθ w' ιγ' λθ' = $29\frac{2}{3}\frac{1}{13}\frac{1}{39}$ for $29\frac{10}{13}$, and so on. A less orthodox method (found in later manuscripts) was to affix two accents, e.g. ζ" = $\frac{1}{7}$. Diophantus uses, instead of the accent, a sign which Tannery printed as ×, e.g. $\gamma^{\times} = \frac{1}{3}$.

An ordinary proper fraction could be expressed in various ways. It consists of a certain multiple of a particular aliquot part (i.e. 1 divided by the denominator). Hence the numerator could be represented by the ordinary numeral, and the denominator by the accented numeral meaning the aliquot part. Thus in Archimedes $\overline{\iota}$ oa' = ten seventy-oneths ($\frac{10}{71}$), and $\overline{,a\omega\lambda\eta}$ $\overline{\theta}$ $\iota a'$ = $1838\frac{9}{11}$. The objection to this method is that, but for the context, $\overline{\iota}$ oa' would naturally mean $10\frac{1}{71}$. Sometimes we find the numerator written in full along with the accented letter representing the denominator; thus δύο $\mu\epsilon'$ = 2/45ths (Aristarchus of Samos). A method less open to misconstruction was the following. Just as δ^{os} was an abbreviation for τέταρτος, so δ^{ov} was used as an abbreviation for 'one-fourth' (ordinarily written as δ') and then made a kind of substantive (with case-termination appended), to which was added the numeral representing the numerator; thus $\delta^{\omega\nu}\varsigma$ is literally 'six fourths', and similarly ν $\kappa\gamma^{\omega\nu}$ = $\frac{50}{23}$, $\rho\kappa a^{\omega\nu}{,}a\omega\lambda\delta'L'$ = $1834\frac{1}{2}$/121 (Diophantus).

Ambiguity was also avoided by two other methods. The first predominates in the *Geometrica* and other works attributed to Heron. Here the accented numeral representing the submultiple corresponding to the denominator is written twice (as if to show that there are more than one of these submultiples), and the numerator is written as an ordinary numeral; thus ϵ $\iota\gamma'$ $\iota\gamma'$ = $\frac{5}{13}$ths, τὰ ς $\zeta'\zeta'$ = $\frac{6}{7}$. The fractional signification is often emphasized by adding the word λεπτά ('small parts' or fractions) in contrast to the μονάδες (units) of which a whole number consists; cf. μονάδες $\rho\mu\delta$ λεπτὰ $\iota\gamma'$ $\iota\gamma'$ $\sigma\varphi\theta$ = $144\frac{299}{13}$. Sometimes fractions are given in two alternative forms, the normal and the Egyptian: e.g. β γ' $\iota\epsilon'$ ἤτοι β καὶ β $\epsilon'\epsilon'$ = '$2\frac{1}{3}$ $\frac{1}{15}$ or $2\frac{2}{5}$'; ζ L' ι' $\iota\epsilon'$ $o\epsilon'$ ἤτοι μονάδες ζ ϵ' ϵ' γ καὶ β $\epsilon'\epsilon'$ τῶν $\epsilon'\epsilon'$ = '$7\frac{1}{2}$ $\frac{1}{10}$ $\frac{1}{15}$ $\frac{1}{75}$ or $7\frac{3}{5} + \frac{2}{5} \times \frac{1}{5}$'. In the same works ϵ' τὸ ϵ' means 1/5th of 1/5th.

The second method is the most convenient of all. It predominates in Diophantus, and is occasionally used in the *Metrica* of Heron. The method was to write the numerator in the line and the denominator directly above it, which is practically the reverse of our method. In Tannery's edition of Diophantus a line is printed between the numerator and denominator: thus $\overline{\rho\kappa\alpha}^{\iota\varsigma} = \dfrac{121}{16}$. But it is better to omit the line (cf. $\overset{\rho\kappa\eta}{\rho} = \dfrac{100}{128}$ in Kenyon's Papyri ii, No. cclxv. 40, and $\overset{\rho\xi\delta}{\delta} = \dfrac{4}{164}$ in Heron's *Metrica*). Diophantus even writes a submultiple in this way (e.g. $\overset{\phi\iota\beta}{\alpha} = \dfrac{1}{512}$), abandoning the orthodox plan according to which $\dfrac{1}{512}$ would be $\phi\iota\beta'$ simply. The denominator could also be placed above but to the right, like an exponent, e.g. $\iota\epsilon^{\delta} = \dfrac{15}{4}$. Where numbers contain integers and also fractions, they are written much as we write them, first the integer, then the fraction or fractions: e.g. $\alpha\,\gamma^{\lambda} = 1\tfrac{1}{3}$, τo $\text{L}'\ \iota\varsigma^{\lambda} = 370\tfrac{1}{2}\,\tfrac{1}{16}$. Complicated fractions where the numerator and denominator are algebraical expressions or large numbers are expressed by writing the numerator first and interposing $\mu o\rho\acute{\iota}o\upsilon$ or $\dot{\epsilon}\nu\ \mu o\rho\acute{\iota}\omega$ (= 'divided by') between it and the denominator: thus $\overset{\text{Y}}{\text{M}}\ \rho\nu\,.\,\zeta\overline{\lambda}\pi\delta\ \mu o\rho\acute{\iota}o\upsilon$ $\kappa\varsigma\,.\,\beta\rho\mu\delta = 1507984/262144$.

Sexagesimal fractions.

Great interest attaches to the sexagesimal system of fractions (Babylonian in its origin), which was used by the Greeks in astronomical calculations, and is found in full operation in Ptolemy. The circumference of a circle and with it the four right angles subtended by it at the centre are divided into 360 τμήματα ('segments') or μοῖραι ('parts') or, as we should say, 'degrees', each μοῖρα again into 60 parts called (*first*) *sixtieths* or *minutes* (λεπτά), each of these again into 60 *second-sixtieths* (*seconds*), and so on. The diameter was divided into 120 τμήματα ('segments'), each of which was divided into sixtieths, each sixtieth again into sixty parts, and so on. A convenient system was thus available for calculations to any degree of accuracy; any mixed number could be expressed as so many units, so many fractions which we should write as $\frac{1}{60}$, so many of those which we should write as $(\frac{1}{60})^2$, $(\frac{1}{60})^3$, &c. The units, τμήματα or μοῖραι (often denoted by μ°), were written first, then followed a numeral with one accent representing the number of 60ths or minutes, then a numeral with two accents representing second-sixtieths, and so on. Thus $\mu^\circ \beta = 2^\circ$, μοιρῶν μζ μβ′ μ″ = 47° 42′ 40″. Similarly, τμημάτων ξζ δ′ νε″ = 67^p 4′ 55″, where p denotes the 'part' of the diameter. Where there was no unit, or none of any particular denomination of the fractions, the symbol O, signifying οὐδεμία μοῖρα, οὐδὲν ἑξηκοστόν and the like, was used: thus μοιρῶν O α′ β″ O‴ = 0° 1′ 2″ 0‴. The sexagesimal fractions, though no doubt less convenient to work with than our decimals, furnished a speedy way of approximating to the values of surds. Ptolemy says in his Table of Chords, that the chord subtending at the centre an angle of 120° is (τμημάτων) ργ νε′ κγ′

or $103^p\ 55'\ 23''$; since the radius $= 60^p$, this is equivalent to saying that
$$\sqrt{3} = 1 + \frac{43}{60} + \frac{55}{60^2} + \frac{23}{60^3},$$
which again is equivalent to 1·7320509 . . . , and is correct to the seventh decimal place.

PRACTICAL CALCULATION

(α) *The abacus.*

The essence of the abacus is the arrangement of it in columns, vertical or horizontal but generally vertical, each of which is marked off by lines or in some other way, and is allocated to one of the successive denominations of the numerical system in use, i.e., in the case of the decimal system, units, tens, hundreds, thousands, and so on. The number of units of each denomination was shown by means of pebbles, buttons, pegs, or the like, placed in the columns. When in the process of addition or multiplication the number of pebbles contained in one column becomes sufficient to make one or more units of the next higher denomination, the number of pebbles equivalent to the whole number of the higher units is withdrawn from the column in which they are, and pebbles representing the proper number of the higher units are added to the higher column. Similarly, in subtraction, when a number of units of one denomination has to be subtracted, but there are not a sufficient number in the column to subtract from, one pebble is withdrawn from the next higher column and decomposed actually or mentally into the number of the lower units equivalent in value; the addition of this number to the lower column enables the operator to withdraw the number requiring to be subtracted.

THE ABACUS

Unfortunately the details of the Greek abacus have to be inferred from those of the Roman abacus, since the only abaci which have been preserved and can with certainty be identified as such are Roman. There were two kinds of these; in one kind the marks used were buttons or knobs which could be moved up and down in each column but could not be taken out of it, while in the other kind there were pebbles, which could be moved from one column to another. Each column was in two parts, a shorter portion at the top containing one button only, which represented half the number of units necessary to make up one of the next higher denomination, and a longer portion containing one less than half the same number. This enabled the total number of buttons to be economized. After the columns representing integers and containing respectively units, tens, hundreds, thousands, &c., came columns representing fractions; the first contained *unciae* of which there were twelve to the unit, i.e. fractions of $\frac{1}{12}$ (the button in the top portion represented $\frac{6}{12}$ths and there were 5 in the lower portion, each representing $\frac{1}{12}$th), then there might be shorter columns containing respectively fractions of $\frac{1}{24}$th (one button), fractions of $\frac{1}{48}$th (one button), and fractions of $\frac{1}{72}$nd (two buttons).

According to the medieval writer of the so-called *Geometry* of Boëtius, the 'abacus' was a later name for what was previously called 'mensa Pythagorea' in honour of the Master who taught its use. Here the method was to put in the columns, not pebbles, or buttons, but the corresponding *numeral*, which might be written in sand sprinkled over the surface (just as the Greek geometers are said to have drawn their diagrams in sand strewn over boards similarly called ἄβαξ or ἀβάκιον). The figures put in the columns were called *apices*.

That the Greeks did use some kind of abacus is clear from allusions in classical writers to the use of pebbles in calculations, and the assignment of different values to the pebbles according to the position in which they were placed. In the *Wasps* of Aristophanes Bdelycleon tells his father to do an easy sum 'not with pebbles but with fingers', as much as to say 'you don't need pebbles for this sum; you can do it on your fingers'. Herodotus says that, in reckoning with pebbles, as in writing, the Greeks move their hand from left to right, the Egyptians from right to left; apparently therefore the columns were vertical, facing the operator. Diogenes Laërtius attributes to Solon a remark that those who carried weight with tyrants were like the pebbles on a reckoning-board, because they sometimes stood for more, sometimes for less. A character in a fourth-century comedy calls for an abacus and pebbles to do his accounts. More definitely still, Polybius says 'These men are really like the pebbles on reckoning-boards. For these, according to the pleasure of the reckoner, have the value now of a χαλκοῦς ($\frac{1}{8}$th of an obol or $\frac{1}{48}$th of a drachma) and at the next moment of a talent'.

There appear to be three monuments left which may throw light on the form and details of the Greek abacus, but that is all that can be said. The first is the famous 'Salaminian Table', discovered and described by Rangabé (1846); the Table, now broken into two unequal parts, is in the Epigraphical Museum at Athens. Annexed is a representation of it. The size and material of the table (according to Rangabé it is 1·5 metres long and 0·75 metres broad) show that it was no ordinary abacus. It may have been a fixture intended for quasi-public use, such as a banker's or money-changer's table, or it may have been a scoring table for some game like tric-trac or backgammon. There is nothing to show how it was used, but the series of letters are such as might have marked the columns of an abacus. The letters on the three sides are the same except that two of them go no higher than X (1000 drachmae); but the third has ⌷ (5000 drachmae) and T (the talent or 6000 drachmae) in addition; ⊢ is the sign for a drachma, I for an obol ($\frac{1}{6}$th of a drachma), C for $\frac{1}{2}$-obol, T for $\frac{1}{4}$-obol (τεταρτημόριον), and X for $\frac{1}{8}$-obol (χαλκοῦς).

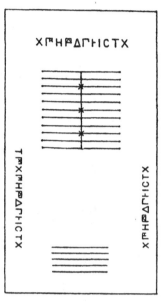

Secondly, the so-called Darius-vase found at Canossa (Canusium), north-west of Barletta, represents a collector of tribute of distressful countenance with a table in front of him having pebbles, or (as some maintain) coins, upon it, and on the right-hand edge, beginning on the side farthest away and written in the direction towards him, the letters MΨHΡΓΟ⟨T, while in his left hand he holds a book in which presumably he enters the receipts. Now M, Ψ(= X), H, and Ρ are of course the initial letters of the words for 10000, 1000, 100, and 10 respectively. Here, therefore, we have a purely decimal system, without the half-way numbers represented by Γ (5) in combination with the other initial letters used in the 'Attic' system. The sign Γ after Ρ seems to be a mistake for Ρ, the older sign for a drachma, O is an obol, ⟨ a $\frac{1}{2}$-obol, and T a $\frac{1}{4}$-obol. The letters form suitable headings for columns in an abacus, but the table is said to be more like an ordinary money-changer's or tax-collector's table such as might be seen anywhere in the East to-day.

Thirdly, there is a monument of the same sort, a so-called σήκωμα (arrangement of measures) which was discovered about sixty years ago. It is of stone, with fluid measures, and has, on the right side, the 'Herodianic' numeral signs ΧⲦΗⲢΔⲄⲏΤΙC, which include those for 500, 50, and 5 drachmae; Ⲁ is the sign for a drachma, Ι, C those for an obol and a ½-obol, while Τ must be some number of obols making part of a drachma, presumably τριώβολον or 3 obols.

When once the alphabetic numerals were definitely established, the Greeks had in truth no need of an abacus; the alphabetic numerals enabled them to carry out the ordinary operations of addition, subtraction, multiplication, &c., in a form little less convenient than ours.

(β) *Addition and subtraction.*

In writing out numbers for the purpose of these operations the Greek would naturally place all the numbers of one denomination above one another, in one vertical row. The following would then be a typical addition-sum:

$$\begin{array}{ll} \,\alpha\,\upsilon\,\kappa\,\delta \;=\; & 1424 \\ {}^{\alpha}\quad\rho\;\;\gamma & 103 \\ \mathsf{M}\,{,}\beta\,\sigma\,\pi\,\alpha & 12281 \\ \overset{\gamma}{\mathsf{M}}\quad\lambda & 30030 \\ \hline \overset{\delta}{\mathsf{M}}\,{,}\gamma\,\omega\,\lambda\,\eta & 43838, \end{array}$$

while a subtraction would be like this:

$$\begin{array}{ll} \overset{\theta}{\mathsf{M}}\,{,}\gamma\,\chi\,\lambda\,\varsigma \;=\; & 93636 \\ \overset{\beta}{\mathsf{M}}\,{,}\gamma\,\upsilon\;\;\theta \;=\; & 23409 \\ \hline \overset{\zeta}{\mathsf{M}}\quad\sigma\,\kappa\,\zeta & 70227 \end{array}$$

ARITHMETICAL OPERATIONS

(γ) *Multiplication*.

The first requisite here was a multiplication table, and this the Greeks are certain to have had from very early times. The Egyptians indeed never seem to have had such a table. They effected their multiplications by successive multiplications by two and addition of products as required. That is, they obtained successively numbers representing twice, four times, eight times, sixteen times the multiplicand, and so on, as far as necessary, and then added such of these products (including once the multiplicand) as gave the product required. Thus, to multiply by 13, they did not take 10 times and 3 times the multiplicand, but once, twice, four times, and eight times it, and then added the products corresponding to 8, 4, and 1, since $8+4+1 = 13$. Similarly, to obtain 25 times a number, they added once and 8 times and 16 times the number $(1+8+16 = 25)$.[1]

[1] There is an exact parallel to the Egyptian practice in the so-called 'Russian peasant's' method of multiplication said to be in use to-day (though I have not seen any confirmation of this). It will be best explained by an example. Suppose that we have to multiply 157 by 83. Write out (1) the multiplier and multiplicand side by side, then (2) under the multiplier write the half of it, or the integer next below the half if it is odd, and under the multiplicand twice it, (3) do the same with the 41 and 314, thus giving 20 and 628, and so on, until we have 1 in the multiplier column. Now strike out all the lines in the multiplicand column which have an *even* number opposite to them in the other column, and add the numbers which are then left in the multiplicand column. This gives 13031, the desired result.

83	157
41	314
20	~~628~~
10	~~1256~~
5	2512
2	~~5024~~
1	10048
	13031

A little consideration shows that the above amounts to an elegant practical method of carrying out what is exactly the Egyptian procedure. We strike out the figures opposite the *even* numbers under the multiplier because, when we take the exact half of the multiplier-number, the new line we put down is the exact equivalent of the preceding line, but where we have taken half the multiplier-number *less* 1,

PRACTICAL CALCULATION

The Greek method of multiplication, as of addition and subtraction, was essentially the same as ours. The multiplicand was written first, and below it was placed the multiplier preceded by ἐπί ('upon' = 'by'). The only difference is that the Greek began by multiplying the highest denominations instead of the lowest. The term in the multiplier containing the highest power of 10 was taken and multiplied successively into the separate terms of the multiplicand in descending order beginning from that containing the highest power of 10 and ending with the units; the results of each successive multiplication were put down in order from left to right. Next the term containing the second highest power of 10 in the multiplier was multiplied into the terms of the multiplicand in the same descending order and the results put down in a fresh line, and so on. The same procedure was followed when either or both of the numbers to be multiplied contained fractions.

Two examples from Eutocius' commentary on Archimedes will make the whole matter clear.

(1)   ,ατνα   1351
    ἐπὶ ,ατνα   ×1351

ρ λ ε MMM,α	1000000	300000	50000	1000
λ θ α MMM,ετ	300000	90000	15000	300
ε α MM,ε,βφν		50000	15000	2500	50
,ατνα			1000	300	50	1

ὁμοῦ   ρπβ
M,εσα   *together* 1825201.

the new line is *not* equivalent to the preceding one, as we have omitted once the figure in the multiplicand-column, so that we must leave it in, to be included in the final addition.

MULTIPLICATION. DIVISION 31

(2) ,γιγL'δ'   3013½ ¼ (= 3013¾)
ἐπί ,γιγL'δ'   × 3013½ ¼

	ᾱ ͞γ	͞γ			
MM,θ,αφψν	9000000	30000	9000	1500	750
ΜρλεβL'	30000	100	30	5	2½
,θλθαL'L'δ'	9000	30	9	1½	½ ¼
,αφεαL'δ'η'	1500	5	1½	¼	⅛
ψνβL'L'δ'η' ις'	750	2½	½ ¼	⅛	1/16

ὁμοῦ M̄,βχπθ ις'   together 9082689 1/16.

A difficulty could only arise if the numbers to be multiplied (either or both) contained unusually high powers of 10, in which case it would require thought to settle the denomination of the product. Both Archimedes and Apollonius gave rules by which this could be done. For instance, Archimedes in his *Sand-reckoner* deals with the geometrical progression 1, a, a^2 ... a^n, where a is any number, and proves in effect that $a^m . a^n = a^{m+n}$.

Theon of Alexandria, in his commentary on Ptolemy's *Syntaxis*, shows how to multiply two expressions containing sexagesimal fractions. The successive fractions are treated as separate denominations like the powers of 10 in the above examples, and the procedure is the same, *mutatis mutandis*.

(δ) *Division*.

Here again the Greek method does not differ substantially from ours. Theon of Alexandria explains a case of long division which is specially interesting in that an expression containing sexagesimal fractions (minutes and seconds) has to be divided by another like expression. The problem is to divide 1515 20' 15" by 25 12' 10", and

Theon's working can be shown by the aid of modern notation, as follows:

Divisor.	Dividend.			Quotient.
25 12′ 10″	1515	20′	15″	First term 60
25 . 60 =	1500			
Remainder	15 =	900′		
Sum		920′		
12′ . 60 =		720′		
Remainder		200′		
10″ . 60 =		10′		
Remainder		190′		
25 . 7′ =		175′		Second term 7′
		15′ =	900″	
Sum			915″	
12′ . 7′ =			84″	
Remainder			831″	
10″ . 7′ =			1″ 10‴	
Remainder			829″ 50‴	Third
25 . 33″ =			825″	term 33″
Remainder			4″ 50‴ =	290‴
12′ . 33″ =				396‴
(too great by)				106‴

Thus the quotient is something less than 60 7′ 33″. It will be observed that, in view of the number of separate denominations, Theon makes his multiplications in three parts and increases correspondingly the number of subtractions.

(ε) *Extraction of the square root.*

We can now see how the problem of extracting the square root of a number would be attacked. The number

EXTRACTION OF SQUARE ROOT

would first be arranged according to denominations, units, tens, hundreds, &c., as before. The first term of the square root would be some number of tens or hundreds or thousands, &c., and would have to be found by trial like the first term of the quotient in a long division. If A is the number the square root of which is required, while a represents the first term or denomination in the square root, it would be necessary to use the identity $(a+x)^2 = a^2 + 2ax + x^2$, and to find x so that $2ax + x^2$ might be less than $A - a^2$, i.e. we have to divide $A - a^2$ by $2a$, allowing for the fact that not only must $2ax$ (where x is the quotient) but also $2ax + x^2$ be less than $A - a^2$. The highest possible value of x satisfying the condition would be found by trial. If that value were b, the further quantity $2ab + b^2$ would have to be subtracted from the first remainder $A - a^2$, and from the second remainder thus left a third term or denomination in the square root would have to be found in like manner; and so on. That this was the procedure is proved by a simple case given by Theon in his commentary on Ptolemy's *Syntaxis*. We have here to find the square root of 144. The highest possible denomination (or power of 10) in the square root is 10; 10^2 subtracted from 144 leaves 44, which must contain not only twice the product of 10 and the next term of the square root but also the square of the next term itself. Now twice 1.10 itself produces 20, and the division of 44 by 20 suggests 2 as the next term of the square root; this turns out to be the exact number required, since $2.20 + 2^2 = 44$.

The same procedure is illustrated by Theon's explanation of Ptolemy's method of extracting square roots according to the sexagesimal system of fractions. The problem is to find approximately the square root of 4500 μοῖραι or *degrees*, and a geometrical figure is used which proves beyond all doubt the essentially Euclidean basis of the whole method.

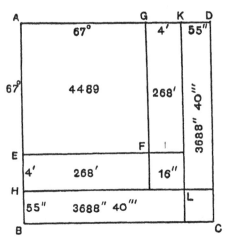

The following arithmetical representation of the purport of the passage, when looked at in the light of the figure, will make the matter clear. Ptolemy has first found the integral part of the square root to be 67. Now $67^2 = 4489$, so that the remainder is 11. Suppose now that the rest of the square root is expressed by means of sexagesimal fractions, and that we may therefore write

$$\sqrt{(4500)} = 67 + \frac{x}{60} + \frac{y}{60^2},$$

where x, y have yet to be found. Thus x must be such that $2 \cdot 67x/60$ is somewhat less than 11, or x must be somewhat less than $\frac{11}{2} \cdot \frac{60}{67}$ or $\frac{330}{67}$, which is itself greater than 4. On trial it turns out that 4 will satisfy the conditions of the problem, namely that $\left(67 + \frac{4}{60}\right)^2$ must be less than 4500,

EXTRACTION OF SQUARE ROOT

so that a remainder will be left by means of which y can be found.

Now this remainder is $11 - \dfrac{2 \cdot 67 \cdot 4}{60} - \left(\dfrac{4}{60}\right)^2$, which reduces to $\dfrac{7424}{60^2}$. We have therefore to find y so that $2\left(67 + \dfrac{4}{60}\right)\dfrac{y}{60^2}$ may be somewhat less than $\dfrac{7424}{60^2}$, that is to say, $8048y$ must be somewhat less than $7424 \cdot 60$. Thus y is approximately equal to 55.

We have then to subtract $2\left(67 + \dfrac{4}{60}\right)\dfrac{55}{60^2} + \left(\dfrac{55}{60^2}\right)^2$, or $\dfrac{442640}{60^3} + \dfrac{3025}{60^4}$, from the remainder $\dfrac{7424}{60^2}$ above found. If we subtract $\dfrac{442640}{60^3}$ from $\dfrac{7424}{60^2}$, we obtain $\dfrac{2800}{60^3}$ or $\dfrac{46}{60^2} + \dfrac{40}{60^3}$; but Theon does not go farther and subtract the remaining $\dfrac{3025}{60^4}$; he merely remarks that the square of $\dfrac{55}{60^2}$ approximates to $\dfrac{46}{60^2} + \dfrac{40}{60^3}$.

III

PYTHAGOREAN ARITHMETIC

PYTHAGORAS was born at Samos about 572 B.C. In 532, or thereabouts, in order to escape the rule of Polycrates, tyrant of Samos, he migrated to southern Italy and founded a society at Croton. Then, when the brotherhood, having involved itself in politics, became the subject of plots and attacks, notably by one Cylon, Pythagoras moved to Metapontium, where he died at a great age (some say 75, others 80).

The early philosophers who were nearly contemporary with Pythagoras recognize his wide learning. Heraclitus, who was hostile to him, admits that he, most of all men, cultivated 'inquiry'. Empedocles was his enthusiastic admirer, calling him a man of prodigious knowledge and the greatest master of skilled arts of every kind. Later, Herodotus speaks of him as 'the most able philosopher among the Greeks'.

The first evidence of his work in mathematics is apparently the statement of Aristotle, in his separate book *On the Pythagoreans*, of which only fragments survive, that 'Pythagoras, the son of Mnesarchus, first worked at mathematics and arithmetic, and afterwards, at one time, condescended to the wonder-working practised by Pherecydes'. Elsewhere Aristotle says (*Metaph. A.* 5) that, 'in the time of Leucippus and Democritus and before them, the so-called Pythagoreans applied themselves to the study of mathematics and were the first to advance that science; insomuch that, having been brought up in it, they thought that its principles must be the principles of all existing things'.

The extent of Pythagoras' own share in the mathematical discoveries of his school will always remain uncertain, for he left no written treatises (so it is said), nor apparently was any account of the Pythagorean doctrines written by any one before Philolaus. Aristotle himself evidently felt the difficulty, since he never attributes any philosophical doctrines to the founder of the school but always refers to 'the Pythagoreans' or 'the so-called Pythagoreans'.

Nevertheless, there is good reason to believe the testimony of Proclus that it was Pythagoras who 'transformed the study of geometry into a liberal education, examining the principles of the science from the beginning'; and we can accept the statement of Aristoxenus that Pythagoras attached supreme importance to the study of arithmetic, 'which he advanced and took out of the region of commercial utility'.

We may take it as certain that Pythagoras himself discovered that musical harmonies depend on numerical ratios, the octave representing the ratio 2 : 1 in length of string at the same tension, the fifth 3 : 2 and the fourth 4 : 3. This capital discovery must have created a deep impression; it would confirm in a striking manner (if it was not what actually suggested) the theory that 'all things are numbers'. This theory appears in different forms. Philolaus said that all things *have* numbers, and that it is this fact which enables them to be known. The earliest form of the theory, however, maintained that things *are* numbers. Aristotle, in one place, says of the Pythagoreans that 'they constructed the whole heaven out of numbers, but not of *monadic* numbers, since they suppose the numbers to have *magnitude*'. It is tempting to suppose that this idea arose from observation of the heavens, which would show that each constellation consists of a certain *number* of stars forming a certain geometrical figure. This would agree with what we learn from Aristotle, who has the interesting remark that one Eurytus, an ancient Pythagorean, 'settled what is the number of what object (e.g. this is the number of a man, that of a horse), and imitated the shapes of living things by pebbles after the manner of those who bring numbers into the forms of triangle or square' (what we call 'figured' numbers).

DEFINITIONS OF THE UNIT AND OF NUMBER

The Pythagoreans connected the unit in arithmetic and the point in geometry by saying that the unit is a 'point without position' (στιγμὴ ἄθετος), and a point is 'a unit having position' (μονὰς θέσιν ἔχουσα). Aristotle observes that the One is reasonably regarded as not itself being a number, since the measure is not measures, but the measure or the One is the beginning (or principle) of number. According to Iamblichus, Thymaridas (an ancient Pythagorean) defined the unit as 'limiting quantity' (περαίνουσα ποσότης) or, as we might say, 'limit of fewness', while some Pythagoreans called it 'the confine between number and parts', i.e. that which separates multiples from submultiples.

The first definition of number is attributed to Thales, who defined it as a 'collection of units' (μονάδων σύστημα), a definition almost identical with Euclid's, namely 'the multitude made up of units'. The Pythagoreans similarly 'made number out of one'; some of them called it 'a progression of multitude beginning from a unit and a regression ending in it'. Eudoxus defined a number as a 'determinate multitude' (πλῆθος ὡρισμένον).

CLASSIFICATION OF NUMBERS

The distinction between *odd* and *even* numbers no doubt goes back to Pythagoras. A Philolaus fragment says that 'number is of two special kinds, odd and even, with a third, even-odd, arising from a mixture of the two; and of each kind there are many forms'. Nicomachus gives the following as the Pythagorean definition of odd and even. 'An *even* number is that which admits of being divided by one and the same operation into the greatest and the least parts, greatest in size but least in number (i.e. into *two* halves) ... while an *odd* number is that which cannot be so divided, but is only divisible into two unequal parts.' Another ancient definition given by Nicomachus says that 'an *even* number is that which can be divided both into two equal parts and into two unequal parts (except the fundamental dyad, which can only be divided into two equal parts), but, however it is divided, must have its two parts *of the same kind* without share in the other kind (i.e. the two parts are either both odd or both even); while an *odd* number is that which, however divided, must in any case fall into two unequal parts, and those parts always belonging to the two *different* kinds respectively (i.e. one being odd and the other even)'.

How far the early Pythagoreans carried the subdivision of Philolaus' three kinds of number, the odd, the even, and the combination of the two, even-odd, we do not know. But they are not likely to have advanced beyond the point of view of Plato and Euclid, which is much the same. Plato has the terms 'even-times even' (ἄρτια ἀρτιάκις), 'odd-times odd' (περιττὰ περιττάκις), 'odd-times even' (ἄρτια περιττάκις), and 'even-times odd' (περιττὰ ἀρτιάκις), by

40   PYTHAGOREAN ARITHMETIC

which he simply means the product of even and even, odd and odd, odd and even, and even and odd factors respectively. Similarly, according to Euclid, an 'even-times even' number is 'a number measured by an even number according to an even number' (i.e. an even number of times), and an 'even-times odd' number 'a number measured by an even number according to an odd number'. Euclid does not seem, any more than Plato, to have distinguished between an 'even-times odd' number and an 'odd-times even' number, or to have troubled about the fact that, as defined, the classes 'even-times even' and 'even-times odd' are not mutually exclusive; for 24, which is 6 times 4 or 4 times 6, is also 8 times 3.

The Neo-Pythagoreans met this point by a new classification. With them the 'even-times even' number is that which has its half even, the half of the half even, and so on until unity is reached; in short, it is a number of the form 2^n. The 'even-odd' number (ἀρτιοπέρισσος, in one word) is the other extreme, a number which can only be halved once, and then leaves an odd number as quotient, i.e. a number of the form $2(2n+1)$. Intermediate is the third class, the 'odd-even' (περισσάρτιος), a number which can be halved twice or more times successively, but then leaves an odd number as quotient, i.e. a number of the form $2^{n+1}(2m+1)$.

Prime or *incomposite* numbers (πρῶτος καὶ ἀσύνθετος) and *secondary* or *composite* numbers (δεύτερος καὶ σύνθετος) are distinguished in a fragment of Speusippus based upon works of Philolaus. Thymaridas the Pythagorean called a *prime* number *rectilinear*, since it can only be set out in one dimension (the only measure of it, except the number itself, being 1). Alternative terms were *euthymetric* and *linear*. Strictly speaking, the prime number should have

'PERFECT' AND 'FRIENDLY' NUMBERS 41

been defined as that which is rectilinear *only*. With Euclid a prime number is 'that which is measured by a unit alone', a composite number 'that which is measured by some number'; numbers *prime to one another* are numbers 'measured by a unit alone as common measure', and numbers *composite to one another* are those 'measured by some number as common measure'. Euclid, as well as Aristotle before him, admitted 2 as a prime number; the Pythagoreans not only excluded 2 from the category of primes, but for them 2, the dyad, was not a number at all, but only the principle of the even, as the unit was the principle of number.

'PERFECT' AND 'FRIENDLY' NUMBERS

'Perfect' numbers in the sense of Euclid's definition (Eucl. VII, Def. 22) appear there for the first time. A 'perfect' number is a number which is equal to (the sum of) all its own parts (i.e. all its factors including 1), e.g.

$$6 = 1+2+3,$$
$$28 = 1+2+4+7+14,$$
$$496 = 1+2+4+8+16+31+62+124+248.$$

Nicomachus knew of these perfect numbers and one other, 8128, and he remarks that they are found in 'ordered' fashion, there being one among the tens, one among the hundreds, and one among the thousands, and that they terminate alternatively in 6 and 8. They do all terminate in 6 or 8 (as can easily be proved by means of the general expression for them established by Euclid, namely $(2^n-1)2^{n-1}$, where 2^n-1 is a prime number), but not alternately, for the fifth and sixth perfect numbers terminate in 6, and the seventh and eighth in 8; and there is no such 'order' in their succession as Nicomachus supposes. The perfect numbers so far discovered, in addition to the four known to Nicomachus, are:

fifth, $2^{12}(2^{13}-1) = 33\ 550\ 336$
sixth, $2^{16}(2^{17}-1) = 8\ 589\ 869\ 056$
seventh, $2^{18}(2^{19}-1) = 137\ 438\ 691\ 328$
eighth, $2^{30}(2^{31}-1) = 2\ 305\ 843\ 008\ 139\ 952\ 128$
ninth, $2^{60}(2^{61}-1) = 2\ 658\ 455\ 991\ 569\ 831\ 744\ 654$
                              $692\ 615\ 953\ 842\ 176$
tenth, $2^{88}(2^{89}-1)$
eleventh, $2^{106}(2^{107}-1)$
twelfth, $2^{126}(2^{127}-1)$.

The first five perfect numbers were known in the fifteenth century; the first eight were calculated by Jean Prestet (d. 1670), Fermat (1601-65) having stated that $2^{31}-1$ is prime, a fact proved later by Euler. The remaining four were discovered in the period from 1886 to the present date.

Two numbers are 'friendly' when each is the sum of all the aliquot parts of the other, e.g. 284 and 220 (for $284 = 1+2+4+5+10+11+20+22+44+55+110$, while $220 = 1+2+4+71+142$. Iamblichus refers the discovery of such numbers to Pythagoras, since, the parts of either number producing the other, the numbers realize Pythagoras' definition of a friend as 'Alter ego'. The subject of 'friendly' numbers was taken up by Euler, who discovered no less than sixty-one pairs of them, after Descartes and van Schooten had found only three.

While with Euclid and later writers the 'perfect' number was the kind of number described above, the Pythagoreans (we are told) made 10 the perfect number. Being the sum of the first four numbers 1, 2, 3, 4, the number 10 formed the 'set of four (numbers)' called *tetractys*. This set of numbers includes the numbers out of which are formed the ratios corresponding to the musical intervals, namely 4 : 3 (the fourth), 3 : 2 (the fifth), and 2 : 1 (the octave). Such virtue was attached to the τετρακτύς that it was for the Pythagoreans their 'greatest oath' and was alternatively called 'Health'. It also gives, when graphically represented by points in four lines one below the other,

.

. .

. . .

. . . .

a triangular number. Hence Lucian's story that Pythagoras once told some one to count, and, when he had said 1, 2, 3, 4, Pythagoras interrupted 'Do you see? What you take for 4 is 10, a perfect triangle, and our oath'.

Speusippus observes further that 10 contains in it the 'linear', 'plane', and 'solid' varieties of number, for 1 is a point, 2 a line, 3 a triangle, and 4 a pyramid. This is easily seen by placing the right number of dots in the proper positions. We are thus brought to the theory of 'figured' numbers, which seems to go back to Pythagoras himself. It is clear that the oldest Pythagoreans were acquainted with the formation of triangular and square numbers by means of dots or pebbles; and we judge from the account of Speusippus' book *On the Pythagorean Numbers*, which was founded on works of Philolaus, that the latter dealt with linear numbers, polygonal numbers, and plane and solid numbers of all sorts.

FIGURED NUMBERS

(a) *Triangular Numbers.*

It was probably Pythagoras who discovered that the sum of any number of successive terms of the series of natural numbers is a triangular number. This is seen diagrammatically thus:

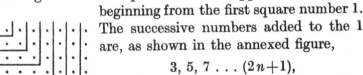

In general, $1+2+3+\ldots+n = \tfrac{1}{2}n(n+1)$ is a triangular number with side n.

(b) Square numbers and gnomons.

It is easy to see that, if we have a number of dots forming and filling up a square (say 16, as in the accompanying figure), the next higher square, the square of 5, can be formed by adding rows of dots round two sides of the original square as shown. The number of dots so added is $2.4+1 = 9$. This process of forming successive squares can be applied throughout, beginning from the first square number 1. The successive numbers added to the 1 are, as shown in the annexed figure,

$$3, 5, 7 \ldots (2n+1),$$

that is to say, the successive odd numbers. The method of formation shows that the sum of any number of successive terms of the series of odd numbers 1, 3, 5, 7 ... (starting from 1) is a square, and in fact $1+3+5+\ldots+(2n-1) = n^2$, while the addition of the next odd number $2n+1$ makes the next higher square, $(n+1)^2$. All this was known to Pythagoras. The odd numbers successively added were called *gnomons*. Aristotle speaks of gnomons placed round 1 which now produce different figures every time (oblong figures, each dissimilar to the preceding one), now preserve one and the same figure (squares); the latter is the case with the gnomons now in question.

The history of the term *gnomon* is interesting. The gnomon was originally the upright stick which cast shadows on the plane or hemispherical surface of a sundial to mark the hours; and 'marker' or 'pointer' is the literal meaning of the word. It connoted perpendicularity; hence we find Oenopides of Chios calling a perpendicular a straight line drawn 'gnomon-wise' (κατὰ γνώμονα). On a like ground the term is used in Theognis to describe an instrument for drawing right angles, like a carpenter's square. The transition was natural to the figure which remains over in a square when a smaller square is cut out of it (or the figure which, as Aristotle says, when added to a square preserves the shape and makes a larger square). In Euclid, Book II, it has a slightly wider meaning, covering not only the gnomon which is part of a square, but the figure which is similarly related to any parallelogram. Later still, Heron 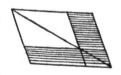 of Alexandria defines a gnomon as that which, when added to anything, number or figure, makes the whole similar to that to which it is added.

(c) *Gnomons of the polygonal numbers.*

In accordance with the generalized meaning of the word 'gnomon', all the polygonal numbers, like triangular and square numbers, are formed by adding gnomons successively to 1, which is potentially the first polygonal number of any form. The method of formation in the case of the pentagonal and hexagonal numbers is shown in the accompanying figures, and it is seen that the successive gnomons are in the case of the pentagon 4, 7, 10 ... with

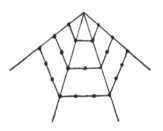 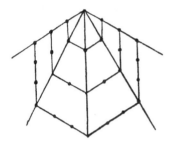

the common difference 3, and in the case of the hexagon 5, 9, 13 ... with the common difference 4. In general, the successive gnomonic numbers for any polygonal number, say, of n sides have $n-2$ for their common difference.

(d) *Right-angled triangles with sides in rational numbers.*

Pythagoras was aware that, while $3^2+4^2=5^2$, any triangle with its sides in the ratio of the numbers 3, 4, 5 is right angled. It would be natural to inquire whether there were other squares besides 5^2 which are the sum of two squares, or, in other words, to seek other sets of three integral numbers which can be made the sides of right-angled triangles; and here we have the beginning of the *indeterminate analysis* which culminated in the work of Diophantus. Since the sum of any number of successive terms of the series 1, 3, 5, 7 ... is a square, it is only necessary to pick out of this series the odd numbers which are themselves squares; for if we take one of these, say 9, the addition of this square to the sum of the numbers which precede it in the series (which sum is a square) makes the square number which is the sum of the terms of the series up to and including 9. It would be natural to seek a formula which should enable all sets of three

RIGHT-ANGLED TRIANGLES IN NUMBERS

numbers of this kind to be written down, and such a formula is actually attributed to Pythagoras. This formula amounts to the statement that, if m be any odd number,

$$m^2 + \{\tfrac{1}{2}(m^2-1)\}^2 = \{\tfrac{1}{2}(m^2+1)\}^2.$$

Pythagoras would presumably arrive at this rule of formation in this way. Observing that the gnomon put round the figure representing n^2 is $2n+1$, he would have to make $2n+1$ a square.

If we suppose that $2n+1 = m^2$,

we obtain $n = \tfrac{1}{2}(m^2-1)$,

and therefore $n+1 = \tfrac{1}{2}(m^2+1)$.

It follows that

$$m^2 + \{\tfrac{1}{2}(m^2-1)\}^2 = \{\tfrac{1}{2}(m^2+1)\}^2.$$

Another formula devised for the same purpose is attributed to Plato, namely

$$(2m)^2 + (m^2-1)^2 = (m^2+1)^2.$$

This is not equivalent to the result of multiplying Pythagoras' formula by 4 throughout, because m need not here be odd, as it must in Pythagoras' formula. But it may have been obtained, like the other, by consideration of gnomons in a figure. Consider the square with n dots in its side in relation both to the next smaller square, $(n-1)^2$, and to the next larger, $(n+1)^2$. Then n^2 exceeds $(n-1)^2$ by the gnomon $2n-1$, but falls short of $(n+1)^2$ by the gnomon $2n+1$. Therefore the square $(n+1)^2$ exceeds the square $(n-1)^2$ by the sum of the two gnomons $2n-1$ and $2n+1$, which is $4n$.

That is, $4n + (n-1)^2 = (n+1)^2$,

and, substituting m^2 for n in order to make $4n$ a square, we have the Platonic formula
$$(2m)^2 + (m^2-1)^2 = (m^2+1)^2.$$
The formulae of Pythagoras and Plato supplement each other. Euclid's solution of the problem (Eucl. X, Lemma following Prop. 28) is more general, amounting to the following.

If AB be a straight line bisected at C and produced to D, then (Eucl. II. 6)
$$AD \cdot DB + CB^2 = CD^2,$$
which we may write thus,
$$uv = c^2 - b^2,$$
where
$$u = c+b, \quad v = c-b,$$
and consequently
$$c = \tfrac{1}{2}(u+v), \quad b = \tfrac{1}{2}(u-v).$$

Now, says Euclid, in order that uv may be a square, both u and v must either be squares, or 'similar plane numbers', and further they must be either both odd or both even in order that b and c may be whole numbers. 'Similar plane numbers' are of course numbers which are the product of two factors proportional in pairs, as $mp \cdot np$ and $mq \cdot nq$, or mnp^2 and mnq^2. Provided, therefore, that these numbers are both even or both odd,
$$m^2 n^2 p^2 q^2 + \{\tfrac{1}{2}(mnp^2 - mnq^2)\}^2 = \{\tfrac{1}{2}(mnp^2 + mnq^2)\}^2,$$
which is the formula required.

(e) *Oblong numbers.*

Just as the sum of any number of terms of the series of natural numbers 1, 2, 3 . . . is a triangular number, and the sum of any number of terms of the series of odd

OBLONG NUMBERS

numbers 1, 3, 5 ... is a square, so the sum of any number of terms of the series of even numbers, 2, 4, 6 ... is an 'oblong' number (ἑτερομήκης) with sides or factors differing by 1. In fact the sum of the series

$$2+4+6\ldots+2n=n(n+1).$$

It cannot be doubted that the earliest Pythagoreans, if not Pythagoras himself, discovered this fact. They would also observe that the 'oblong' number is double of a triangular number. These facts would be brought out by taking two dots representing 2, and then placing round them, as successive gnomons, dots representing the even numbers 4, 6 ... thus:

The successive oblong numbers are

$$1.2=2, \quad 2.3=6, \; 3.4=12, \ldots n(n+1) \ldots$$

No two of these numbers are 'similar', for the ratio $n : (n+1)$ is different for all different values of n. As Aristotle says, the addition of each of these gnomons (4, 6, 8 ...) changes the shape of the figure every time.

It should be noted that the Greek word for 'oblong', ἑτερομήκης, would literally cover any number which is the product of any two unequal numbers, and it is used by Plato and Aristotle in this sense; it came, however, to be restricted to numbers the factors of which differ by 1, i.e. numbers of the form $n(n+1)$. In Theon of Smyrna and Nicomachus a number of the form $m(m+n)$ where $n > 1$ is called 'prolate' (προμήκης).

It is obvious that any oblong number $n(n+1)$ is the sum of two equal triangular numbers,

and that any square number is the sum of two successive triangular numbers. The annexed diagrams illustrate these facts; the second is proved by the identity

$$\tfrac{1}{2}n(n-1)+\tfrac{1}{2}n(n+1)=n^2.$$

Another theorem connecting triangular numbers and squares is to the effect that 8 times any triangular number $+1$ makes a square. The theorem is quoted by Plutarch and assumed by Diophantus; it may easily go back to the Pythagoreans. In our notation

$$8 \cdot \tfrac{1}{2}n(n+1)+1 = 4n(n+1)+1 = (2n+1)^2.$$

The truth of this theorem also would be obvious from a figure made up of dots in the usual way. The annexed figure shows a square with side 7 divided into four equal 'oblong' numbers with a single dot in the middle, and each of the four 'oblongs' gives two equal triangular numbers.

Speusippus and Philippus of Opus (fourth century B.C.) are said to have written on polygonal numbers; and Hypsicles, who wrote about 170 B.C., is twice mentioned by Diophantus as the author of a 'definition' of a polygonal number.

PYTHAGOREAN ARITHMETIC

THE THEORY OF PROPORTION AND MEANS

In an often-quoted passage of his 'summary' Proclus says that Pythagoras discovered the theory of *proportionals* and the construction of the cosmic figures (the five regular solids); for it seems now to be generally agreed that we should read 'proportionals' (τῶν ἀνάλογον) instead of the variant 'irrationals' (τῶν ἀλόγων). The theory of means in particular was developed very early in the Pythagorean school in connexion with arithmetic and the theory of music. We are told that in Pythagoras' time there were three means, the arithmetic, the geometric, and the subcontrary, and that the name of the third, the subcontrary, was changed by Archytas and Hippasus to 'harmonic'. A fragment by Archytas *On Music* defines the three as follows: we have the *arithmetic* mean when, of three terms, the first exceeds the second by the same amount as the second exceeds the third; the *geometric* mean when the first is to the second as the second is to the third; the '*subcontrary* which we call *harmonic*' when the three terms are such that, by whatever part of itself the first exceeds the second, the second exceeds the third by the same part of the third. That is, if b is the harmonic mean between a and c, and if $a = b + \dfrac{a}{n}$, then $b = c + \dfrac{c}{n}$, whence in fact

$$\frac{a-b}{b-c} = \frac{a}{c}, \text{ or } \frac{1}{c} - \frac{1}{b} = \frac{1}{b} - \frac{1}{a}.$$

Philolaus is said to have called the cube a 'geometrical harmony' because it has 12 edges, 8 angles, and 6 faces, and 8 is, in harmonics, the mean between 12 and 6.

Iamblichus, after Nicomachus, mentions a special 'most perfect proportion', consisting of four terms and called 'musical', which was discovered by the Babylonians and first introduced into Greece by Pythagoras. It was, he says, used by Aristaeus of Croton, Timaeus of Locri, Philolaus and Archytas (among other Pythagoreans), and finally by Plato in his *Timaeus*. The proportion is

$$a : \frac{a+b}{2} = \frac{2ab}{a+b} : b,$$

a particular case being $12 : 9 = 8 : 6$. The two middle terms are the arithmetic and harmonic means between the extremes.

The theory of means was further developed in the school, seven others being added from time to time to the first three, making ten in all. The fourth, fifth, and sixth are credited, at least in part, to Archytas and Hippasus, or alternatively to Eudoxus; the last four are said to have been added by two later Pythagoreans, Myonides and Euphranor.

The ten means are described by Nicomachus and Pappus, whose accounts only differ as regards one of the ten. They need not all be set out here. Their nature will be understood from the following explanation. If $a > b > c$, the following formulae show the first three means, the arithmetic, geometric, and harmonic.

Formula. Equivalent.

(1) $\dfrac{a-b}{b-c} = \dfrac{a}{a} = \dfrac{b}{b} = \dfrac{c}{c}$ $a + c = 2b$ (arithmetic).

(2) $\dfrac{a-b}{b-c} = \dfrac{a}{b}$ $ac = b^2$ (geometric).

(3) $\dfrac{a-b}{b-c} = \dfrac{a}{c}$ $\dfrac{1}{a} + \dfrac{1}{c} = \dfrac{2}{b}$ (harmonic).

The next three are obtained by varying the right-hand

side in these formulae and equating $(a-b)/(b-c)$ to c/a, c/b, and b/a respectively. Three more are obtained by changing the left-hand side to $(a-c)/(a-b)$ and equating this to a/c, b/c, and a/b respectively; and a tenth arises from equating $(a-c)/(b-c)$ to b/c.

Geometric means.

Plato speaks in the *Timaeus* of the geometric means between two squares and two cubes, observing that between two 'planes' one such mean suffices, but to connect two 'solids' two means are necessary. This is equivalent to saying that, if p^2, q^2 are two square numbers,

$$p^2 : pq = pq : q^2,$$

while, if p^3, q^3 be two cube numbers,

$$p^3 : p^2q = p^2q : pq^2 = pq^2 : q^3,$$

the one mean in the first case being pq, and the two means in continued proportion in the second case being p^2q, pq^2. Nicomachus quotes the substance of Plato's remark as a 'Platonic theorem', and adds in explanation the equivalent of Eucl. VIII. 11, 12; but the theorem no doubt goes back to the Pythagoreans.

Boëtius has preserved in his *De inst. musica* a proof by Archytas of an interesting theorem about geometric means, namely that, if we have two numbers in the ratio known as ἐπιμόριος or *superparticularis*, i.e. in the ratio of $n+1$ to n, there can be no number which is a mean proportional between the two numbers. The theorem is Prop. 3 in the *Sectio Canonis* attributed to Euclid, and Archytas' proof is substantially identical with that in the *Sectio* (see pp. 136–7, *post*). This fact creates a presumption that there existed, at least as early as the date of Archytas (say 430–365 B.C.), an *Elements of Arithmetic* in the form which we call Euclidean.

THE IRRATIONAL

The subject of irrationals in general was for the Greeks a part of geometry rather than arithmetic, and necessarily so, because, for want of notation, an irrational of any sort could only be denoted by a straight line or a combination of lines. This is illustrated by Euclid's Book X on irrationals simple and compound; these are always irrational *straight lines*, and the whole treatment of the subject is geometrical. The first discovery of the existence of the irrational must, however, have been made as the result of arithmetical considerations or reasoning with numbers. It is certain that it was made with reference to the diagonal of a square in relation to its side, that is to say, the first irrational or incommensurable to be discovered was the equivalent of what we write as $\sqrt{2}$. The discovery can hardly have been made by Pythagoras himself, but it was certainly made in his school. The approximate date can only be conjectured. According to Plato's *Theaetetus*, Theodorus of Cyrene was the first to prove the irrationality of what we write as $\sqrt{3}, \sqrt{5} \ldots \sqrt{17}$; and we may infer that the irrationality of $\sqrt{2}$ had already been proved before Theodorus' time. Theodorus was Plato's teacher in mathematics, and the irrationality of $\sqrt{2}$ is unmistakably alluded to in the *Republic* as a thing well known, for there Plato speaks of 'the rational diameter of 5' (by '5' he means the square having 5 for its side) as being 7 ($= \sqrt{49}$), in contradistinction to the 'irrational diameter of 5' which is what we should write as $\sqrt{50}$, the 'rational diameter' (7) being an approximation to this. Again, there is a well-attested title of a work by Democritus (born 470 or 460 B.C.) περὶ ἀλόγων γραμμῶν καὶ ναστῶν, *On irrational straight lines and solids* (atoms); and it is difficult to resist

INCOMMENSURABILITY OF DIAGONAL

the conclusion that the irrationality of $\sqrt{2}$ was discovered before Democritus' time. The traditional proof of it, given by Aristotle and repeated in a proposition interpolated in Euclid, Book X, is by a *reductio ad absurdum* showing that, if the diagonal of a square is commensurable with its side, it will follow that one and the same number is both odd and even. The proof is substantially as follows.

Suppose the diagonal AC of a square to be commensurable with AB the side; and let $\alpha : \beta$ be their ratio expressed in the least possible numbers.

Then $\alpha > \beta$, and therefore α is necessarily > 1.

Now $AC^2 : AB^2 = \alpha^2 : \beta^2$,

and, since $AC^2 = 2AB^2$, $\alpha^2 = 2\beta^2$.

Hence α^2, and therefore α, is even.

Since $\alpha : \beta$ is a ratio in its lowest terms, it follows that β must be *odd*.

Let $\alpha = 2\gamma$; therefore $4\gamma^2 = 2\beta^2$ or $2\gamma^2 = \beta^2$, so that β^2, and therefore β, is *even*.

But β was also odd: which is impossible.

Therefore the diagonal AC cannot be commensurable with the side AB.

ALGEBRAIC EQUATIONS

(a) *The indeterminate equations of the second degree*
$$2x^2 - y^2 = \pm 1.$$

Not only did the Pythagoreans prove the incommensurability of $\sqrt{2}$ with 1, but they showed how to find any number of successive approximations to the value of $\sqrt{2}$ by finding any number of integral solutions of the above equations. The pairs of values of x, y were called 'side-' and 'diameter-' (diagonal-) 'numbers' respectively and, as the values increase, the ratio of y to x approximates more and more closely to $\sqrt{2}$. Theon of Smyrna explains the formation of the series of 'side-' and 'diameter-' numbers thus. We will denote the successive 'diameters' by $d_1, d_2 \ldots$ and the corresponding 'sides' by $a_1, a_2 \ldots$ Theon begins by making $a_1 = 1, d_1 = 1$ (these values satisfy the above equation with the positive sign). The second 'side' and 'diameter' are formed from the first, the third from the second, and so on, according to the following scheme:

$$a_2 = a_1 + d_1, \quad d_2 = 2a_1 + d_1,$$
$$a_3 = a_2 + d_2, \quad d_3 = 2a_2 + d_2,$$
$$\cdots \cdots \cdots \cdots$$
$$a_{n+1} = a_n + d_n, \quad d_{n+1} = 2a_n + d_n.$$

Since $a_1 = d_1 = 1$, it follows that

$$a_2 = 1+1 = 2, \quad d_2 = 2 \cdot 1+1 = 3,$$
$$a_3 = 2+3 = 5, \quad d_3 = 2 \cdot 2+3 = 7,$$
$$a_4 = 5+7 = 12, \quad d_4 = 2 \cdot 5+7 = 17,$$

and so on.

Theon states, with reference to these numbers, the general proposition that

$$d_n^2 = 2a_n^2 \pm 1,$$

and observes (1) that the signs alternate as successive d's and a's are taken, $d_1^2 - 2a_1^2$ being equal to -1, $d_2^2 - 2a_2^2$ equal to $+1$, $d_3^2 - 2a_3^2$ equal to -1, and so on, while (2) the sum of the squares of *all* the d's will be double of the sum of the squares of all the a's. (If the number of successive terms in each series is finite, the number must be even.)

The properties stated depend on the truth of the identity

$$(2x+y)^2 - 2(x+y)^2 = 2x^2 - y^2;$$

for, if x, y be numbers which satisfy one of the two

equations $2x^2-y^2=\pm 1$, the formula (if true) gives us two higher numbers, $(x+y)$ and $(2x+y)$, which satisfy the other equation.

Not only is the identity true, but we know how it was proved; for Proclus tells us, in his *Commentary on the Republic of Plato*, that Euclid proved it graphically in the second Book of the *Elements*, namely in II. 10, a proposition to the effect that, if AB is bisected in C and produced to D, then

$$AD^2+DB^2=2AC^2+2CD^2.$$

If $AC=CB=x$, and $BD=y$, this proposition gives

$$(2x+y)^2+y^2=2x^2+2(x+y)^2,$$
or $$(2x+y)^2-2(x+y)^2=2x^2-y^2,$$

which is the formula required.

The general property is easily proved algebraically thus:

$$\begin{aligned}d_n^2-2a_n^2 &= (2a_{n-1}+d_{n-1})^2-2(a_{n-1}+d_{n-1})^2\\ &= 2a_{n-1}^2-d_{n-1}^2\\ &= -(d_{n-1}^2-2a_{n-1}^2)\\ &= +(d_{n-2}^2-2a_{n-2}^2), \text{ in like manner;}\end{aligned}$$

and so on.

As the proof of the property is contained in Eucl. II. 10, it is a fair inference that that theorem is Pythagorean, and may even have been formulated for the specific purpose.

(β) *The ἐπάνθημα ('bloom') of Thymaridas.*

Thymaridas of Paros, an ancient Pythagorean, is famous as the author of a rule for solving a certain set of n simultaneous simple equations connecting n unknown quantities; so well known was it that it went by the name of the 'flower' or 'bloom' of Thymaridas. The rule is stated in general terms, but the substance is pure algebra. The known quantities in the equations are ὡρισμένα (determinate), the unknown ἀόριστα (undetermined): cf. Diophantus' definition of his unknown as an 'undefined or undetermined number of units'.

The rule states in effect that, if we have the following n equations connecting the n unknowns $x, x_1 \ldots x_{n-1}$,

$$x+x_1+x_2+\ldots+x_{n-1}=s,$$
$$x+x_1=a_1,$$
$$x+x_2=a_2,$$
$$\cdot\quad\cdot\quad\cdot\quad\cdot$$
$$x+x_{n-1}=a_{n-1},$$

the solution for x is

$$x=\frac{(a_1+a_2+\ldots+a_{n-1})-s}{n-2}$$

Iamblichus, our informant, goes on to show that other types of equations can be reduced to this, so that the rule does not 'leave us in the lurch' in those cases either. He gives as an instance the indeterminate problem represented by the following three equations connecting four unknown quantities:

$$x+y=a\,(z+u),$$
$$x+z=b\,(u+y),$$
$$x+u=c\,(y+z).$$

From these equations we obtain

$$x+y+z+u=(a+1)(z+u)=(b+1)(u+y)=(c+1)(y+z).$$

If now x, y, z, u are all to be integers, $x+y+z+u$ must contain $a+1, b+1, c+1$ as factors. If L be the least common multiple of $a+1$, $b+1$, $c+1$, we can put

$x+y+z+u=L$, and we obtain from the above equations, in pairs,

$$x+y = \frac{a}{a+1}L,$$

$$x+z = \frac{b}{b+1}L,$$

$$x+u = \frac{c}{c+1}L,$$

while $x+y+z+u=L$.

These equations are of the type to which Thymaridas' rule applies, and, since the number of unknown quantities is 4, $n-2$ is in this case 2, and

$$x = \frac{L\left(\frac{a}{a+1}+\frac{b}{b+1}+\frac{c}{c+1}\right)-L}{2}$$

The numerator is integral, but it may be an odd number, in which case we must substitute $2L$ for L as the value of $x+y+z+u$.

Iamblichus has the particular case where $a=2$, $b=3$, $c=4$. L is thus $3 \cdot 4 \cdot 5 = 60$, and the numerator of the above expression becomes $133-60$, or 73, an odd number; hence we take $2L$ or 120 in place of L, and so obtain $x=73$, $y=7$, $z=17$, $u=23$.

Iamblichus goes on to apply the method to the equations

$$x+y = \frac{3}{2}(z+u),$$

$$x+z = \frac{4}{3}(u+y),$$

$$x+u = \frac{5}{4}(y+z),$$

which give
$$(x+y+z+u) = \frac{5}{2}(z+u) = \frac{7}{3}(u+y) = \frac{9}{4}(y+z),$$
whence $x+y+z+u = \frac{5}{3}(x+y) = \frac{7}{4}(x+z) = \frac{9}{5}(x+u).$

In this case we take L, the least common multiple of 5, 7, 9, or 315, and put
$$x+y+z+u = L = 315,$$
so that
$$x+y = \frac{3}{5}L = 189,$$
$$x+z = \frac{4}{7}L = 180,$$
$$x+u = \frac{5}{9}L = 175;$$
therefore
$$x = \frac{544-315}{2} = \frac{229}{2}.$$

In order that x may be integral, we take $2L$, or 630, instead of L, and the solution is $x=229$, $y=149$, $z=131$, $u=121$.

(γ) *Equation* $xy = 2(x+y)$..

It would appear that the Pythagoreans considered the equivalent of this equation, for in the *Theologumena Arithmetices* it is observed that 16 is the only square the area of which is numerically equal to its perimeter. This corresponds to the solution $x=y=4$. If we write the equation in the form $(x-2)(y-2) = 4$, we see that the integral solutions are obtained by equating $x-2$, $y-2$ to the respective factors of 4. These factors are (2, 2) and (4, 1). The only possible values of x, y are thus found to be (4, 4) or (3, 6). Therefore, besides the square, the only rectangle having the property in question is 3.6.

MANUALS OF 'ARITHMETIC'

The treatises on the Pythagorean theory of numbers which have survived are the work of Neo-Pythagoreans, though there is little in their content that does not go back to the immediate successors of Pythagoras. The first is the *Introductio arithmetica* of Nicomachus. Nicomachus of Gerasa, a city in Judaea beyond Jordan, flourished about A.D. 100. Of his life nothing is known. He wrote other works, one of which, the *Enchiridion Harmonices* or *Handbook of Harmony*, has survived. A *Theologumena Arithmetices*, or treatise on the theology or the mystic properties of numbers, is also attributed to him; this is not extant in its original form, but it cannot be doubted that the compilation with the same title which has come down to us, and which was edited by Ast along with the *Introductio* in 1817, contains extracts from the original work of Nicomachus as it does fragments from Speusippus and Anatolius, Bishop of Laodicea (A.D. 270). An *Introductio geometrica* is also mentioned by Nicomachus himself, who says in one place, with regard to certain solid numbers, that they were specially treated in it; but it does not necessarily follow that the treatise was his own.

The *Introductio arithmetica* with which we are here concerned deals in great part with the same subjects as the arithmetical Books (VII–IX) of Euclid's *Elements*. But the treatment is wholly different. In Euclid numbers are represented by straight lines with letters attached. Since a straight line can represent any number, the notation is as general as, though less concise than, our algebraical notation; and Euclid's proofs are general and scientific. In Nicomachus numbers are no longer denoted by straight lines, so that, when undetermined numbers have to be expressed or distinguished, it is necessary to use ordinary language, which makes enunciations cumbrous and hard to follow. Nicomachus has little or nothing in the shape of proofs in the proper sense of the term. As a rule, he merely states a proposition, and then illustrates it by means of particular numbers. Mathematically speaking, therefore, Nicomachus' treatise (unlike Euclid's) has little scientific value. Indeed Nicomachus seems to have been a philosopher rather than a mathematician. His object was apparently to write a popular treatise which should arouse in the beginner an interest in the theory of numbers by making him acquainted with the more obvious properties which they had been proved to possess. But he himself was more interested in the mystic properties of numbers which appealed to the philosopher; hence his high-flown and rhetorical language when he is stating even the most obvious relations of numbers. It is difficult to account for the success of the work unless on the assumption that it was at first read by philosophers rather than mathematicians (Pappus evidently despised it), and afterwards became popular at a time when there were no mathematicians left, but only philosophers who incidentally took an interest in mathematics. But a success it was. This is shown by the fact that it was translated into Latin by Apuleius of Madaura (born about A.D. 125), and again by Boëtius, and that commentaries on it were written by Iamblichus, Heronas, Asclepius of Tralles, Johannes Philoponus, and Proclus. Its vogue is further attested by the allusion in Lucian's *Philopatris*, 'You calculate like Nicomachus', a remark which, in view of its context, seems like a jibe rather than a compliment. Both the *Introductio* and the commentary of Iamblichus have been edited in the Teubner series, the

INTRODUCTIO OF NICOMACHUS

former by Hoche and the latter by Pistelli. We have now an elaborate English translation by Martin Luther D'Ooge, edited, with studies in Greek arithmetic, by F. E. Robbins and L. C. Karpinski (University of Michigan Studies, vol. xvi, 1926).

Book I of the *Introductio* begins with generalities, partly philosophical, on the relation of the four subjects of the *quadrivium*, arithmetic, geometry, astronomy, and music. Then (c. 7) we have definitions of number, even and odd, next (cc. 8–10) the distinctions between three kinds of even numbers and (cc. 11–13) three kinds of odd numbers. Incidentally c. 13 explains the method of Eratosthenes' *sieve* (κόσκινον), a device for finding prime numbers. We set out the series of odd numbers beginning with 3,

$$3, 5, 7, 9, 11, 13, 15, 17, 19, 21, 23, 25 \ldots$$

Now 3 is a prime number, but multiples of it are not; the multiples forming part of the series, namely 9, 15, 21 ... are obtained by passing over two numbers at a time beginning from 3; we strike out these multiples as not being prime. Similarly 5 is a prime number, but, by passing over *four* numbers at a time beginning from 5, we obtain multiples of 5, namely 15, 25 ...; these again we strike out for a like reason. In general, if n be a prime number, we find its multiples in the series by passing over $n-1$ terms at a time, and we strike out these multiples. When we have gone far enough with this process, the numbers which are still left will be primes. It is clear, however, that this primitive method would become useless as soon as we reached numbers of any considerable size. The same c. 13 gives the rule for finding out whether two numbers are prime to one another; it is the method of Eucl. VII. 1, equivalent to our rule for finding the greatest common measure, but is described in general terms. The next chapters (cc. 14–16) discuss *perfect* numbers and distinguish them from *over-perfect* (ὑπερτελής) and *deficient* (ἐλλιπής) numbers. As regards Nicomachus' knowledge of perfect numbers see above (p. 41). An 'over-perfect' number is a number such that the sum of all its aliquot parts is greater, and a 'deficient' number a number such that the sum of all its aliquot parts is less, than the number itself.

Having considered numbers by themselves, Nicomachus next considers numbers in relation to one another. One number may be greater than, equal to, or less than, another. If it is greater, it may be greater, says Nicomachus, in one of five different ways according to the particular ratio (greater than 1) that it bears to it; and the reciprocals of these ratios give five corresponding different ways in which one number may be less than another.

1. The greater number may be a *multiple* (πολλαπλάσιος) of the lesser, in which case the lesser is a *submultiple* (ὑποπολλαπλάσιος) of the greater. Particular multiples are the *double* (διπλάσιος), *triple* (τριπλάσιος), &c.

2. If a, b are the greater and lesser numbers respectively, we may have

$$a = \left(1 + \frac{1}{n}\right)b \text{ or } \frac{n+1}{n}b, \text{ so that } b = \frac{n}{n+1}a,$$

where n is any integer. These ratios are called respectively ἐπιμόριος or *superparticularis* (Boëtius) and ὑπεπιμόριος (*subsuperparticularis*). There were special names for the particular cases where n is 2, 3, &c.; when n is 2, the ratio is ἡμιόλιος (*sesquialter*), when n is 3 ἐπίτριτος (*sesquitertius*, &c.), and the reciprocals are the same with ὑπ(ό) or *sub* prefixed. ἐπίτριτος means of course 'one-third on', and so with the other cases.

3. If $a = \left(1+\dfrac{m}{n}\right)b$ $(m < n)$, and therefore $b = \dfrac{n}{n+m}a$, the ratios were called ἐπιμερής (*superpartiens*) and ὑπεπιμερής (*subsuperpartiens*) respectively.

Two subdivisions were distinguished in which

(1) $m = n-1$, so that $m/n = \dfrac{2}{3}, \dfrac{3}{4}$, &c.

(2) $m < n-1$, e.g. $m/n = \dfrac{2}{5}, \dfrac{3}{7}$, and the like.

4 and 5. In these cases the greater number contains, not *once* the number, but a *multiple* of it, plus a fraction. That is, $a = \left(p+\dfrac{1}{n}\right)b$, or $a = \left(p+\dfrac{m}{n}\right)b$, and there are subdivisions of the latter according as $m = n-1$ or $< n-1$. The word for the ratio $p+\dfrac{1}{n}$ is πολλαπλασιεπιμόριος, and that for $p+\dfrac{m}{n}$ is πολλαπλασιεπιμερής, and so on.

In c. 23 Nicomachus illustrates the fact that, if a, b, c be three numbers in geometrical progression with one of the above ratios as common ratio, three terms formed as follows, namely

$$a,\ a+b,\ a+2b+c,$$
or
$$c,\ c+b,\ c+2b+a,$$

are in geometrical progression with one or other of the above types of ratio as common ratio. Suppose, for example, that the original terms are a, na, n^2a. The first transformation gives three terms $a, a(n+1), a(n+1)^2$ with common ratio $n+1$; the second gives three terms $n^2a, n(n+1)a, (n+1)^2a$, the common ratio being $(n+1)/n$; and so on.

66 PYTHAGOREAN ARITHMETIC

Book II begins with a similar transformation. If a, b, c be three numbers in ascending geometrical progression with any common ratio other than 1, and if we form three numbers
$$a, \quad b-a, \quad c+a-2b,$$
we have three more terms in geometrical progression, but with a less common ratio. E.g. if the original terms are a, na, n^2a with n as common ratio, the transformation gives $a, (n-1)a, (n-1)^2a$, three terms in geometrical progression but with $n-1$ as common ratio; and by applying the process the necessary number of times we shall arrive at numbers in the ratio of equality.

C. 6 is preliminary to the subject of polygonal numbers and their respective gnomons. The numbers are shown graphically, with a's for dots, as above explained. There are no general proofs, but triangular numbers, squares, pentagonal, hexagonal, and heptagonal numbers are exhibited, and it is shown that the corresponding gnomons are the successive terms after 1 in the arithmetical progressions which have 1 for the first term and the numbers 1, 2, 3, 4, 5 respectively for common difference (cf. pp. 44-6 above). That is, the successive gnomons for triangles are 2, 3, 4 ..., for squares 3, 5, 7 ..., for pentagons 4, 7, 10 ..., and so on. By the aid of modern notation we may generalize thus. The gnomons for polygonal numbers of a sides are
$$1+a-2, \; 1+2(a-2), \; 1+3(a-2) \ldots$$
and an a-gonal number with n in its side is
$$1+\{1+(a-2)\}+\{1+2(a-2)\}+\ldots+\{1+(n-1)(a-2)\}$$
$$=n+\tfrac{1}{2}n(n-1)(a-2).$$

C. 12 mentions that a square is the sum of two consecutive triangular numbers, and that, if we add to any polygonal number a certain triangular number, we obtain a polygonal number with one more side. In fact, an a-gonal number of side n plus the triangular number of side $n-1$ gives the $(a+1)$-gonal number of side n, for

$$n+\tfrac{1}{2}n(n-1)(a-2)+\tfrac{1}{2}n1(n-)=n+\tfrac{1}{2}n(n-1)\{(a+1)-2\}.$$

In c. 13 Nicomachus passes to the first *solid* number, the *pyramid*. The base of the pyramid may be a triangular number, a square, or any polygonal number, but Nicomachus only mentions the first triangular pyramids 1, 4, 10, 20, 35, 56, 84, and explains (c. 14) the formation of pyramids on square bases. We may generalize as follows. An a-gonal number with n in its side is, as we have seen,

$$n+\tfrac{1}{2}n(n-1)(a-2).$$

It follows that the pyramidal number with that polygonal number for base is

$$1+2+3+\ldots+n+\tfrac{1}{2}(a-2)\{1.2+2.3+\ldots+(n-1)n\}$$
$$=\tfrac{1}{2}n(n+1)+\tfrac{1}{2}(a-2)\cdot\tfrac{1}{3}(n-1)n(n+1).$$

In c. 14 Nicomachus speaks of pyramids *once, twice, or thrice truncated* (κόλουρος, δικόλουρος, τρικόλουρος). These are pyramids in which we have cut off (*a*) the unit at the top, (*b*) the unit and the next layer, (*c*) the unit and the next two layers respectively.

Other solid numbers are classified (cc. 15–17). Cubes are the products of three equal numbers; a *scalene* solid number is the product of three numbers all unequal. There were a variety of other names according to the relations between the factors: e.g. a *beam* (δοκίς) or *column* (στηλίς) has a square base while the height is greater than the side of the square, but in a *tile* (πλινθίς) the base is a square while the other edge is less than the side of the square. Cubes the last digit of which is the same as the last digit of the side (the sides and cubes in these cases end in 1, 5, or 6) are called *spherical* or *recurrent* (ἀποκαταστατικοί); the squares of the same numbers end in the same digits and are called *circular* (κυκλικοί).

Sum of series of cube numbers.

In c. 20 we have the interesting statement that, if we set out the series of odd numbers

1, 3, 5, 7, 9, 11, 13, 15, 17, 19 ...,

the first, namely 1, is a cube; the sum of the next *two*, 3+5, is a cube; the sum of the next *three*, 7+9+11, is a cube; and so on. It is easy to deduce from these facts the summation of any number of terms of the series of natural cubes. In general, n^3 is the sum of n successive odd numbers beginning with a certain number determined by the fact that the number of terms preceding it in the above series is

$$1+2+3+ \ldots +(n-1) = \tfrac{1}{2}(n-1)n.$$

Thus $1^3+2^3+3^3+ \ldots +n^3$ is the sum of $\tfrac{1}{2}(n-1)n+n$, that is, $\tfrac{1}{2}n(n+1)$, terms of the series of odd numbers.

Therefore

$$1^3+2^3+3^3+ \ldots +n^3 = \tfrac{1}{4}n(n+1)[2+\{\tfrac{1}{2}n(n+1)-1\}2]$$
$$= \{\tfrac{1}{2}n(n+1)\}^2.$$

Nicomachus does not give this formula, but it was known to the Roman *agrimensores*, and it is not likely that Nicomachus was unaware of it. That it was a Greek discovery may fairly be inferred from the fact that al-Karkhī, the Arabian algebraist, who mainly followed Greek models, gives in his algebra entitled *al-Fakhrī* a proof by means of a figure with gnomons drawn about squares in the traditional Greek fashion.

Let AB be the side of a square AC; let

$$AB = 1+2+ \ldots +n = \tfrac{1}{2}n(n+1),$$

and along BA measure $BB'=n$, $B'B''=n-1$, $B''B'''=n-2$, and so on.

Draw the squares on AB', AB'' ... showing the gnomons which are the differences between the squares, as in the diagram.

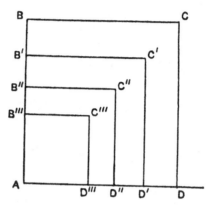

Now (gnomon $BC'D$) $= BB' \cdot BC + DD' \cdot C'D'$
$$= BB' (BC + C'D').$$

But $BC = \tfrac{1}{2} n (n+1)$, and
$$C'D' = 1+2+3+ \ldots +(n-1) = \tfrac{1}{2} n (n-1).$$

Therefore (gnomon $BC'D$) $= n \cdot n^2 = n^3$.
Similarly (gnomon $B'C''D'$) $= (n-1)^3$, and so on.

Therefore $1^3 + 2^3 + 3^3 + \ldots + n^3 =$ the sum of the gnomons down to the gnomon about the small square at A which has 1 for its side, *plus* that small square itself.

That is, $1^3 + 2^3 + 3^3 + \ldots + n^3 =$ (square AC)
$$= \{\tfrac{1}{2} n (n+1)\}^2.$$

In cc. 21-9 Nicomachus deals with arithmetical and geometrical progressions and with means, namely the arithmetic, geometric, and harmonic, and the others already mentioned (pp. 52-3). Various properties are mentioned in the first three cases; the rest are treated summarily.

If a, b, c be in arithmetical progression,
$$b^2 - ac = (a-b)(b-c) = (a-b)^2 = (b-c)^2.$$
We notice that, if the terms are written in the form $d+a, a, a-d$, this property may be written
$$a^2 = (a+d)(a-d) + d^2;$$
and this would seem to be the origin of the *regula Nicomachi* stated by one Ocreatus (O'Creat?) in the twelfth or thirteenth century to be used for finding the square of a number between 5 and 10. If $a+d=10$ and $a > d$, the theory is that a^2 is easier to arrive at through the formula than by direct squaring, because the multiplications on the right hand are easier, though there are two of them.

In c. 24 Nicomachus states certain obvious relations between three terms in descending geometrical progression and the common ratio. The same chapter cites the 'Platonic theorem' about the number of means in continued proportion between two square numbers and two cube numbers respectively, namely one in the first case and two in the second. Finally in c. 25 Nicomachus observes that, if $a > b > c$, then $a/b <=> b/c$ according as a, b, c are in arithmetical, geometrical, or harmonical progression.

The book by Theon of Smyrna purporting to be a manual of mathematical subjects such as a student would require to enable him to read Plato includes an arithmetical section treating the elementary theory of numbers on much the same lines as Nicomachus, though less systematically. It contains, however, two important things which are not found in Nicomachus. One is the account of the 'side-' and 'diameter-numbers' invented by the Pythagoreans for the purpose of finding the successive integral solutions of the equations $2x^2-y^2=\pm 1$ (see pp. 55-7 above). The other is a statement to the effect that, if m^2 is a square number, either m^2 or m^2-1 is divisible by 3, and again either m^2 or m^2-1 is divisible by 4. This is equivalent to saying that a square number cannot be of any of the forms $3n+2$, $4n+2$, or $4n+3$. Theon says further that, if m^2 is a square number, one of the following alternatives must hold:

(1) $\dfrac{m^2-1}{3}$, $\dfrac{m^2}{4}$ both integral (e.g. $m^2=4$),

(2) $\dfrac{m^2-1}{4}$, $\dfrac{m^2}{3}$ both integral (e.g. $m^2=9$),

(3) $\dfrac{m^2}{3}$, $\dfrac{m^2}{4}$ both integral (e.g. $m^2=36$),

(4) $\dfrac{m^2-1}{3}$, $\dfrac{m^2-1}{4}$ both integral (e.g. $m^2=25$).

We can hardly doubt that these discoveries were Pythagorean. The truth of the statements can easily be verified. Since any number must have one of the forms

$$6k,\ 6k\pm 1,\ 6k\pm 2,\ 6k\pm 3,$$

any square must have one or other of the forms

$$36k^2,\ 36k^2\pm 12k+1,\ 36k^2\pm 24k+4,\ 36k^2\pm 36k+9,$$

and the consideration of these forms separately gives what is required to prove the facts stated.

Iamblichus, born at Chalcis in Coele-Syria, lived in the first half of the fourth century A.D. He was a pupil of Anatolius and Porphyry, and wrote nine books on the Pythagorean Sect with the following titles: I. On the Life of Pythagoras; II. Exhortation to Philosophy; III. On mathematical science in general; IV. On Nicomachus' *Introductio arithmetica*; V–VII. On arithmetical science in physics, ethics, and theology respectively; VIII. On the Pythagorean geometry; IX. On the Pythagorean music. The first four of these books survive and are accessible in modern editions. We are here concerned with the Commentary on Nicomachus. This is an elaborate work, amounting almost to a revised edition with certain things added. Those which deserve notice are as follows.

A square is represented as a racecourse (δίαυλος) formed by the series of natural numbers beginning from 1 as the *start* (ὕσπληξ), going up to n as the turning-point (καμπτήρ), and then returning through $n-1$, $n-2$, &c., to 1, the *goal* (νύσσα). The sum is clearly made up of two triangular numbers with sides n, $n-1$ respectively, and is therefore equal to $\frac{1}{2}n(n-1)+\frac{1}{2}n(n+1)=n^2$.

Another proposition is of greater interest. Take any three consecutive numbers the greatest of which is divisible by 3. Add them, and we have a number consisting of a certain number of units, a certain number of tens, a certain number of hundreds, and so on. The numbers of each denomination are the *digits* of the number as we write it. Add the digits; this gives a smaller number. Add the digits of the smaller number, and so on. Then, says Iamblichus, the final result will always be the number 6. Take, e.g., the numbers 10, 11, 12; the sum is 33; add the digits and we have 6. Take again the numbers 994, 995, 996; the sum is 2985; the sum of the digits of this number is 24, and the sum of the digits of 24 is again 6.

Iamblichus, thirdly, gives the notice about the solution of the set of any number of simultaneous equations between as many unknowns which went by the name of the *Epanthema* of Thymaridas (see pp. 57–8 above).

IV

THE EARLIEST GREEK GEOMETRY. THALES

THE 'SUMMARY' OF PROCLUS

WE shall often, in the course of this history, have to refer to the so-called 'summary' of Proclus. This is contained in a few pages (65–70) of Proclus' *Commentary on Euclid, Book I*, and it reviews in the briefest possible outline the course of Greek geometry from the earliest times to Euclid, with special reference to the evolution of the Elements. It has often been called the 'Eudemian' summary, on the assumption that it is an extract from the great *History of Geometry* in four Books by Eudemus, the pupil of Aristotle. But a perusal of the actual summary suffices to show that it cannot have been written by Eudemus himself, though the earlier portion down to a certain sentence was probably based on material drawn from Eudemus' *History*. The sentence in question marks a break in the narrative.

'Those', it runs, 'who have compiled histories bring the development of this science to this point. Not much younger than these is Euclid, who put together the Elements, collecting many of the theorems of Eudoxus, perfecting many others by Theaetetus, and bringing to irrefragable demonstration the propositions which had only been somewhat loosely proved by his predecessors.'

Since Euclid was later than Eudemus, this could not have been written by Eudemus; on the other hand, the description of 'those who have compiled histories' suits Eudemus perfectly. There is, however, no such difference

74   THE EARLIEST GREEK GEOMETRY. THALES

in style between the two portions of the summary as to suggest different authorship. The author, whoever he was, speaks of the question of the origin of the Elements in a way in which no one would be likely to write who was not later than Euclid. It was clearly his object to trace the growth not so much of geometry in general as of the Elements in particular. Hence he pays the most attention to the geometers who had written Elements, or who had contributed to their content; he omits to mention important discoveries in geometry (or only mentions them in parenthesis, as it were) when they had no particular bearing on the Elements. Thus he speaks of Hippocrates of Chios as a famous geometer for the particular reason that he was the first person of whom it was recorded that he wrote Elements, and there is no reference to his other achievements beyond the addition to his name, for the purpose of identification, of the words 'the discoverer of the quadrature of the lune'. There is no mention of Democritus, who was a distinguished mathematician. Now we know that Plato was an opponent of Democritus, never once mentions him, and is said to have wished to burn his writings. A Platonist therefore might have deliberately omitted to mention Democritus; but Eudemus would not be likely to do so. Was the author of the 'summary' Proclus himself, the Platonist? This seems unlikely because (1) the style is not such as to point to Proclus as the author, and (2) Proclus would hardly have spoken in the detached way that the author does of Euclid and the great book which was the subject of his whole commentary: 'Not much younger than these is Euclid who compiled the Elements...' 'This man lived in the time of the first Ptolemy....' On the whole it would seem most probable that Proclus took the body of the summary from

SUPPOSED EGYPTIAN ORIGIN

a compendium made by some writer (later than Eudemus) who based the first portion of it, directly or indirectly, upon notices in Eudemus' *History*.

Greek tradition is unanimous in tracing the beginnings of geometry to Egypt. Revenue was raised there by the taxation of land, and, when the flooding of the Nile obliterated the boundaries between holdings, it was necessary to restore them, or to determine the taxable area independently of them, by calculation depending on measurements. Or again, as Herodotus says, the river would sweep away a portion of a plot, and, when the owner applied for a corresponding reduction of tax, surveyors had to be sent down to certify what the reduction in area had been; 'this', in his opinion, 'was the origin of geometry, which then passed into Greece'. Heron of Alexandria, Diodorus Siculus, and Strabo all have the same story.

Thales, we are told, went to Egypt and brought geometry thence into Greece. But what he learnt in Egypt was a few geometrical facts rather than geometry. For, while the Egyptians had practical rules for measuring with more or less accuracy (1) certain areas such as squares, triangles, trapezia, and even circles, (2) the solid content of measures of corn, &c., of different shapes, there is no trace of any attempt by them to give a proof of any rule; they had no idea of geometry as a demonstrative science.

Geometry in this sense was the creation of the Greeks. No one before them had thought of proving such a thing as that the two base angles of an isosceles triangle are equal; the idea was an inspiration unique in the history of the world, and the fruit of it was the creation of mathematics as a science.

76 THE EARLIEST GREEK GEOMETRY. THALES

We can form a fair estimate of what Thales was in a position to learn from Egypt.

The most important available source of information about Egyptian mathematics is still the Papyrus Rhind, written probably about 1650 B.C. but copied from an original of the time of Amenemhet III (Twelfth Dynasty), who reigned from 1849 to 1801 B.C. The geometry in this 'guide for calculation' is rough mensuration. The cases which concern us here are the following.

(1) The area of a rectangle is of course given as the product of the sides.

(2) The area of a triangle is given as half the *tp-r*, which means the side of the triangle taken as base, multiplied by what is called the *mryt*. Nothing is said of the shapes of the triangles drawn in the Papyrus. They look like *isosceles* triangles with bases rather small in comparison with the other sides. As they stand on the page, they lie flat, as it were; the bases are vertical and the perpendicular height (not drawn) would, if drawn, be horizontal, i.e. parallel to the line of script. The difficulty is to decide what is meant by *mryt*. To judge by the figures alone, as drawn, it might be a 'side' of the triangle other than the base. But in that case, unless the 'sides' were equal, the formula would give two results according to the particular 'side' taken; and, apart from this, it is hardly likely that even the Egyptians of the time would use a formula so inexact.

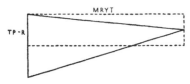

That *mryt* actually means what we call the 'height' is now, I think, made probable by the considerations urged in a review (1926) by Mr. Battiscombe Gunn of the edition of the Papyrus by Professor T. Eric Peet (1923). The text uses the expression 'to give its (the triangle's) rectangle' or 'to cause it to be rectangular'. This would be done by drawing perpendiculars to the base (1) from its upper extremity, (2) from its middle point, and completing the rectangle by drawing through the vertex a parallel to the base. Now *mryt* apparently means a *quay*, and that would be the appearance of the rectangle drawn as described, since it would have the effect of (as it were) levelling up the sloping side of the triangle to the horizontal. The above figure shows the triangle, its base, 'its rectangle', and the *mryt*. The same question arises as to

(3) The formula for the area of a parallel-trapezium (described as a 'truncated triangle'), namely $\frac{1}{2}(a+c) \cdot b$, where a, c are the base and the opposite side respectively, while b is the *mryt*. In the case taken the figure seems to have been intended to be isosceles, and the base and the opposite side (6, 4 respectively) are short relatively to the *mryt* (20), so that the angles at the base are not far short of being right angles. But, as the formula would be inexact if the *mryt* were one of the non-parallel sides, there is the same reason here as in the case of the triangle for taking *mryt* to mean the perpendicular height.

(4) Certain inscriptions on the Temple of Horus at Edfu, which belong to the reign of Ptolemy XI, Alexander I (107–88 B.C.), refer to the assignment of plots of land to the priests. From portions of these inscriptions published by Lepsius we gather that $\frac{1}{2}(a+c) \cdot \frac{1}{2}(b+d)$ was a formula in use for the area of a quadrilateral of any form in which (a, c) and (b, d) are pairs of opposite sides. Some of the quadrilaterals are evidently trapezia with the non-parallel sides equal; but others are not, although they are commonly not far from being rectangles or isosceles trapezia. Examples are '16 to 15 and 4 to $3\frac{1}{2}$ make $58\frac{1}{8}$',

i.e. $\frac{1}{2}(16+15) \times \frac{1}{2}(4+3\frac{1}{2}) = 58\frac{1}{8}$; '$9\frac{1}{2}$ to $10\frac{1}{2}$ and $24\frac{1}{2}$ $\frac{1}{8}$ to $22\frac{1}{2}$ $\frac{1}{8}$ make $236\frac{1}{4}$'; '22 to 23 and 4 to 4 make 90', and so on. Triangles are not made the subject of a separate formula, but are regarded as cases of quadrilaterals in which one side (e.g. d) is made zero; the formula for the triangle then becomes $\frac{1}{2}(a+c) \cdot \frac{1}{2}b$, e.g. the triangle 5, 17, 17 is described as a figure with sides '0 to 5 and 17 to 17', the area accordingly being taken as $\frac{1}{2}(0+5) \times \frac{1}{2}(17+17) = 42\frac{1}{2}$.

We come now to (5) the mensuration of circles in the Rhind Papyrus. If d is the diameter, the area is given as $\{(1-\frac{1}{9})d\}^2$ or $\frac{64}{81}d^2$. As the area is in fact $\frac{1}{4}\pi d^2$, this means that the value of π is put at $\frac{256}{81}$ or $(\frac{16}{9})^2$, which is 3·16 very nearly. The use of this value is illustrated in the measurements of containers with circular bases, which are in fact right cylinders. The content is taken to be the product of the area of the base and the height. That is, if d is the diameter of the base and h the height, the content is found to be $(\frac{8}{9}d)^2 \cdot h$. Then, curiously enough, this result is multiplied by $1\frac{1}{2}$. This makes it appear that the base was multiplied by $1\frac{1}{2}$ times the height. It is now explained (*vide* Peet's edition) that this is not so; what happens is that, when the content has been found in cubic cubits, the result is multiplied by $1\frac{1}{2}$ in order to express it in terms of a measure of capacity called the *khar* which was $\frac{2}{3}$rds of a cubic cubit. For example, Problem No. 41 is about a circular container of diameter 9 and height 10. The above formula gives $8^2 \cdot 10 = 640$, and the text adds 'Its half is now added to it: it becomes 960. ⟨This is⟩ its content in *khar*'. With this case should be compared the measurement of a mass of corn (also cylindrical) found in one of the Kahun Papyri. The figure shows a circle with $1365\frac{1}{3}$ inside it, representing the content of the container, and with 12 and 8 written above and to the left of the circle respectively. The calculation is done in this way. 12 is taken, then $1\frac{1}{3}$ times it, making 16. The 16 is squared, producing 256, and finally the 256 is multiplied by $\frac{2}{3}$ of 8, which gives $1365\frac{1}{3}$. Schack-Schackenburg showed that the problem is that of finding the content of a cylinder of diameter 12 and height 8 cubits. The formula here used is $(\frac{4}{3}d)^2 \cdot \frac{2}{3}h$, which is seen to be the same thing as the above formula $(\frac{8}{9}d)^2 \cdot h$ multiplied by $1\frac{1}{2}$, for both reduce to $\frac{32}{27}d^2h$. That is to say, the method of the Kahun Papyrus is to work out the result directly in *khar* instead of first working out the volume in cubic cubits and then multiplying by $1\frac{1}{2}$ in order to turn it into *khar*.

(6) More important geometrically are certain calculations with reference to the proportions of pyramids (Nos. 56-9 of the Rhind Papyrus) and a monument (No. 60). A certain relation called *seqet*, literally 'that which makes the nature', i.e. which determines the form or the proportions of the pyramid, has to be found from two lines distinguished in the figure, namely

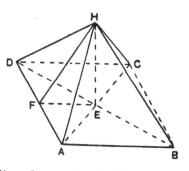

(1) *ukha-thebt*, which is some line drawn in the base, and (2) *pir-em-us* or *per-em-us*, 'height' (a word from which the name πυραμίς may have been derived). It is now established beyond doubt that *ukha-thebt* means the side of the square base and *pir-em-us* the vertical height of the pyramid. If the diagonals of the base meet in E, and EF be drawn parallel to a side (say AB), and if FH be joined, then $FE=$ half the side of the base, and $HE=$ the height of

80   THE EARLIEST GREEK GEOMETRY. THALES

the pyramid. Now the *seqet* is the ratio $\dfrac{\frac{1}{2}\,ukha\text{-}thebt}{pir\text{-}em\text{-}us}$, which is therefore the ratio of FE to EH, or what we call the *co-tangent* of the angle of slope of the faces of the pyramid. The correctness of this view is confirmed by the actual results obtained in Problems No. 56 and Nos. 57–9. In No. 56 the *seqet* is $\frac{18}{25}$; this is the co-tangent of an angle of 54° 14′ 46″, which agrees fairly well with the measurements by Flinders Petrie of the angle of slope of the lower half of the southern stone pyramid at Dahshur, namely (at the mean) 55° 1 for the lower and 54° 31′ for the higher part of the faces. In Nos. 57–9 the *seqet* is $\frac{3}{4}$, the co-tangent of an angle of 53° 7′ 48″, while 53° 10′ ± 4′ is the mean figure given by Flinders Petrie for the slope of the faces of the second pyramid of Gizeh.

In the case of the monument *seqet* has the same meaning, but the shape is different, tapering more than the pyramid, and the lines in it determining the *seqet* are called by different names. The exact form of the monument is uncertain; it might be, say, like an obelisk or a cone (Professor Peet inclines to the latter); the *seqet* is the ratio of half the *senti* to the *qay en ḥeru*. Now *senti* means 'foundation' or 'base', so that half of it is either the radius or half the length of the side of the base, while *qay en ḥeru* is the 'vertical length' or height. The *seqet* in Problem No. 60 is $\frac{1}{4}$, which is the co-tangent of an angle of 75° 57′ 50″, and this again is near enough to the figure of 76° given by Flinders Petrie as the 'characteristic angle' of a mastaba-tomb such as those at Saqqara and Medum.[1]

The measurements in the Rhind of the *seqet* of pyramids naturally connect themselves with the story of Thales' discovery of a method of finding the height of a pyramid.

[1] See further note on Egyptian mathematics in Appendix.

THALES

THALES of Miletus, whose mother was of Phoenician origin, lived about 624–547 B.C. He was in the year 582 declared one of the Seven Wise Men, and no wonder. The others were not philosophers, but mostly shrewd men of affairs with a turn for legislation; Thales stands by himself. His genius was many-sided. Statesman, engineer, man of business, philosopher, mathematician, and astronomer, he covered almost the whole field of human thought and activity. As a statesman he gave the wisest counsel to his compatriots, as when he dissuaded the Milesians from listening to Croesus' proposals for an alliance, and when he advised the Ionian city-states to combine, for their mutual protection, in a federation, with Teos as capital. As engineer, he is said to have enabled Croesus' army to cross the river Halys by first making an artificial channel and diverting the stream into it, and then directing it back to its old course. He even condescended, on occasion, to trade, as when he undertook to show how easy it was to get rich. Foreseeing that in a certain season there would be a great crop of olives, he obtained control over all the oil-presses in the neighbourhood, paying only a small consideration when there was nobody bidding against him, and then charging what he liked when the demand for the accommodation became urgent, with the result that he made a large fortune.

He combined, too, a rare humour with profound knowledge of human nature. Many stories confirm this. 'Know thyself' is his maxim. Asked how we shall lead the best and most righteous lives, he replied, 'By refraining from doing what we blame in others'. To the question what was the strangest thing he had ever seen he answered, 'An aged tyrant'. When some one asked him what he would take for a certain discovery he had made in astronomy, he said, 'It will be sufficient reward for me if, when telling it to others, you will not claim it as your own discovery but will say it was mine.'

We are here concerned with Thales' mathematics and astronomy. All the traditions about his mathematics, except that recording his definition of a 'number', relate to his geometry. They are as follows:

(a) *Measurement of height of pyramid.*

Thales evoked general admiration by showing how to calculate the height of a pyramid by means of shadows. There are two versions of the story. The earliest is that of Hieronymus, a pupil of Aristotle, who says that Thales observed the length of the shadow of a pyramid at the particular moment when our shadows are of the same length as ourselves. The later version (Plutarch) says that he set up a stick at the end of the shadow of the pyramid and, having thus conceived two (similar) triangles, argued that the height of the pyramid is to the length of the stick as the shadow is to the shadow.

The method described in the first version, leading to the simpler calculation, seems the more probable. Thales would probably observe that, when one object throws a shadow of length equal to its own height, other objects do so also; he would probably convince himself of this by induction, after actual measurement in a number of cases. The inference with regard to the pyramid would then be obvious. But even if his method was the more general one, it required no more knowledge of the properties of similar triangles than was involved in the use of the *seqet* by the Egyptians; his solution is in fact a *seqet* calculation like that in the Rhind problem No. 57, where, given the base and the *seqet*, we have to calculate the height. The *seqet* in Thales' case is, of course, the ratio of the length of the shadow of the stick to that of the stick itself, which would be obtained by measurement. The only difficulty would be to measure or estimate the length of the shadow of the pyramid, i.e. the distance from the apex of its shadow to the centre of its base.

(b) *Geometrical theorems.*

The following are the general theorems in elementary geometry attributed to Thales:

(1) that a circle is bisected by its diameter (=Eucl. I, Def. 17);

(2) that the angles at the base of an isosceles triangle are equal (=Eucl. I. 5);

(3) that, if two straight lines cut one another, the vertically opposite angles are equal (=Eucl. I. 15);

(4) that, if two triangles have two angles equal to two angles respectively and one side to one side (namely that adjoining the equal angles or that subtending one of the equal angles), the triangles are equal in all respects (=Eucl. I. 26).

Further (5), Pamphile says that Thales 'was the first to describe on a circle a triangle (which shall be) right-angled, and that he sacrificed an ox (on the strength of the discovery)'. This must apparently mean that Thales discovered that the angle in a semicircle is a right angle (=Eucl. III. 31).

Thales is said to have demonstrated (1), but only to have stated (2), while Eudemus is quoted as saying that he discovered (3) but did not prove it scientifically, and that he must have known (4) because it was necessary to his method of finding the distance of ships from the shore. The dictum that he demonstrated that a circle is bisected by its diameter (a fact which Euclid states as a definition) need not be taken too literally. Thales may have observed rather than proved the fact, which might perhaps have been suggested to him by the appearance of certain drawings, which he saw on monuments, &c., in Egypt, of circles divided into sectors by two, four, or six diameters.

As regards (4) we are not told how Thales measured the distance of ships from the shore. Several suggestions have been made as to his method. The first supposes him to have been on the top of a tower on the sea-shore and to have used similar triangles. If B be the base of the tower, C the ship, and A the eye of the observer vertically above 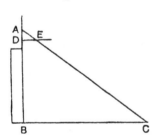 B, then ABC is a right-angled triangle. If a small right-angled triangle be drawn, as ADE, so that AD lies along AB and AE along AC, while ADE is the right angle, the two right-angled triangles are similar. In these triangles AB is known, while AD, DE can be measured. Then, since the triangles are similar, $CB : BA = ED : DA$. If, therefore, $AD = l, DE = m, AB = h$, BC is found to be hm/l. The objection to this solution is that it does not suit Eudemus' description, since it does not depend directly on the theorem of Eucl. I. 26.

Tannery favoured the hypothesis of a solution on lines followed by the Roman *agrimensor* Marcus Junius Nipsus in his *fluminis varatio*. To find the distance from A to an inaccessible point B. Measure from A any length AC in a direction at right angles to AB. Bisect AC at D. From

DISTANCE OF SHIP FROM SHORE

C and on the side of AC away from B draw CE at right angles to AC, and let E be the point on CE which is in a straight line with B, D. Then clearly, by Eucl. I. 26, $CE = AB$, and CE can be measured, so that AB is known. The objection to this solution is that, as a rule, it would be difficult to get a sufficient amount of free and level space for the construction and measurement.

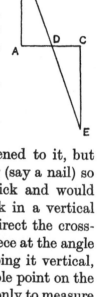

The following seems to me to be the easiest possible solution depending on Eucl. I. 26. An observer on the top of the tower had only to use a rough instrument consisting of a straight stick and a cross-piece fastened to it, but capable of being turned about the fastening (say a nail) so that it could form any angle with the stick and would stay where it was put. Now fix the stick in a vertical position (by means of a plumb-line) and direct the cross-piece to the ship. Next, leaving the cross-piece at the angle so found, turn the stick round, while keeping it vertical, until the cross-piece points to some accessible point on the shore. Mentally noting this point, we have only to measure the straight line drawn to it from the base of the tower, which, by Euclid, I. 26, is equal to the distance of the ship. It appears that this method is actually found in many practical geometries of the first century of printing and had therefore long been known. There is a story that one of Napoleon's engineers won the imperial favour by quickly measuring, in this way, the breadth of a stream which blocked the progress of the army.

There is more difficulty about the statement of Pamphile (5) implying that Thales first discovered that the angle in a semicircle is a right angle. The matter is further confused

86   THE EARLIEST GREEK GEOMETRY. THALES

by an addition of Diogenes Laërtius to the citation from Pamphile: 'others, however, including Apollodorus the "calculator", say that it was Pythagoras'. The reference of Pamphile to a sacrifice evidently brought to Diogenes' mind the distich of Apollodorus about the sacrifice by which Pythagoras celebrated the discovery of his famous proposition, and Diogenes forgot for the moment that the latter was quite a different theorem. We may, therefore, ignore Diogenes' addition to the story.

The following is the dilemma: (1) Euclid proves (in III. 31) that the angle in a semicircle is a right angle by means of the proposition (I. 32) that the sum of the angles of any triangle is equal to two right angles; the proof is well known. But we are distinctly told by Proclus, on the authority of Eudemus, that the first to *discover*, as well as to give a general proof of the fact, that the angles of any triangle are together equal to two right angles were the Pythagoreans. It is, therefore, hardly permissible to suppose that Thales used Euclid's method of proof. On the other hand (2), if Thales proved in some other way that the angle in a semicircle is a right angle, he could hardly have failed to see the obvious deduction that the sum of the angles of a *right-angled* triangle is equal to two right angles. For, if BAC be a right angle in a semicircle, and A be joined to the centre O, we have two isosceles triangles OAB, OAC, and by Thales' own proposition (= Eucl. I. 5) the base angles in each are equal, that is, the angles OAB, OBA are equal, and the angles OAC, OCA are equal; therefore the sum of the angles OAB, OAC is equal to the sum of the angles OBA, OCA. The former sum is known

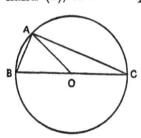

to be a right angle; therefore the sum of the angles OBA, OCA is also a right angle, and the sum of all four (which make up the three angles of the triangle BAC) is equal to two right angles. And again, if it is known that the angles of a right-angled triangle are together equal to two right angles, it is an easy inference that the angles of *any* triangle are together equal to two right angles. For we have only to divide the latter triangle into two right-angled triangles by drawing a perpendicular from a vertex to the opposite side.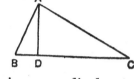
Thus, if, in the triangle ABC, AD is perpendicular to BC, the sum of all the angles of the two right-angled triangles ABD, ADC is known to be four right angles. Now the angles of the triangle ABC are together equal to the sum of all the angles in the two triangles ABD, ADC less the two angles ADB, ADC. The latter two angles make up two right angles; therefore the sum of the angles of the triangle ABC is equal to four right angles less two right angles, i.e. to two right angles.

In view of these difficulties, it seems possible that Thales' argument was of a more primitive kind making no assumption about the sum of the angles of even a right-angled triangle. No doubt, in the infancy of geometry, all sorts of diagrams would be drawn, and lines in them, by way of experiment, in order to see whether any property could be detected by mere inspection. We may imagine Thales drawing what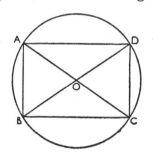
we call a rectangle, a figure with four right angles (which, it would be found, could be drawn in practice), and then

drawing the two diagonals as AC, DB in the annexed figure. The equality of the pairs of opposite sides would leap to the eye, and could be verified by measurement. Thales might then argue thus. Since, in the triangles ADC, BCD, the two sides AD, DC, are equal to the two sides BC, CD respectively, and the included angles (being right angles) are equal, the triangles are equal in all respects. Therefore the angle ACD (i.e. OCD) is equal to the angle BDC (or ODC), whence it follows (by the converse of Eucl. I. 5 known to Thales) that $OC=OD$. Starting from the equality of BA, CD it would be proved in the same manner that $OD=OA$. Hence OA, OD, OC (and OB) are all equal, and a circle described with O as centre and OA as radius passes through B, C, D also. Now AOC, being a straight line, is a diameter of the circle, and therefore ADC is a semicircle. The angle ADC is an 'angle in the semicircle', and by hypothesis it is a right angle. The construction amounts to circumscribing a circle about the right-angled triangle ABC, which seems to answer well enough to Pamphile's phrase about 'describing on a circle a triangle which is right-angled'.

It seems probable that the equality of the sum of the angles of a triangle to two right angles was first discovered with reference to a right-angled triangle, and by an elementary argument of the above kind. A rectangle would be drawn and one of its diagonals inserted. It would then be immediately assumed that the two triangles into which the diagonal divides the rectangle are equal in all respects, and hence the sum of the angles of either triangle is equal to half the sum of the four angles of the rectangle, and therefore to the half of four right angles. The transition to any triangle could then be made, as above shown, by dividing it into two right-angled triangles.

It is true that Geminus says that the 'ancients' investigated each species of triangle separately, first the equilateral, then the isosceles, and afterwards the scalene triangle, whereas later geometers proved the property generally for *any* triangle. But we need not take this too seriously. Aristotle, in the *Posterior Analytics*, observes that, if one should prove separately for each sort of triangle, equilateral, isosceles, and scalene, that its angles are together equal to two right angles, either by one proof or by different proofs, he does not yet know that the *triangle* in general has the property, except in a sophistical sense, even if he knows that no other sort of triangle exists besides those specified. For he does not know it of the triangle *qua* triangle or *notionally* of a triangle in general. It may well be that Geminus was misled into taking for a historical fact what Aristotle only gives as a hypothetical illustration.

(c) *Thales as astronomer.*

Thales was the first Greek astronomer. Every one knows the story of his falling into a well when star-gazing, and being rallied by 'a clever and pretty maidservant from Thrace' (as Plato has it) for being so eager to know what goes on in the heavens that he could not see what was in front of him, nay at his very feet. There is good evidence that Thales predicted a solar eclipse which took place on 28 May 585 B.C. We can conjecture the basis of this prediction. The Babylonians had, as the result of observations continued for centuries, discovered the period of 223 lunations after which eclipses recur; and Thales had doubtless heard of this either directly or through the Egyptians as intermediaries. He can hardly have known the *cause* of eclipses.

According to Eudemus, Thales discovered 'that the

90   THE EARLIEST GREEK GEOMETRY

period of the sun with reference to the solstices is not always the same', which is taken to mean that he observed the inequality of the lengths of the four astronomical seasons, the four parts of the 'tropical' year as divided by the solstices and equinoxes. Diogenes Laërtius mentions written works by him *On the Solstice* and *On the Equinox*.

Thales used the Little Bear as a means of finding the pole, and advocated the Phoenician practice of sailing by the Little Bear instead of by the Great Bear as the Greeks did. A *Nautical Astronomy* is attributed by some to Thales and by others to Phocus of Samos.

FROM THALES TO PYTHAGORAS

The history of geometry between the times of Thales and Pythagoras is blank except that the Proclus-summary gives the name of one person who occupied himself with geometry; the name is variously given as Ameristus, Mamercus (in Friedlein's edition of Proclus), Mamertinus (Suidas), and Mamertius (in Heiberg's edition of Heron). Whichever name is right, Suidas and Proclus agree in calling him a brother of Stesichorus the poet (c. 630–550 B.C.).

Suidas says that ANAXIMANDER (born about 611/10 B.C.) introduced the *gnomon* (or sun-dial with needle vertical) and 'generally set forth a sketch or outline of geometry'. It seems likely that the word 'geometry' is here used in its primitive sense of 'land-measurement', and that the reference is, not to a geometrical work, but to the famous map of the inhabited earth. For Anaximander was the first person to draw such a map. The Egyptians drew maps, but only of particular districts; Anaximander boldly planned out the whole world with the circumference of the earth and sea. On the *gnomon* (sun-dial), which he is said to have introduced into Greece, Anaximander, so we are told, showed the solstices, the times, the seasons, and the equinoxes. According to Herodotus, the Greeks learnt the use of the gnomon from the Babylonians.

Anaximander put forward some very original and daring hypotheses in astronomy. According to him the earth is like a tambourine, a short cylinder with two circular bases (on one of which we live), and of depth equal to one-third of the diameter of either base. It is suspended freely in the middle of the universe, without support, being kept in equilibrium by virtue of its equidistance from the extremities and from the other heavenly bodies all round. The sun, moon, and stars are enclosed in opaque rings of compressed air concentric with the earth and filled with fire; what we see is the fire shining through vents (like gas-jets, as it were). The sun's ring is 27 or 28 times, and the moon's ring 19 times, as large as the earth, i.e. the sun's and moon's distances are estimated in terms (as we may suppose) of the radius of the circular face of the earth; the fixed stars and the planets are nearer to the earth than the sun and moon. This is the first recorded speculation on sizes and distances.

Anaximander, like Thales, is also credited with having constructed a sphere to represent the heavens.

V
PYTHAGOREAN GEOMETRY

ABOUT fifty years separated Thales and Pythagoras. With Pythagoras geometry became for the first time a scientific subject pursued for its own sake. 'Pythagoras', says the Proclus-summary, 'transformed the study of geometry into a liberal education, examining the principles of the science from the beginning and probing the theorems through and through in a purely intellectual manner'. Favorinus says that he 'used definitions on account of the mathematical nature of the subject'. We conclude that Pythagoras first laid down certain principles (including definitions), and then built up an ordered sequence of propositions. 'A figure and a platform, not a figure and sixpence'; this was the Pythagorean motto, meaning that each new theorem sets up a platform from which to ascend to the next, and so on.

A comparatively early authority, Callimachus (about 250 B.C.), is quoted by Diodorus as having said that Pythagoras discovered some geometrical theorems himself and was the first to introduce others from Egypt into Greece. Five lines quoted by Diodorus (*minus* a few words) also form part of a longer fragment in the Oxyrhynchus Papyri, though the text is still uncertain. The verses tell us about the cup bequeathed by Bathycles, an Arcadian, to be given to the best of the Seven Wise Men, and how it was first brought to Thales by Bathycles' son, who 'by a happy chance found the old man scraping the ground and drawing the figure discovered by the Phrygian Euphorbus, who was the first to draw even scalene triangles and a circle ...'. Euphorbus is of course Pythagoras, who claimed to have been Euphorbus in one of his various incarnations. And, in spite of the anachronism, the figure discovered by Euphorbus is presumably the theorem of the square on the hypotenuse. The rest is uncertain. After the word 'circle' (κύκλον) Diodorus has ἑπταμήκη, 'seven-lengthed', which, if correct, can hardly be taken to mean anything else but the circle including seven orbits, i.e. the zodiac circle, which embraces the independent circles of the sun, moon, and planets. But this leaves the words 'even scalene triangles' high and dry, as it were. It would be more natural if the reading were such as to enable us to connect the circle with the scalene triangle, e.g. if the circle were the circle circumscribing the scalene triangle.[1] If Thales actually circumscribed a circle about a right-angled triangle, as the citation from Pamphile suggests, it would be most appropriate that the Pythagoreans should generalize the problem and show how to circumscribe a circle about *any* scalene triangle.

We proceed to set out the propositions in geometry which are definitely attributed to the Pythagoreans, including those associated with the name of Pythagoras himself.

(a) *The sum of the angles of any triangle is equal to two right angles.*

As we have seen (p. 88), it is likely enough that this was first discovered with reference to the particular case

[1] Diodorus' reading of the line in question is καὶ κύκλον ἑπταμήκη δίδαξε νηστεύειν, which does not scan. The Papyrus, in place of ἑπταμήκη, has what looks like ἐπ̃ (with λ above), and the rest of the line apparently blank. Diels reads καὶ κύκλον ἕ⟨λικα⟩ κἠδίδαξε νηστεύειν. I should like to suggest καὶ κύκλον ἔμπλην instead of καὶ κύκλον ἕ⟨λικα⟩, the word ἔμπλην meaning 'next to' or 'close by', which seems a possible description (in a poem) of a circle *circumscribed* about a triangle.

of a right-angled triangle, after which the extension of the theorem to any triangle would be made by dividing the triangle, by a perpendicular drawn from a vertex to the opposite side, into two right-angled triangles. All that we are told, however, is that Eudemus attributed the discovery of the general theorem to the Pythagoreans and gave their proof of it. This proof, as elegant as that of Euclid, depends, equally with his, on the properties of parallels, which must therefore have been known to the authors of it. It is as follows:

Let ABC be any triangle, and through A draw DAE parallel to BC.

Then, since BC, DE are parallel, the alternate angles DAB, ABC are equal.

Similarly, the alternate angles EAC, ACB are equal.

Therefore the sum of the angles ABC, ACB is equal to the sum of the angles DAB, EAC.

Add to each sum the angle BAC; therefore the sum of the three angles ABC, ACB, BAC, i.e. the three angles of the triangle, is equal to the sum of the three angles DAB, BAC, CAE, i.e. to two right angles.

We need not hesitate to credit the Pythagoreans with the more general propositions about the angles of any polygon, namely (1) that, if n be the number of the sides or angles, the interior angles of the polygon are together equal to $2n-4$ right angles, and (2) that the exterior angles of the polygon (being the supplements of the interior angles respectively) are together equal to four right angles. The propositions are interdependent, and Aristotle twice quotes the latter. The Pythagoreans also discovered that the only three regular polygons the angles of which, if placed together round a common point as vertex, will just fill up the space (four right angles) round the point are the equilateral triangle, the square, and the regular hexagon.

(β) *The 'Theorem of Pythagoras'* (=Eucl. I. 47).

Tradition is unanimous in referring to Pythagoras the discovery of the theorem of the square on the hypotenuse; but the documentary evidence is far from conclusive. Callimachus speaks of the 'figure' discovered by Pythagoras, and the distich of Apollodorus the 'calculator' or mathematician (date uncertain) says, 'When Pythagoras discovered that famous proposition on the strength of which he offered a splendid sacrifice of oxen'. Unfortunately neither author says what the proposition he refers to actually was. But Plutarch, Athenaeus, Diogenes Laërtius, and Porphyry all connect the story of the sacrifice with the theorem of the square on the hypotenuse, though Plutarch, in giving Apollodorus' verses, expresses doubt whether the proposition referred to is that theorem or a certain problem of 'applying an area', while in another passage he says that the occasion of the sacrifice was the solution of the problem 'given two (rectilineal) figures, to apply [he should rather have said 'construct'] a third which shall be equal to the one and similar to the other'. Vitruvius, however, a century or so before Plutarch, definitely connected the sacrifice with the discovery that the particular triangle 3, 4, 5 is right angled. Proclus will not commit himself to a definite opinion; he says, 'If we listen to those who wish to recount ancient history, we may find some of them referring this theorem (Eucl. I. 47) to Pythagoras and saying that he sacrificed an ox in celebration of his discovery. But, for my part, while I admire *those who* first observed the truth of the theorem, I marvel more at the writer of the Elements, not only because he confirmed it by a most lucid demonstration, but because he compelled assent to the still more general theorem in the sixth book by the irrefutable arguments of science'. It is difficult for us to be more positive than Proclus was; but for myself I like to believe that, so far as the general theorem and the proof of it are concerned, the commonly accepted tradition is right.

Some knowledge, however, of the property of right-angled triangles can be traced long before the date of Pythagoras. The Egyptians indeed do not seem to have had it, for, although they knew that $3^2+4^2 = 5^2$, there is nothing in their mathematics, so far as known to us, to suggest that they knew that the triangle (3, 4, 5) is right-angled (T. Eric Peet, *The Rhind Mathematical Papyrus*, p. 32). On the other hand, it would appear that practical use was made of the theorem of the square on the hypotenuse, as early as (say) 2000 B.C., by the Babylonians. The evidence for this is the text of certain Babylonian tablets containing mathematical problems which have just recently (1928-9) been interpreted for the first time by O. Neugebauer, W. Struve, and others. Two of the problems are: to calculate the length (1) of a chord of a circle from its *sagitta* and the diameter of the circle, and (2) of the *sagitta* from the chord and the diameter. If c be the chord, a its *sagitta*, and d the diameter of the circle, the formulae intended to be used are evidently $c = \sqrt{\{d^2-(d-2a)^2\}}$ and $a = \frac{1}{2}\{d-\sqrt{(d^2-c^2)}\}$, and it is not possible to account for these formulae except on the assumption that they were based, in some form or other, on the theorem of Pythagoras. In the particular case $a = 2$, $c = 12$, $d = 20$, and the property used is $20^2 = 16^2+12^2$, equivalent to $5^2 = 4^2+3^2$.

Again, there are those who credit the Indians with the

discovery of the theorem. The claim is mainly based on the *Āpastamba-Śulba-Sūtra*, which is thought to be at least as early as the fourth or fifth century B.C. A feature in this work is the construction of right angles by means of stretched cords in the ratios of the sides of certain right-angled triangles in rational numbers. Seven such triangles are used, which however reduce themselves to four, namely (3, 4, 5), (5, 12, 13), (8, 15, 17), and (12, 35, 37).

One of these triangles (5, 12, 13) was known as early as the eighth century B.C., while yet another (7, 24, 25) appears in the *Baudhāyana Ś.-S.*, which is supposed to be earlier than Āpastamba. Hence the Indians knew that five distinct triangles in rational numbers a, b, c such that $a^2+b^2=c^2$ are right angled. Yet, strangely enough, Āpastamba says, with reference to the seven triangles which he mentions, 'so many *recognizable* constructions are there', as if he knew of no other rational right-angled triangles. But Āpastamba does also state the equivalent of Eucl. I. 47 in general terms, though without proof, and bases on it constructions for finding the square equal to (1) the sum, (2) the difference, of two given squares. He also recognizes the truth of the theorem for an isosceles triangle, and even gives a construction for $\sqrt{2}$ or the length of the diagonal of a unit-square; he in fact constructs a line which is $\left(1+\frac{1}{3}+\frac{1}{3.4}-\frac{1}{3.4.34}\right)$ times the side. This approximation to $\sqrt{2}$ is no doubt derived from the consideration that $2.12^2 = 17^2-1$, but the author does not betray any knowledge of the fact that this approximate value is not exact.

The Indians, therefore, knew empirically of the property of right-angled triangles and stated it generally. But they gave no indication of any proof; their statement appears to have been the result of an imperfect induction from a very small number of cases of right-angled triangles in rational numbers known to them. This is in great contrast to what is attributed to Pythagoras, which includes the discovery of a general formula for finding an unlimited number of rational right-angled triangles.

Assuming that, as Vitruvius says, Pythagoras began with the triangle (3, 4, 5), the next step would be to seek for other similar cases. An experiment may have been made 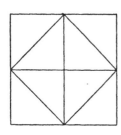 with an isosceles right-angled triangle, and the mere drawing of a figure would indicate the property in this case. If the middle points of the sides of a square be joined in order as in the annexed diagram, we have a square inside the original square and obviously half as large. But the original square is equal to the square on the diagonal of the inner square; therefore the square on the diagonal is equal to twice the square on the side.

In suggesting a possible method by which the general theorem was first proved we have a choice between two different lines of proof. One would be to represent the three squares in a figure, and to show how the two are equal to the one; this would be after the manner of Euclid, Book II. The other would be to use proportions after the manner of Euclid, Book VI.

If the first method is preferred, no better suggestion can be made than that of Bretschneider and Hankel. The first of the subjoined figures, which is like that of Euclid, II. 4, represents a larger square of side $(a+b)$ and two smaller squares of sides a, b respectively, with the two complementary rectangles (a, b). Dividing each complementary rectangle into two equal right-angled triangles by drawing the diagonal c, we then dispose the four triangles within another square of side $(a+b)$ as shown in the second figure. Deducting the four right-angled triangles

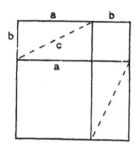 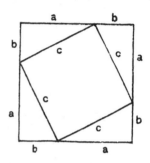

(a, b, c) from the larger square in each figure, we have as remainders, (1) in the first figure, the squares on a and b, and (2) in the second figure, the one square on c. Therefore the sum of the squares on a, b is equal to the square on c.

The proof by proportion might take different forms. Let ABC be a triangle right angled at A. Draw AD perpendicular to BC. Then the triangles DBA, DAC are similar to the triangle ABC and to one another.

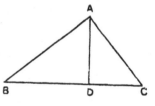

Now (1) it follows from the theorems of Eucl. VI. 4 and 17 that
$$BA^2 = BD \cdot BC,$$
and
$$AC^2 = CD \cdot BC.$$
Therefore
$$BA^2 + AC^2 = BC^2.$$

Alternatively (2) it would be seen that, in the similar triangles DBA, DAC, ABC, the corresponding sides opposite to the right angle in each case are BA, AC, BC.

The triangles are therefore in the duplicate ratio of those sides, and so are the squares on the latter. Therefore the squares are proportional to the corresponding triangles.

But two of the triangles, DBA, DAC, together make up the third triangle ABC.

Therefore the same is true of the corresponding squares,

or $$BA^2 + AC^2 = BC^2.$$

It must not be overlooked that the Pythagorean theory of proportion was only applicable to commensurable quantities. This would be no obstacle to the use of proportions in such a proof so long as the existence of the incommensurable remained undiscovered. But, when once the incommensurable was discovered, it would be necessary, pending the appearance of a new theory of proportion applicable to incommensurable as well as to commensurable magnitudes, to invent new proofs independent of proportions in place of those in which proportions were used. Now it will be noticed that the first of the above proofs by proportion shows that the square on BC is equal to the sum of two rectangles, and this is *precisely* what Euclid proves in his proposition I. 47. It appears probable therefore that Euclid found the proposition proved by means of proportions and, by a stroke of genius, gave the proof a different form in order to get the proposition into Book I in accordance with his general arrangement of the *Elements*.

(γ) *Application of areas and geometrical algebra.*

For want of the necessary notation the Greeks had no algebra in our sense. They were obliged to use geometry as a substitute for algebraical operations; and the result is that a large part of their geometry may appropriately be called 'geometrical algebra'. One of the two main

APPLICATION OF AREAS

methods at their disposal was the 'application of areas' (the other being the method of proportions). We have it on the authority of Eudemus, cited by Proclus, that the method of 'application of areas' (παραβολὴ τῶν χωρίων), their *exceeding* (ὑπερβολή), and their *falling-short* (ἔλλειψις), was the discovery of the Pythagoreans. The method is fundamental in Greek geometry, and gives the geometrical solution of the equivalent of algebraic equations of a degree not higher than the second.

The simplest case is 'application' pure and simple, as in Eucl. I. 44, 45: To apply to a given straight line as base a parallelogram containing a given angle and equal in area to a given triangle or rectilineal figure. This is equivalent to the operation of finding x where $ax = bc$, i.e. of dividing the product bc by a.

The general case where the applied area 'exceeds' or 'falls short' is enunciated thus: To apply to a given straight line a parallelogram equal to a given rectilineal figure and (1) *exceeding* or (2) *falling-short* by a parallelogram similar to a given parallelogram. In the accompanying figures

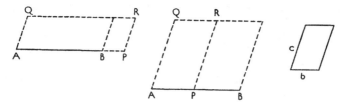

the parallelograms AR are *applied* to the straight line AB, but in the first figure the base AP overlaps AB and the parallelogram *exceeds* [the parallelogram on AB itself] by the parallelogram BR, while in the second figure the base AP falls short of AB and the parallelogram AR *falls short* by the parallelogram BR. The problem is, given AB, to draw the figure so that the parallelogram AR shall be equal to a given area (C, say), while the excess or defect BR is similar to a given parallelogram. What has in effect to be done is to determine the *size* of the excess or defect BR (its *shape* is determined by the given figure to which it has to be similar); that is to say, to determine one of the sides BP, PR in such a way that, when the figure is completed, the parallelogram AR may be equal to C. Let the ratio of BP to PR be that of b to c, and suppose that $BP = x$. Let $AB = a$. Then $AP = a \pm x$ and $PR = \dfrac{c}{b}x$. Now the area of the required parallelogram is $m \cdot AP \cdot PR$, that is, $m(a \pm x)\dfrac{c}{b}x$, where m is a certain constant depending on the size of the given angle BPR (actually the sine of that angle). Hence the equation to be solved is

$$m(a \pm x)\frac{c}{b}x = C.$$

In the case of defect, corresponding to the negative sign, the possibility of a solution is subject to a certain condition. Euclid actually proves the necessary condition in that case, and gives the geometrical solution of the two cases (VI. 27–9).

The cases arising most commonly in Greek geometry are simpler cases in which the parallelogram to be applied is a rectangle and the excess or defect is a square. The corresponding equation is then of the form

$$(a \pm x)x = b^2.$$

To solve this equation we should first, if necessary, change the sign throughout so as to make the term in x^2 positive, then add $\tfrac{1}{4}a^2$ on both sides so as to make the left side a complete square. We have then on the right hand

APPLICATION OF AREAS

$\frac{1}{4}a^2 \pm b^2$, and we equate the square root of this to the side of the complete square on the left hand. The Greek geometrical procedure was the exact equivalent, as we see from the particular case solved by Euclid in II. 11. We have to divide AB at G so that $AB \cdot BG = AG^2$. If $AB=a, AG=x$, this is equivalent to

$$a(a-x) = x^2$$

or $$x^2 + ax = a^2.$$

Euclid bisects AD, the side of the square on AB, at E, and joins EB. Then, after producing EA to F so that $EF = EB$, he makes AG equal to AF.

Now $EB^2 = \frac{1}{4}a^2 + a^2$
$= (x + \frac{1}{2}a)^2$, from above.

And $EF = EB$; therefore $EF = x + \frac{1}{2}a$, so that $AF = x$, which is therefore found.

The solutions of the cases

$$(a \pm x)x = b^2$$

are connected with Eucl. II. 5, 6. These propositions are in the form of theorems. But suppose e.g. that, in the figure of II. 5, $AB=a, BD=x$.

Then

$(a-x)x$ = rectangle AH
= gnomon NOP.

If, then, the area of the gnomon ($=b^2$, say) is given, we have the equation

$$ax - x^2 = b^2,$$

or $$x^2 - ax = -b^2.$$

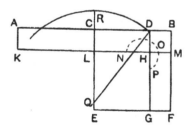

To solve this equation, we add $\frac{1}{4}a^2$ to both sides and equate ($\frac{1}{2}a-x$) to the square root of ($\frac{1}{4}a^2-b^2$). (For a real solution, therefore, b^2 must not exceed $\frac{1}{4}a^2$.)

The geometrical equivalent is this. Bisect AB at C and draw CQ at right angles to AB and of length equal to b. Then, with Q as centre and $\frac{1}{2}a$ as radius, draw a circle. If $\frac{1}{2}a > b$, the circle will cut CB in some point D.

By construction, $CD^2 = QD^2 - QC^2 = \frac{1}{4}a^2 - b^2$; and, by the equation, this is equal to $(\frac{1}{2}a-x)^2$. Thus by finding D we have found CD or $\frac{1}{2}a-x$, and x, or DB, is determined.

It is important to note that Apollonius employs the terminology of 'application of areas' to describe the fundamental properties of the three conics. These properties are equivalent to the following Cartesian equations referred to axes which are in general oblique:

$$y^2 = px \quad \text{(the parabola)},$$

$$y^2 = px + \frac{p}{d}x^2 \text{ (the hyperbola)},$$

$$y^2 = px - \frac{p}{d}x^2 \text{ (the ellipse)},$$

where d is the diameter of reference and p the corresponding 'parameter'. This is the origin of the names which were applied to the three conics for the first time by Apollonius himself: *parabola* = 'application', *hyperbola* = 'exceeding', *ellipse* = 'falling-short'.

The problem of Eucl. II. 14 is dependent on I. 44, 45, and is the equivalent of the solution of the pure quadratic $x^2 = A$, or the extraction of the square root.

The whole of Euclid's Book II, with the section of Book I from Prop. 42 to the end, may be said to deal with the transformation of areas (or the sums or differences of areas) of rectilineal figures into equivalent areas of different

THE IRRATIONAL

shape or composition by means of 'application' and the use of the theorem of I. 47. A characteristic of Book II is the use of the *gnomon*, which is essentially Pythagorean. Pythagorean, too, are the theorems of II. 9, 10, which are not only very useful in geometry, but were specifically used for the purpose of proving the property of the successive 'side'- and 'diameter'-numbers (cf. pp. 55-7 above).

The quantitative comparison of areas could be made by means of proportions, the other main method employed in the geometrical algebra. The ratio of one area to another (or of the content of one solid figure to that of another) could be expressed as a ratio between straight lines, and such ratios could be compounded or otherwise manipulated to any desired extent.

(δ) *The irrational*.

The discovery of the incommensurable by the Pythagoreans was bound to cause a great sensation, the more so as it would immediately be seen to throw doubt on so much of the Pythagorean proofs of theorems in geometry as rested on their (arithmetical) theory of proportion. To avoid the *impasse*, it would be necessary to seek proofs on other lines where possible; but geometry undoubtedly suffered a serious set-back pending the discovery by Eudoxus (408-355 B.C.) of the new theory of proportion applicable to commensurable and incommensurable magnitudes alike. In the meantime the position was so inconvenient that we can understand a desire on the part of the inner circle of the Pythagoreans that the discovery should not bécome known to the profane. This may, perhaps, account for the legend that the first of the Pythagoreans (whether it was Hippasus or another) who made it public perished at sea for his impiety, or (according to another version) was banished from the community and had a tomb erected for him as if he were dead.

(ε) *The five regular solids.*

Proclus, speaking of Pythagoras, says in parenthesis that, in addition to a theory of proportionals, he discovered 'the putting-together of the cosmic figures', i.e. the five regular solids. Next we have the story that Hippasus 'was a Pythagorean, but, owing to his having been the first to publish the (construction of the) sphere from the twelve pentagons (i.e. the inscribing of the dodecahedron in a sphere) perished by shipwreck for his impiety, but received credit for the discovery although it really belonged to HIM, for it is thus that they refer to Pythagoras, and they do not call him by his name'.

Now in what sense, if at all, did Pythagoras or the Pythagoreans discover the 'putting-together' of the five regular solids? Some light is thrown on this question by the procedure of Plato in the *Timaeus*. Plato there shows how to construct the regular solids in the elementary sense of putting triangles and pentagons together to form their faces. He forms a square from four isosceles right-angled triangles, and an equilateral triangle from three pairs of triangles which are the halves of equal equilateral triangles cut into two by bisecting one of the angles. Then he forms solid angles by putting together (1) squares three by three, (2) equilateral triangles three by three, four by four, and five by five respectively; the first figure so formed is a cube with eight solid angles, the next three are the tetrahedron, octahedron, and icosahedron respectively. The fifth figure, the dodecahedron, has pentagonal faces, and Plato forms solid angles by putting together equal

THE FIVE REGULAR SOLIDS

pentagons three by three; the result is a regular solid with twelve faces and twenty solid angles.

There is nothing in all this that would be beyond Pythagoras or the Pythagoreans, provided that the construction of the regular pentagon was known to them; the method of formation of the solids agrees well with the known fact that the Pythagoreans put angles of certain regular figures together round a point and showed that only three of such angles would fill up the space in one plane round the point. Moreover, there is evidence of the existence of dodecahedra in very early times. Thus a regular dodecahedron of Etruscan origin discovered on Monte Loffa (Colli Euganei, near Padua) in 1885 is held to date from the first half of the first millennium B.C. It is possible, therefore, that Pythagoras or the Pythagoreans had actually seen such a dodecahedron.

As regards the regular pentagon, we may observe that the construction of it in Euclid, Book IV, depends on the construction of a certain isosceles triangle, which again depends on the problem of cutting a straight line 'in

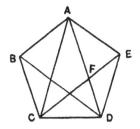 

extreme and mean ratio'. The last-named problem is certainly Pythagorean, being a simple case of 'application of areas'. The construction of the regular pentagon was, therefore, well within the powers of the Pythagoreans. It may have been evolved in some such way as this. Suppose the pentagon constructed, as $ABCDE$. Join AC, AD, CE. Now the Pythagoreans knew the theorems about the sum of the angles of any triangle and the sum of the interior angles of any polygon. They would, therefore, see that each angle, as ABC, of the regular pentagon is $\frac{6}{5}$ths of a right angle. It follows that, in the isosceles triangle BAC, the angle BAC is $\frac{2}{5}$ths of a right angle. So is the angle EAD, and so is the angle ECD, for the like reason. Subtracting the two angles BAC, DAE from the angle BAE (which is $\frac{6}{5}$ths of a right angle), we see that the angle CAD is $\frac{2}{5}$ths of a right angle. So is the angle ACE, for the like reason. It follows that, in the isosceles triangle ACD, each of the base angles is double of the vertical angle.

Again, if AD, CE meet in F, CFD is an isosceles triangle because the angle CFD, being equal to the sum of the angles CAF, ACF, is $\frac{4}{5}$ths of a right angle, and is therefore equal to the angle CDF. Hence $CD = CF = AF$.

Moreover, the triangles ACD, CDF are equiangular and therefore similar;

therefore $\qquad AC : CD = CD : DF$
or $\qquad AD : AF = AF : FD$,

that is, AD is divided at F in extreme and mean ratio.

If, therefore, AD is given, F can be found, and we can construct the regular pentagon on CD as base.

The interest of the Pythagoreans in the regular pentagon is further attested by the 'triple interwoven triangle' or pentagram, i.e. the star-pentagon which, according to Lucian and the scholiast to the *Clouds* of Aristophanes, was used by the Pythagoreans as a symbol of recognition between members of the same school and was called by them Health. I have drawn the star-pentagon separately, and the close connexion between the two figures could hardly be missed.

That the dodecahedron is inscribable in a sphere would probably be inferred from a consideration of its regular formation, without constructing the sphere and finding the centre of it in the scientific manner of Eucl. XIII. 17, where the relation between an edge of the dodecahedron and the radius of the sphere is also found: an investigation probably due to Theaetetus. For, according to Suidas, Theaetetus was the first to 'write upon' (or 'construct') the five regular solids, which probably means that he was the first to construct them theoretically and to investigate their relations with the circumscribing spheres and with one another. The scholium No. 1 to Euclid's Book XIII says of that Book that it is about 'the five so-called Platonic figures, which however do not belong to Plato, three of the five being due to the Pythagoreans, namely the cube, the pyramid, and the dodecahedron, while the octahedron and the icosahedron are due to Theaetetus'. This may well have been the case.

(ζ) *Pythagorean astronomy*

In astronomy Pythagoras realized that the earth (and no doubt each of the other heavenly bodies also) is spherical in form, and he also knew that the sun, moon, and planets have movements of their own in a sense opposite to that of the daily rotation. So far as we know, however, he kept the earth in the centre. His successors in the school (one Hicetas of Syracuse and Philolaus are alternatively credited with this innovation) deposed the earth from its place in the centre and made it revolve, like the sun, the moon, and the planets, and an assumed additional body, the 'counter-earth', round the 'central fire', wherein was located the governing principle which directs the movement and activity of the universe. The counter-earth, which accompanies the earth and revolves in a smaller orbit, is not seen by us because the hemisphere of the earth on which we live is turned away from the counter-earth. As the Pythagoreans, according to Aristotle, considered that eclipses of the moon occur owing to the interposition, between it and the sun, sometimes of the earth, sometimes of the counter-earth, the latter may have been invented in order to account for the frequency of lunar eclipses as compared with solar.

SUMMARY.

It may be useful to summarize the contribution of Pythagoras and the Pythagoreans to geometry. With them geometry became a scientific subject studied for its own sake; indeed for Pythagoras geometry was science itself ('geometry was called by Pythagoras "inquiry", ἱστορία'). Pythagoras explored the first principles, starting with definitions, and built upon them a logically connected system.

The positive achievements of the Pythagoreans in geometry were these:

1. They knew the properties of parallels, and used them to prove generally that the sum of the angles of any triangle is equal to two right angles. They deduced the well-known theorems about the sums of (1) the exterior, (2) the interior, angles of any polygon.

2. The transformation of areas of rectilineal figures, and the sums and differences of such areas, into equivalent areas of different shapes, in the manner of Eucl. I. 42–7 and Eucl. II, was their creation. To this end they invented the powerful method of 'application of areas', the main

constituent of the geometrical algebra by which they effected the geometrical equivalent of addition, subtraction, division, extraction of the square root, and finally (with the help of the theorem of the square on the hypotenuse and proportions) the complete solution of the general quadratic equation $x^2 \pm pq \pm q = 0$ so far as it has real roots.

3. They had a theory of proportion pretty fully developed, though it was only applicable to commensurable magnitudes, being presumably a numerical theory on the lines of Euclid, Book VII. They were aware of the properties of similar figures; for Plutarch attributes to Pythagoras himself the solution of the problem of describing a rectilineal figure similar to one given figure and equal in area to another, and this implies a knowledge of the theorem that similar rectilineal figures are in the duplicate ratio of corresponding sides. Much of the content of Euclid, Book VI, must, therefore, have been known to the Pythagoreans.

4. They had discovered, or were aware of the existence of, at least three of the regular solids. There is no reason to doubt that they could construct a regular pentagon in the manner of Eucl. IV. 10, 11.

5. They discovered the existence of the incommensurable in at least one case, that of the diagonal of a square in relation to its side; and they also devised a method of obtaining closer and closer approximations to the value of $\sqrt{2}$ in the form of numerical fractions, by means of the series of 'side'- and 'diameter'-numbers which are the successive solutions of the equations

$$2x^2 - y^2 = \pm 1,$$

for which purpose they used the theorems of Eucl. II. 9, 10.

VI
PROGRESS IN THE ELEMENTS DOWN TO PLATO'S TIME

FOR the period from Pythagoras to Plato the summary of Proclus gives us little more than a list of names of persons distinguished in geometry. After Pythagoras, he says,

Anaxagoras of Clazomenae dealt with many questions in geometry, and so did Oenopides of Chios, who was a little younger than Anaxagoras.... After them came Hippocrates of Chios, the discoverer of the quadrature of the lune, and Theodorus of Cyrene, both of whom became distinguished geometers; Hippocrates indeed was the first person of whom it is recorded that he actually compiled Elements. Plato, who came next to them, caused mathematics in general and geometry in particular to make a very great advance, owing to his own zeal for these studies; indeed every one knows that he filled his writings with mathematical discourses and strove on every occasion to arouse enthusiasm for mathematics in those who took up philosophy. At this time, too, lived Leodamas of Thasos, Archytas of Taras, and Theaetetus of Athens, by whom the number of theorems was increased and a further advance was made towards a scientific grouping of them.

The special object of the writer was clearly the limited one of tracing progress in the Elements as distinct from geometry in general. The name of Hippias of Elis who belongs to the period in question is omitted, doubtless because the curve afterwards called the *quadratrix*, his special discovery, belongs to higher geometry. There is no mention of Democritus, perhaps for a like reason. Hippocrates of Chios comes into the list, but mainly because he was supposed to have been the first writer of Elements; the fact that he squared certain lunes is only mentioned parenthetically, and nothing is said of his reduction of the problem of doubling the cube to that of finding two mean proportionals in continued proportion between two given straight lines. The particular contributions of the distinguished geometers mentioned to the body of the Elements are also not stated in our passage, though in a later passage Proclus attributes to Oenopides two propositions in Euclid, Book I, presently to be mentioned.

We will now take the names belonging to this period in their order.

ANAXAGORAS (about 500–428 B.C.), born at Clazomenae in the neighbourhood of Smyrna, neglected his possessions, which were considerable, in order to devote himself to science. Being asked what was the good of being born, he replied, 'the investigation of sun, moon, and heaven'. He took up his abode in Athens, where he enjoyed the friendship of Pericles. When Pericles became unpopular shortly before the outbreak of the Peloponnesian War, he was attacked through his friends, and Anaxagoras was accused of impiety for saying that the sun is a red-hot stone and the moon earth. According to one account he was fined five talents and banished; another account says that he was kept in prison and that it was intended to put him to death, but that Pericles obtained his release; he then left Athens and spent the rest of his life at Lampsacus.

In astronomy Anaxagoras has the distinction of being the first to state clearly that the moon has no light of its own, but receives its light from the sun. Solar eclipses he attributed to the interposition of the moon, and lunar eclipses to the interposition, sometimes of the earth, but sometimes of other opaque and invisible bodies which he imagined to exist 'below the moon'. He accounted for the evolution of the universe by a hypothesis which contained some fruitful ideas. First a vortex was set up in a portion of the mixed mass in which 'all things were together', by Mind. The rotation, beginning at the centre, gradually spread further outwards. This resulted, first, in the separation of two great masses, the 'aether' on the outside (consisting of the rare, hot, light, dry), and on the inside the 'air' (consisting of the opposite categories). From the 'air' were next separated clouds, water, earth, and stones. As the result of the circular motion, all the heaviest things collected in the centre, and from these elements, when consolidated, the earth was formed. After this, in consequence of the violence of the whirling motion, the surrounding fiery aether tore stones away from the earth and kindled them into stars. These whirlings-off are of precisely the same kind as the theory of Kant and Laplace assumed for the formation of the solar system.

We can believe that Anaxagoras was a good mathematician, but all that we are told is that, when in prison, he tried to write or prove (ἔγραφε) the squaring of the circle, that he held that magnitudes are divisible without limit ('in the small there is no smallest but only smaller and smaller *ad infinitum*'), and that he wrote on perspective (the art of painting on a plane surface in such a way as to make some things appear to be in the background and others in the foreground).

Of OENOPIDES of Chios no personal details seem to be recorded save that he was a little younger than Anaxagoras. He was primarily an astronomer. Eudemus' *History of Astronomy* credited him with the discovery of the 'cincture of the zodiac circle', which must be taken to mean the obliquity of the ecliptic. This is no doubt what the writer of the pseudo-Platonic *Erastae* had in mind when he represented Socrates as going into the school of Dionysius and finding two lads disputing a certain point, something about Anaxagoras or Oenopides, he was not certain which; but they appeared to be drawing circles and imitating certain inclinations by placing their hands at an angle.

Oenopides also estimated the period of a Great Year (the period after which the heavenly bodies resume their relative places). He is said to have put the Great Year at 59 years, and the length of the year itself at $365\frac{22}{59}$ days. This Great Year clearly had reference to the sun and moon only.

In geometry Proclus says that Oenopides was the first to investigate the problem of Eucl. I. 12 (to draw a perpendicular to a given straight line from a given point outside it), because he thought it useful for astronomy; only Oenopides used the archaic name for perpendicular, 'gnomon-wise' (κατὰ γνώμονα), the gnomon being perpendicular to the horizon. Now perpendiculars had, of course, been drawn before, e.g. by the aid of a set-square; it may therefore be that Oenopides was the first to construct them theoretically, in the manner of Euclid, by means (ultimately) of the ruler and compasses only.

Proclus further cites Eudemus as attributing to Oenopides the discovery of the problem of Eucl. I. 23 (on a given straight line, and at a given point on it, to construct a rectilineal angle equal to a given rectilineal angle).

DEMOCRITUS, both as mathematician and as physicist, may be said at last to have come into his own. In the theory of atoms common to him and Leucippus he is the forerunner of the most modern researches into that subject. In mathematics (as we learn from the *Method* of

Archimedes discovered only twenty-four years ago) he enunciated the important propositions that the volume of a pyramid on any polygonal base is one-third of that of the prism with the same base and height, and that the volume of a cone is similarly one-third of that of the cylinder with the same base and height, though the discovery of a scientific proof of these propositions was reserved for Eudoxus.

Democritus came from Abdera and, according to his own account, was young when Anaxagoras was old. Apollodorus placed his birth in Ol. 80 ($=460$–457 B.C.), while, according to Thrasylus, he was born in Ol. 77. 3 ($=470/69$ B.C.), being one year older than Socrates. He lived to a great age, 90 according to Diodorus, 100, 104, 108, or 109 according to other accounts. There was no subject to which he did not contribute, from mathematics and physics on the one hand to ethics and poetics on the other; he even went by the name of 'Wisdom'. Sharing out his patrimony with two brothers, he took a smaller share because it was in cash, and he wanted it to spend in travelling for the purpose of study. He is said to have visited Egypt, Persia, and Babylonia, where he consorted with priests and magi; some say he went to India and Aethiopia also. He cared nothing for fame: 'I came to Athens and no one knew me.' His sole thought was for philosophy and science: 'he would rather find out one single explanation of cause than win the throne of Persia.' Plato ignored him and is said to have wished to burn all his works. Aristotle, however, in *De generatione et corruptione*, pays a warm tribute to his genius; no one, he says, had observed anything about change and growth except superficially, whereas Democritus seemed to have thought of everything.

A long list of his works is given by Diogenes Laërtius on the authority of Thrasylus. In astronomy he wrote, among other things, a book *On the Planets* and another called *The Great Year* or *Astronomy*, with a *Parapegma* (calendar). Censorinus describes Democritus' Great Year as being 82 (LXXXII) years including 28 intercalary months; but, as Callippus' cycle of 76 years contained the same number of intercalary months, it is probable that LXXXII is an incorrect reading for LXXVII (77). Democritus also wrote a geographical and nautical survey or description of the inhabited earth, in which he made an innovation. Whereas Anaximander, Hecataeus, and Damastes of Sigeum all made the inhabited earth round, with Greece in the middle and Delphi in the centre of Greece, Democritus was the first to make it elongated, with length $1\frac{1}{2}$ times its breadth.

Except for the reference in Archimedes' *Method* and another by Plutarch presently to be mentioned, we have no information about Democritus' mathematics beyond the titles of works classed as mathematical. These include:

1. *On a difference of opinion, or on the contact of a circle and a sphere.*
2. *On geometry.*
3. *Geometricorum* (? I, II).
4. *Numbers.*
5. *On irrational lines and solids* (atoms).

In the first title I have translated the reading γνώμης ('opinion'), as Cobet's reading γνώμονος (gnomon) gives no sense in this connexion. Democritus evidently discussed in the treatise the nature of the contact between a circle and a tangent to it, and between a sphere and a tangent plane. I formerly suggested to read γωνίης instead of γνώμης, making 'On a difference in an *angle*', in order to bring the alternative titles into closer relation, for nothing would be more natural than that Democritus should discuss the supposed 'angle' which a circle and a tangent to it make at the point of contact. The Greeks (as well as the mathematicians of the thirteenth to the seventeenth century) were much exercised by this particular 'angle' (as at one time they thought it), which they called by the name 'horn-like'. The idea of it even survives in one place in Euclid (III. 16), where he proves that it is less than any rectilineal angle whatever. But it is not really necessary to alter the reading γνώμης (opinion) if we read 'difference of opinion' in the light of a genuine fragment of Democritus preserved by Sextus Empiricus:

There are two kinds of γνώμη (opinion or knowledge), the one genuine, the other dark or blind; to the latter belong sight, hearing, smell, taste, touch; the former (the genuine) is entirely distinct from these. When the dark or blind kind can no longer see, hear, smell, taste, or touch a thing in any further minuteness of detail, but finer penetration ⟨is yet required, then the genuine kind comes to the rescue, since it possesses a more refined organ of apprehension⟩.

(That is, the mind's eye can still see when sense-perception has given out.) Now this is just the sort of distinction to which we should expect Democritus to appeal if he were discussing such a popular objection to the mathematician's view of the nature of the contact of a circle and its tangent as we know that Protagoras put forward 'to confute the geometers' (*vide* Aristotle, *Metaph.* B. 2. 997 b 35), namely, that no such straight lines or circles as the geometer assumes exist in nature, and we shall find that a material ruler does not touch a material circle in one point only. We can imagine Democritus replying to this

DEMOCRITUS

effect: 'It is true that, owing to the imperfection of our instruments, we cannot *draw* a mathematical straight line touching a mathematical circle, and hence we cannot *visualize* the tangent touching the circle at one point. But, none the less, we can see this with the eye of the *mind*; and we *know*, by force of demonstration, that it cannot be otherwise.' We have the less difficulty in believing that the object of the treatise 'On a difference of opinion . . .' was precisely the refutation of such a popular fallacy, since we know that Democritus broached other matters in geometry involving the consideration of infinitesimals.

Plutarch quotes, on the authority of Chrysippus, a dilemma put by Democritus with regard to parallel sections of a cone as follows:

If a cone were cut by a plane parallel to the base [meaning a plane very close to the base], what must we think of the surfaces forming the sections? Are they equal or unequal? For, if they are unequal, they will make the cone irregular as having many indentations, like steps, and unevennesses; but if they are equal, the sections will be equal, and the cone will appear to have the property of the cylinder, and to be made up of equal, not unequal, circles: which is very absurd.

The words '*made up* of equal, not unequal, circles' show that Democritus had already conceived the notion of a solid being the sum of an infinite number of plane laminae, parallel to one another, infinitely thin, and infinitely near together: an idea which Archimedes afterwards used in his *Method*. This idea may also be at the root of Democritus' discovery that the volume of a pyramid is one-third of that of the prism with the same base and height. By drawing three diagonals in the three parallelogrammic faces of a prism with triangular bases, as in Eucl. XII. 7, we divide the prism into three pyramids, and Euclid's proposition proves that the three pyramids are equal to one another. Now the figure shows that the pyramids, two and two, have equal bases and heights. It is, therefore, only necessary to prove that triangular pyramids on equal bases and of the same height are equal. This is a particular case of a more general proposition proved by Euclid in XII. 5 by the rigorous 'method of exhaustion' invented by Eudoxus. Democritus may have argued thus. If the two pyramids be cut respectively by planes parallel to the base and dividing the heights in the same ratio, the corresponding sections are equal. Therefore the pyramids contain the same infinite number of equal plane sections or infinitely thin laminae; therefore the pyramids are equal in content.

Seeing that Democritus came so near to the subject of infinitesimals, there is nothing surprising in his having written on irrational lines and solids. Lest it should be supposed that his physical theory of atoms would be an obstacle to his holding that mathematical magnitudes, e.g. lines, are divisible *ad infinitum*, we have the evidence of Simplicius that, according to Democritus, his atoms were in a mathematical sense divisible *ad infinitum*; as a mathematician, therefore, he would obviously have nothing to do with 'indivisible lines', as is indeed clear from a scholium to Aristotle's *De Caelo* which says that some, as e.g. Leucippus and Democritus, believe in indivisible bodies and others, like Xenocrates, in indivisible lines.

HIPPIAS of Elis, the famous sophist already mentioned (pp. 9, 112), was nearly contemporary with Socrates and Prodicus and was probably born about 460 B.C. Chronologically his place would be here, but as, from the mathematical side, he is only known for his discovery of a higher curve, afterwards called the *quadratrix*, which was used, first for trisecting any angle, and later for squaring the circle, his achievement will come more appropriately in the next chapter (VII) on higher problems.

HIPPOCRATES of Chios would appear to have been in Athens for a considerable portion of the second half of the fifth century, say from 450 to 430 B.C. Why he came to Athens and how he was induced to stay there is told in the story of his capture by pirates on a voyage taken in the course of his business as a merchant (cf. pp. 8, 74 above). He is famous in geometry for three things.

(1) He was, so far as we know, the first to compile a book of Elements.

(2) In an attempt to square the circle he showed how to square certain classes of 'lunes'.

(3) He was the first to observe that the problem of doubling the cube can be reduced to that of finding two mean proportionals in continued proportion between two given straight lines, with the result that the problem was ever afterwards attacked in the latter form.

In the absence of any information about the contents of his Elements, it is fortunate that we possess in the Commentary of Simplicius on the *Physics* of Aristotle considerable extracts from the account of Hippocrates' quadratures of 'lunes' which was contained in Eudemus' *History of Geometry*. We are able to gather from these extracts some valuable indications of the progress made in the Elements down to the time when Hippocrates wrote on the quadratures in question.

The occasion for the disquisition of Simplicius on Hippocrates is a remark by Aristotle that there is no obligation on the exponent of any subject to refute every fallacy that may have been put forward, but only those which their authors purport to base upon the admitted principles forming the foundation of the subject in question. 'Thus', says Aristotle, 'it is for the geometer to refute the (supposed) quadrature of a circle by means of segments, but it is not the business of the geometer to refute the argument of Antiphon' (the latter being held to have infringed one or other of the admitted principles of geometry). It is clear that by the quadrature 'by means of segments' in this passage Aristotle means the same thing as what he elsewhere (*Soph. El.*) calls 'the fallacy of Hippocrates or the quadrature by means of the *lunes*'.

(a) *Quadratures of lunes.*

Simplicius first refers to Alexander's account of the quadratures effected by Hippocrates and the fallacy into which Alexander fell; he then passes to Eudemus' account and concludes very properly that we must pay more regard to Eudemus' version because he was 'nearer the times'.

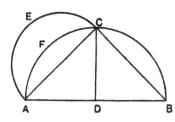

According to Alexander, Hippocrates first squared the lunes formed outside a circle by the semicircles described outwards on the sides of a *square* inscribed in the circle, and then straightway assumed that he had squared the lunes similarly left outside the circle by semicircles described on the sides of a regular hexagon inscribed in the circle. The latter lunes are quite different from the former, and Hippocrates was too good a mathematician to be capable of making such a false inference.

The quadrature in the case of the square is as follows. In the above figure AC is the side of a square inscribed in a circle, and AB the diagonal; AEC is a semicircle described on AC as diameter.

Now $$AB^2 = 2AC^2.$$

And circles (and therefore semicircles) are to one another as the squares on their diameters;
therefore (semicircle ACB) = 2(semicircle AEC),
so that (quadrant $AFCD$) = (semicircle AEC).

Subtract the common part, the segment AFC, and we have (lune $AECF$) = $\triangle ADC$.

Simplicius observes that, while quoting Eudemus 'word for word', he has added a few things from Euclid's *Elements* for clearness' sake, and to aid the memory, in view of the summary and condensed style adopted by Eudemus. Hence there has been controversy as to whether certain sentences come from the original extract from Eudemus or were added by Simplicius. The matter has, however, been very closely studied, and we can now feel fairly satisfied how much we may attribute to Eudemus. I shall reproduce so much of the passage and no more.

Eudemus begins with some general remarks to the effect that the quadratures of lunes, which were considered to belong to an uncommon class of propositions on account of the close relation of lunes to the circle, were first investigated by Hippocrates, whose exposition was thought to be in order.

He began by laying down, as the first proposition useful for his purpose, the theorem that similar segments of circles have the same ratio to one another as the squares on their bases have. And this he proved by first showing that the squares on the diameters have the same ratio as the circles. For, as the circles are to one another, so also are similar segments of them, since similar segments are those which are the same part of the circles respectively. . . .

Simplicius goes on to illustrate 'part': 'as for instance a semicircle is similar to a semicircle, and a third part of a circle to a third part'. This has led some to think that in this last sentence τμήματα (properly 'segments') is used in the sense of 'sectors', since a segment which is a third part of a circle is not easy to visualize, while a *sector* which is a third part of a circle is easily drawn. But the word τμήματα can hardly mean two different things in two consecutive sentences, and it clearly means 'segments' in the preceding sentence and also in the first proposition quoted from Hippocrates. It is therefore best to treat the illustration of a half and one-third of the circle as an unnecessary addition, and to take 'the same part of the circles respectively' as being the phraseology of the definition of proportion then in use (cf. Eucl. VII, Def. 20).

The important fact with regard to the Elements which emerges from the above is that Hippocrates was the first to prove (whatever his method was) that the areas of circles are as the squares on their diameters. He also had a correct idea of similar segments, however he defined them.

Eudemus next gives Hippocrates' *first* quadrature of a lune. This lune had for its outer circumference the
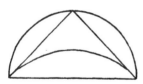
circumference of a semicircle. The inner circumference was arrived at in this way. Joining the extremities of the diameter to the middle point of the arc of the semicircle, we have an isosceles triangle the sides of which cut off segments from the semicircle. We then draw on the diameter of the semicircle a segment similar to one of those segments. It follows that the segment on the diameter is equal to the sum of the two smaller segments. Adding to both the

portion of the isosceles triangle above the arc of the segment described on the diameter, we prove that the 'lune' is equal to the triangle.

Hippocrates' *second* quadrature was that of a lune in which the outer circumference is greater than a semicircle, and which is constructed as follows. $ABDC$ is a parallel-trapezium in which the three sides BA, AC, CD are all equal, and the longer parallel side BD is such that $BD^2 = 3BA^2$. A circle is described about $ABDC$, and a segment of a circle is described on BD which is similar to one of the segments cut off by BA, AC, CD.

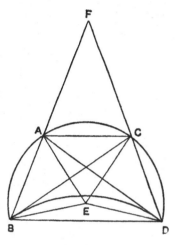

We have first to prove that the segment $BACD$ is greater than a semicircle. Now the angle BAC is greater than a right angle, for, since $BD > AC$, the straight lines BA, DC will, if produced, meet and form an isosceles triangle FAC, so that the angle FAC is less, and the angle BAC greater, than a right angle.

Therefore $\qquad BC^2 > BA^2 + AC^2$.

Therefore $\qquad BC^2 + CD^2 > BA^2 + AC^2 + CD^2$, or BD^2.

Hence the angle BCD is less than a right angle, and the segment $BACD$ is greater than a semicircle.

Eudemus does not give the actual quadrature, but it is obvious.

Since $\qquad BD^2 = 3BA^2$,

(segment on BD) = 3 (segment on BA)
$\qquad\qquad$ = (sum of segments on BA, AC, CD).

Add to each the area between the straight lines BA, AC, CD and the circumference of the segment on BD; therefore (trapezium $BACD$) = (the lune).

Hippocrates' *third* quadrature relates to a lune in which the outer circumference is less than a semicircle. The construction is important, and part of it must be given in Hippocrates' own words:

'*Let there be a circle with diameter AB, and let its centre be K. Let CD bisect BK at right angles, and let the straight line EF be*

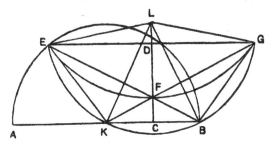

so placed between CD and the circumference that it verges towards B [i.e. will, if produced, pass through B], *while its length is such that the square on it is $1\frac{1}{2}$ times the square on (one of) the radii.*'

Hippocrates next draws EG parallel to AB, joins KE, KF, produces KF to meet EG in G, and joins BF, BG.

'*It is then manifest*', he says, '*that BF produced will pass through E* [since EF verges towards B], *and BG will be equal to EK.*

This being so, I say that the trapezium $EKBG$ can be comprehended in a circle.'

He then circumscribes a segment of a circle about $EKBG$ and draws another segment circumscribing EFG. Now 'each of the segments [on] EF, FG is manifestly similar to each of the segments [on] EK, KB, BG'. [This is clear because $EK = KB = BG$, and all the segments contain equal angles, namely an angle equal to the supplement of the angle EGK.]

The lune $EKBGFE$ is then proved equal to the sum of the triangles BFG, BFK, KFE.

For $\qquad EF^2 = \tfrac{3}{2} EK^2$;

therefore (segment on EF) = $\tfrac{3}{2}$(segment on EK),

or $\quad$ 2(segment on EF) = 3(segment on EK)
$\qquad\qquad\qquad$ = (sum of segments on EK, KB, BG).

Now (the lune) = segments on EK, KB, BG + (trapezium $EKBG$) − (segment EFG).

But the segment EFG is the sum of the triangle EFG and the segments on EF, FG; the sum of the segments on EK, KB, BG, and the sum of those on EF, FG, being equal, cancel out in the equation (see the preceding step); and it follows that the lune is equal to the difference between the trapezium $EKBG$ and the triangle EFG, i.e. to the sum of the triangles BFG, BFK, KFE.

It remains to prove that the segment $EKBG$ is less than a semicircle, or that the angle EKF is obtuse.

The proof purports to be given in Hippocrates' own words, but the text is confused, and it is assumed without proof that $BK^2 > 2BF^2$.

This may be proved thus. Since the isosceles triangles EKB, KFB have one angle (FBK) common, they are similar;

therefore $\qquad EB : BK = BK : BF$,

or $\qquad\qquad EB \cdot BF = BK^2$;

that is, $EF \cdot FB + BF^2 = BK^2 = \tfrac{2}{3} EF^2$, by construction.

It follows that $BF < EF$, and therefore that $EB > 2FB$.

Hence $BK^2 > 2FB^2$, or $EK^2 > 2KF^2$.

Now, says Hippocrates,
$$EF^2 = \tfrac{3}{2}EK^2$$
$$= EK^2 + \tfrac{1}{2}EK^2$$
$$> EK^2 + KF^2,$$
so that the angle EKF is obtuse.

The most remarkable thing in the above construction is the assumption of the solution of the problem 'To place between the circumference of the semicircle AEB and the straight line CD a straight line EF *verging* to B and such that the square on EF is $1\tfrac{1}{2}$ times the square on KB'.

This is one of the problems called by the Greeks νεύσεις (*vergings*, *inclinationes*). It may be expressed in the alternative form, To draw a straight line BFE meeting CD and the circumference of the semicircle AEB in F, E respectively, and such that the square on EF is equal to a given area, in this case $\tfrac{3}{2}BK^2$.

Let us see what this amounts to. Suppose the line BFE drawn as required, and suppose that $BF = x$, $BK = a$. If we could find the proper length of BF, or the proper value of x, the problem would be solved.

Now, since the isosceles triangles EKB, KFB are similar,
$$EB : BK = BK : BF, \quad \text{or} \quad EB \cdot BF = BK^2.$$
But $BF = x$, and $EB = EF + FB = \sqrt{\tfrac{3}{2}} \cdot a + x$.
Therefore $\qquad x(x + \sqrt{\tfrac{3}{2}} \cdot a) = a^2,$
or $\qquad\qquad x^2 + \sqrt{\tfrac{3}{2}} \cdot ax = a^2.$

This is the problem of 'applying to a straight line of length $\sqrt{\tfrac{3}{2}} \cdot a$ a rectangle equal to the square on a and exceeding by a square figure', and would theoretically be solved by the Pythagorean method based on the theorem of Eucl. II. 6. This would be in the power of Hippocrates; but it is possible that in his time, and before the definite restriction of the geometer's permissible instruments to the ruler and compasses, mathematicians may have used the quasi-mechanical method of marking a length equal to $\sqrt{\frac{3}{2}} \cdot a$ on a straight-edge and placing it so that the marked points fell on the circumference and CD respectively, while the edge of the ruler passed through B. This method is perhaps indicated by the fact that Hippocrates first places EF as required and afterwards joins BF.

The *fourth* of Hippocrates' quadratures is that of a certain lune *plus* a circle.

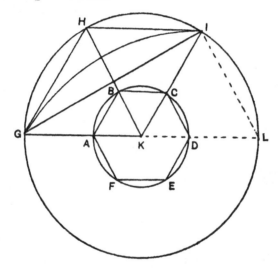

Let there be two circles with common centre K, and such that the square on the diameter of the outer is 6 times the square on the diameter of the inner.

Let $ABCDEF$ be a regular hexagon inscribed in the inner circle. Join $KA, KB, KC,$ and produce them to meet the outer circle in G, H, I. Join GH, HI. Then GH, HI are sides of a regular hexagon inscribed in the outer circle.

On GI draw a segment of a circle similar to the segment cut off by GH or HI.

Now, says Hippocrates, $GI^2 = 3GH^2$.

For, if GL be the diameter of the outer circle, and IL be joined, $GI^2 + IL^2 = 4GK^2 = 4GH^2$, and $IL = GH$;
therefore $\qquad GI^2 = 3GH^2$.
Also $\qquad GH^2 = 6AB^2$, by hypothesis.
Therefore (segment on GI) = 3 (segment on GH)
$\qquad$ = 2 (segment on GH)
$\qquad\quad$ + 6 (segment on AB)
$\qquad$ = (segments on GH, HI) + (all segments in inner circle).

[Add to each side the area between GH, HI, and the arc GI;]
therefore
($\triangle$ GHI) = (lune GHI) + (all segments in inner circle).

Add to each the hexagon $ABCDEF$;
therefore
(lune GHI) + (inner circle) = ($\triangle GHI$) + (hexagon $ABCDEF$).

All the quadratures of Hippocrates as given by Eudemus are perfectly correct. How then came Aristotle to speak of the 'fallacy' in them? There are two possibilities. (1) He may have had before him the same account of Hippocrates' quadratures as that given by Alexander, where the fallacy is obvious. Otherwise (2) the only possible explanation seems to be that a false notion had become current as to what Hippocrates claimed to have done, owing to the use by some ancient author of ambiguous language such as that employed in a sentence in Simplicius which seems to come from Eudemus: 'Thus Hippocrates squared every lune, in so far as (εἴπερ καί) he squared one in which the outer circumference is the arc of a semicircle, one in which it is greater, and one in which it is less, than a semicircle.' The word 'every' in such a statement might, by an unwary reader, have been taken literally, without proper regard to the qualifying phrase.

As it is, Hippocrates' achievement is remarkable, for he discovered and proved the possibility of quadrature in the case of three out of the five species of lunes which alone can be squared by 'plane' methods.

(b) *Reduction of the problem of doubling the cube.*

Hippocrates discovered that the problem of doubling the cube is reducible to that of finding two mean proportionals in continued proportion between two straight lines. That is, he discovered that, if $a : x = x : y = y : b$, then $a^3 : x^3 = a : b$. This shows that he understood the compounding of ratios. Hippocrates may have been led to his result by the analogy of numbers. Plato has in the *Timaeus* the propositions about the number of means in continued proportion required to connect two square numbers and two cube numbers respectively.

(c) *The Elements as known to Hippocrates.*

We have seen that the Pythagorean geometry already contained the substance of Euclid's Books I, II, part of Book IV, and theorems corresponding to a great part of Book VI. There is no direct evidence that the Pythagoreans gave much attention to the geometry of the circle as we find it in Book III. But by the time of Hippocrates the main propositions of Book III were also known and used, as we see from the above account of the quadratures of lunes. Hippocrates also assumes the solution of the problem of circumscribing a circle about a triangle (Eucl. IV. 5) and the property of the hexagon inscribed in a circle (Eucl. IV. 15).

Most remarkable of all is the fact that, according to Eudemus, Hippocrates actually proved the theorem of Eucl. XII. 2 that *circles are to one another as the squares on their diameters*, and afterwards used this to prove that *similar segments of circles are to one another as the squares on their bases*. Unfortunately we are not told how he proved these propositions.

THEODORUS of Cyrene is mentioned by Proclus along with Hippocrates as a celebrated geometer. He is said to have been Plato's teacher in mathematics and to have been visited by Plato in Cyrene; but we gather from the *Theaetetus* of Plato that he had also been in Athens in the time of Socrates. He was distinguished not only in geometry but in astronomy, arithmetic, music, and other educational subjects. There is no reason why we should not take as historical the passage (147 D) in the *Theaetetus* where Theaetetus mentions a discourse in which Theodorus proved that the side of a square containing three square feet is incommensurable with the foot-length, and similarly the sides of squares containing five square feet, seven square feet, and so on up to seventeen square feet, 'at which point for some reason he stopped'. That is to say, Theodorus proved the irrationality of $\sqrt{3}$, $\sqrt{5}$, $\sqrt{7}$ ... up to $\sqrt{17}$. There is no mention of the diagonal of a square in relation to its side, no doubt because the irrationality of $\sqrt{2}$ and the traditional way of proving it were already well known.

It is not known how Theodorus proved the incommensurability of $\sqrt{3}$, $\sqrt{5}$ ... with 1; but his method was no doubt of such a kind that it had to be applied afresh to each case in consequence of the variation of the specific numbers entering into the proofs; and, moreover, it must have been such that, when Theodorus had applied it as far as $\sqrt{17}$, he thought it hardly worth while to proceed to further cases. There seem to be two possibilities. (1) He may have adapted the Pythagorean proof in the case of $\sqrt{2}$, by substituting 3, 5 ... for 2. (2) He may have performed the geometrical equivalent of finding the greatest common measure, and stopped when he found a ratio recurring, and when therefore it would be clear that the process would never end.

THEAETETUS of Athens is associated in the summary of Proclus with Leodamas of Thasos and Archytas of Taras as having 'increased the number of theorems', while Proclus also says that Euclid perfected many of Theaetetus' theorems. Our first information about Theaetetus is contained in Plato's dialogue of that name. He fought for the Athenians in a battle near Corinth in 369 B.C., and was carried home suffering from wounds and dysentery, of which he died. The dialogue, a commemorative tribute to the departed friend, records a conversation between Socrates, Theodorus, and Theaetetus which purports to have taken place some 30 years before, when Theaetetus was quite young (say 18). Theodorus had been discoursing on the irrationality of $\sqrt{3}$, $\sqrt{5}$ ..., and Theaetetus says that the idea then occurred to him of trying to comprehend all these 'square roots' (δυνάμεις), as he calls them, under one term. That is to say, he conceived the idea of generalizing the theory of these 'roots'. This is confirmed by what we learn from a commentary on Euclid, Book X, which survives in Arabic and is attributed to Pappus. We are there told that the theory of irrationals, which began in the school of Pythagoras, was considerably developed by Theaetetus, who had not only distinguished square roots commensurable in length from those which are incommensurable, but had set apart the well-known species of irrational straight lines which we find in Euclid, Book X, the 'medial', the 'binomial', and the 'apotome'; Eudemus is cited as saying that he further associated these irrationals with the three means, the arithmetic, the geometric, and the harmonic respectively. Lastly, a scholiast on Book X definitely ascribes to Theaetetus the discovery of the general theorem of X. 9 (that squares which have not to one another the ratio of a square number to a square number have their sides incommensurable in length).

We infer that Theaetetus laid the foundation of the theory of irrationals as we find it in Euclid, Book X.

He contributed no less notably to Euclid's Book XIII, which (after twelve introductory propositions) is devoted to constructing the five regular solids, circumscribing spheres about them, &c. Suidas says that Theaetetus was the first to write on the five regular solids, while scholium No. 1 to Eucl. XIII adds that, while three of the five were the discovery of the Pythagoreans, namely, the cube, the pyramid, and the dodecahedron, the other two, the octahedron and icosahedron, were discovered by Theaetetus himself. However this may be, Theaetetus was probably the first to construct all five theoretically, and to investigate fully their relations to one another and to the circumscribing spheres, as in Eucl. XIII.

ARCHYTAS of Taras, a Pythagorean, the friend of Plato, flourished in the first half of the fourth century, say 400–365 B.C. Plato made his acquaintance when staying in Magna Graecia, and he is said, by means of a letter, to have saved Plato from death at the hands of Dionysius, tyrant of Syracuse. Statesman and philosopher, he was famous for every sort of accomplishment. He was general of the forces of his city-state for seven years, though ordinarily the law forbade any one to hold the post for more than a year; and he was never beaten. He is said to have been the first to write a systematic treatise on mechanics based on mathematical principles. Vitruvius mentions that, like Archimedes, Ctesibius, Nymphodorus, and Philon of Byzantium, Archytas wrote on machines; two mechanical devices are attributed to him, one a mechanical dove, made of wood, which would fly, the other a rattle which, according to Aristotle, was found useful to 'give to children to occupy them and so prevent them from breaking things about the house (for the young', he adds, 'are incapable of keeping still)'.

Archytas distinguished the four mathematical sciences, arithmetic, geometry, sphaeric (or astronomy), and music. He defined the three means in music, the arithmetic, geometric, and harmonic (a name which, with Hippasus, he substituted for the older name 'subcontrary'). He held that the universe is infinite in extent.

'If', he said, 'I were at the outside, could I stretch my hand or my stick outwards or not? To suppose that I could not is absurd. If I can, that which is outside is either body or space (it makes no difference which). We can in the same way get to the outside of that again, and so on continually. This involves extension without limit. If now what so extends is body, the proposition is proved; but even if it is space, then, since space is that in which body is or can be, and in the case of eternal things we must treat that which potentially is as being, it follows equally that there must be body and space extending without limit.'

In geometry, though Proclus says that Archytas 'increased the number of theorems', only one fragment has come down under his name. This, however, is nothing less than a wonderful construction in three dimensions for finding two mean proportionals in continued proportion between two straight lines, the problem to which Hippocrates had reduced the doubling of the cube. This will be given in the next chapter.

In music Archytas gave the numerical ratios representing the intervals of the tetrachord on three scales, the anharmonic, the chromatic, and the diatonic. He held that sound is due to impact, and that higher tones correspond to quicker motion communicated to the air, and lower tones to slower motion.

In arithmetic (theory of numbers) we have a fragment of Archytas preserved by Boëtius, consisting of a proof of the proposition that there can be no number which is a (geometric) mean between two numbers in the ratio known as ἐπιμόριος or *superparticularis*, that is, the ratio $(n+1) : n$. This is highly interesting because the proof is substantially identical with the proof of the same proposition in the *Sectio Canonis* attributed to Euclid, so that we can appreciate the slight changes of form and expression between the two periods represented, and also because Archytas assumes as known certain propositions which afterwards found their place in Euclid, Book VII.

Let A, B, says Archytas, be the given ratio. [Unlike the *Sectio*, Archytas puts the smaller number first, so that A, B are numbers in the ratio of n to $n+1$.]

Take C, DE the smallest numbers which are in the ratio of A to B. [DE here means $D+E$. In the *Sectio* the corresponding numbers are, as usual, the parts DG, GF of a straight line DF. The step of finding $C, D+E$ as the smallest numbers in the ratio of A to B presupposes Eucl. VII. 33 as applied to two numbers.]

Then DE exceeds C by an aliquot part of itself and of C [i.e. by a number which divides both DE and C: the number is in fact 1, which divides both n and $n+1$. Archytas is using a definition of the ἐπιμόριος ratio like that in Nicomachus, I. 19. 1.]

Let D be the excess [i.e. we suppose $E = C$].

I say that D is not a number, but a unit.

For, if D is a number and an aliquot part of DE, it measures DE; therefore it measures E, that is, C.

Thus D measures both C and DE: which is impossible, since the smallest numbers which are in the same ratio as any numbers are prime to one another [this presupposes Eucl. VII. 22].

Therefore D is a unit; that is, DE exceeds C by a unit.

Therefore no number can be found which is a mean between the two numbers C, DE [for there is no integer intervening].

Therefore neither can any number be a mean between the original numbers A, B which are in the same ratio as C, DE [cf. the more general proposition, Eucl. VIII. 8; the particular inference is a consequence of Eucl. VII. 20, to the effect that the least numbers of those which have the same ratio with them measure the latter the same number of times, the greater the greater and the less the less].

Since this proof presupposes several theorems corresponding to propositions in Euclid, Book VII, the inference is that there already existed, at a date at least as early as Archytas, a treatise of some kind on the Elements of Arithmetic in a form similar to the Euclidean, and containing many propositions afterwards embodied by Euclid in his arithmetical Books.

We can now form an idea of the progress of the Elements down to Plato's time expressed in terms of the content of the *Elements* of Euclid. Book V expounding Eudoxus' great theory of proportion applicable to incommensurable as well as to commensurable magnitudes was of course wanting; but the substance of Books I–IV was practically complete, and the main propositions of Book VI were also known. The Pythagoreans had a theory of proportion which was probably on the lines of Book VII. There existed Elements of Arithmetic partly at all events on the lines of Book VII and containing some propositions of Book VIII. Many of the properties of plane and solid numbers and of similar numbers of both kinds proved in Books VIII and IX must also have been known, while the Pythagoreans had conceived the idea of 'perfect' numbers. The foundations of Books X and XIII had been laid by Theaetetus. The theorems of XI. 1–19 must already have formed part of the Elements; XI. 21 must have been known to the Pythagoreans, and those who were familiar with the theory of plane and solid numbers were in a position to know the properties of parallelepipedal solids, which are the subject of the latter portion of Book XI. Book XII employs the *method of exhaustion*, the orthodox form of which is attributed to Eudoxus. But the theorem of XII. 2 was known to Hippocrates of Chios, and Democritus had enunciated the theorems of XII. 7 Por. and XII. 10 about the volumes of a pyramid and a cone respectively.

VII
SPECIAL PROBLEMS

BEFORE the Elements were complete, the Greeks attacked problems in higher geometry; three such problems, the squaring of the circle, the trisection of any angle, and the doubling of the cube, were rallying points for mathematicians during three centuries at least, and the whole course of Greek geometry was profoundly influenced by this fact, for, in their efforts to find solutions, geometers were irresistibly impelled to seek for new curves and constructions that would serve their turn. Thus, e.g., the investigation of the conic sections seems to have begun with the use of two of them for solving the problem of the two mean proportionals.

The Greeks classified problems and the loci (lines or curves) which served for solving them according to what we should call 'degree'. *Plane* problems were those which could be solved by means of the straight line and circle (these alone are *plane* loci), *solid* problems those which could be solved by the use of one or more conic sections (*solid* loci, as they were called), and finally *linear* problems those which required for their solution what Pappus calls '*linear*' loci, comprising all curves of higher order than conics, such as spirals, *quadratrices*, cochloids (conchoids), and cissoids, or again the various curves included in the class of 'loci on surfaces', by which Pappus seems to mean loci drawn on surfaces, such as the cylindrical helix.

THE SQUARING OF THE CIRCLE

This problem has, perhaps more than any other, had a fascination for inquirers, whether mathematicians or not, throughout the ages. In Egypt, as we have seen, at least as early as 1800 B.C., the area of a circle was taken to be $\frac{64}{81}d^2$, where d is the diameter; this was no doubt a standard value arrived at as the result of repeated attempts at actual measurement; and the approximation (giving 3·16, very nearly, as the value of π) was by no means a bad one.

The first person connected by tradition with the problem in Greece was Anaxagoras, who is said to have worked at it while in prison. Hippocrates of Chios (*vide* the last chapter) squared certain lunes, doubtless in the hope (at least in the first instance) that these investigations might show a way towards the solution of the main problem.

The next important contribution to the subject was that of Antiphon the Sophist, a contemporary of Socrates. He tried the method of inscribing successive regular polygons in a circle. According to some writers he began with a square, according to others with an equilateral triangle, inscribed in the circle. He then drew on each side of the inscribed figure an isosceles triangle with its vertex on the smaller segment cut off by the said side. This gave a regular polygon inscribed in the circle and having twice the number of sides. Proceeding in this way, and doubling continually the number of sides in the inscribed polygon, he thought (so we are told) that 'in this way the area of the circle would be used up, and we should sometime have a polygon inscribed in the circle, the sides of which would, owing to their smallness, coincide with the circumference of the circle'. Since, then, we can make a square equal to any polygon, we shall be in a position to make a square equal to the circle.

ANTIPHON. BRYSON

The idea of Antiphon was at first scouted. Aristotle said it was an error which was even beneath the notice of geometers; geometers were not even called on to refute Antiphon's argument, because it was not based on the admitted principles of geometry. According to Eudemus, the principle which Antiphon's argument violated was the principle that magnitudes are divisible without limit; for, if this is true, Antiphon's process will *never* use up the whole area of the circle, or make the sides of the polygon coincide with the circumference. Yet Antiphon's bold pronouncement was of great significance because it contained the germ of the famous *method of exhaustion* and, ultimately, of the integral calculus; the defect in his statement was little more than verbal, for he had only to put it in the more cautious form of saying, as Euclid does in XII. 2, that, if the process be continued far enough, the segments left over will be together less than any assigned area. The practical use of Antiphon's construction is illustrated by Archimedes' *Measurement of a Circle*, where, by constructing an inscribed regular polygon of 96 sides by Antiphon's method, Archimedes arrives at $3\frac{10}{71}$ as the lower limit to the value of π, while, by circumscribing a similar polygon, he shows that $\pi < 3\frac{1}{7}$. The same construction, starting from a square, was also the basis of Vieta's approximation for $2/\pi$, namely

$$\frac{2}{\pi} = \cos\frac{\pi}{4} \cdot \cos\frac{\pi}{8} \cdot \cos\frac{\pi}{16} \ldots$$
$$= \sqrt{\tfrac{1}{2}} \cdot \sqrt{\tfrac{1}{2}(1+\sqrt{\tfrac{1}{2}})} \cdot \sqrt{\tfrac{1}{2}\{1+\sqrt{\tfrac{1}{2}(1+\sqrt{\tfrac{1}{2}})}\}} \ldots ad\ inf.$$

BRYSON, a pupil of Socrates or of Euclid of Megara, made an attempt at a quadrature which Aristotle calls 'sophistic' and 'eristic' because it was based on principles not special to geometry but applicable equally to other subjects. The commentators speak of Bryson's argument in similar terms, but there is no clear tradition as to what it actually was. All agree that he used both inscribed and circumscribed polygons (or squares), and assumed another polygon (or square) described which is intermediate between the inscribed and circumscribed figures. Then he sought to relate the area of the circle to the intermediate figure, observing that the area of the circle is greater than all inscribed and less than all circumscribed polygons. Perhaps, therefore, we may suppose that he increased the number of sides in the inscribed and circumscribed polygons, much as Antiphon had done with the inscribed polygons, until it was a close enough approximation to the truth to say that, *if* a polygon intermediate between the inscribed and circumscribed polygons could be drawn, the area of the circle would be equal to that of the intermediate polygon. In that case Bryson's idea was useful as suggesting the *compression*, as it were, of inscribed and circumscribed figures into one, so that they will ultimately coincide with the circle and with one another, a characteristic of the method of exhaustion as practised by Archimedes later.

We now come to the geometers who, by means of higher curves, achieved the quadrature or rectification of the circle. The first is HIPPIAS, who invented the curve afterwards known, from its property, as the *quadratrix* (τετραγωνίζουσα). Traditions vary; Pappus says that Dinostratus (a brother of Menaechmus), Nicomedes, and other later geometers used the curve for squaring the circle; according to Proclus, others used 'the *quadratrices* of Hippias and Nicomedes' for trisecting any rectilineal angle. It is possible, therefore, that Hippias first used his curve for the purpose of trisecting any angle or dividing it in any ratio.

(a) The Quadratrix of Hippias.

The *quadratrix* is theoretically constructed as follows: *ABCD* is a square and *BED* a quadrant of a circle with centre *A*. Suppose (1) that a radius of the circle turns uniformly about *A* from the position *AB* to the position *AD*, and (2) that a line remaining always parallel to *AD* moves uniformly and *in the same time* from the position *BC* to the position *AD*.

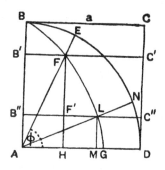

Then in their ultimate positions the moving straight line and the revolving radius will both coincide with *AD*; and at any time during the motion the moving line and the moving radius determine by their intersection a point as *F* or *L*. The locus of these points is the *quadratrix*.

The property of the curve is that

$\angle BAD : \angle EAD = $ (arc *BED*) : (arc *ED*) $= AB : FH$.

In other words, if ϕ is the angle *FAD* made by any radius vector *AF* with *AD*, ρ the length of *AF*, and a the side of the square,

$$\frac{\rho \sin \phi}{a} = \frac{\phi}{\frac{1}{2}\pi}.$$

The curve, when once constructed, clearly enables us not only to trisect any angle but also to divide it in any ratio.

For let *EAD* be the given angle, and let *FH* be divided at *F'* in the given ratio.

Through *F'* draw *B" C"* parallel to *BC* meeting the curve in *L*. Join *AL* and produce it to meet the quadrant in *N*.

Then $\angle NAD : \angle EAD = LM : FH$
$= F'H : FH.$

Therefore $\angle EAN : \angle NAD = FF' : F'H =$ the given ratio.

The application of the *quadratrix* to the rectification of the circle is a more difficult matter because it requires us to know the position of G, the point where the curve meets AD. The difficulties did not escape the Greeks; they are admirably stated by Pappus, after Sporus. Even if we could in practice arrange the two uniform motions so as to take exactly the same time, G is not actually found at all. For, as Sporus says, if in the figure the straight lines CB, BA are made to end their motion together, they will then both coincide with AD itself and will not cut one another any more; yet it was the intersection of the moving straight lines which was supposed to give all the points on the curve. The fact is that we can only *approximate* to the position of G as nearly as we please.

But, assuming G found, we have then to prove the proposition which gives the length of the arc of the quadrant and therefore of the circle itself.

This proposition states that

(arc of quadrant BED) $: AB = AB : AG.$

This is proved in Pappus by *reductio ad absurdum*. The truth of it is easily seen by taking the limit and putting $\phi = 0$ in the equation above (p. 143), whence

$$AG = a / \tfrac{1}{2}\pi ;$$

therefore (arc of quadrant) $= \tfrac{1}{2}\pi a = a^2 / AG.$

(b) *The Spiral of Archimedes.*

We are told that Archimedes used his spiral for squaring the circle; and he does in fact show how to rectify a circle by means of the 'polar subtangent' to the spiral.

THE SPIRAL OF ARCHIMEDES

The spiral is constructed thus. Let a straight line OB revolve in a plane at uniform speed about a fixed point O, starting from the position OA (the 'initial line'); and let a point P on OB also move at uniform speed along OB, starting from O at the same time as OB starts from OA.

The polar equation of the curve is clearly $\rho = a\theta$.

Let the tangent at any point P of the spiral meet at T the straight line drawn from O at right angles to the radius vector OP; then OT is the 'polar subtangent'.

Now in the book *On Spirals* Archimedes proves the equivalent of the fact that, if P be a point on the 'first turn' of the spiral, OT the corresponding 'polar subtangent', and ρ, θ the polar coordinates of P, then OT is equal in length to the arc of the circle with O as centre and $OP\ (=\rho)$ as radius which is subtended by the angle θ at O.

In other words,
$$OT = \rho\theta,$$
or (since $\rho = a\theta$)
$$OT = \rho^2/a.$$

The point P may be on any turn of the spiral (say the nth), in which case θ becomes $2(n-1)\pi + \theta$, where θ is the angle (less than 2π) through which the radius vector has revolved on the nth turn starting from OA, and Archimedes proves in the general case that the subtangent OT is equal to $n-1$ times the complete circumference of the circle with radius OP *plus* the arc of that circle subtended by the angle θ at O.

(c) *Solutions by Apollonius and Carpus.*

Iamblichus, in a commentary on Aristotle's *Categories*, said that Apollonius squared the circle by means of a certain curve 'which he himself called "sister of the cochloid", but which is the same as Nicomedes' curve'. The curve of Nicomedes here mentioned is presumably the *cochloid*, afterwards called the *conchoid*, which is described by Pappus (see *infra*, pp. 150–1); but we know of no curve invented by Apollonius which resembled the cochloid. Apollonius did, however, write a tract on the curve which he called *cochlias*; this is the cylindrical helix, and, as it is possible to use this curve for squaring the circle, Apollonius may quite well have done so.

The same passage of Iamblichus says that Carpus squared the circle by means of a curve 'of double motion'. There is no indication of the nature of this curve.

(d) Ancient approximations to the value of π.

In his *Measurement of a Circle* Archimedes took a circle and inscribed and circumscribed to it regular polygons with 96 sides; then by sheer calculation of the perimeters of the polygons he proved that $3\frac{1}{7} > \pi > 3\frac{10}{71}$. Heron, however, tells us in his *Metrica* that, in another work, Archimedes made another and presumably more accurate calculation. The figures in the Greek text are unfortunately incorrect, the lower limit being given as the ratio of $\mu{,}a\omega o\epsilon$ to $\mu{,}\zeta\upsilon\mu a$, or 211875 : 67441 ($=3{\cdot}141635$), and the higher limit as $\mu{,}\zeta\omega\pi\eta$ to $\mu{,}\beta\tau\nu a$, or 197888 : 62351 ($=3{\cdot}17377$), the lower limit being thus greater than the true value and the higher limit greater than $3\frac{1}{7}$. The best suggestion for correction is perhaps that of J. L. Heiben, who would write $\mu{,}\zeta\upsilon\mu\delta$ for $\mu{,}\zeta\upsilon\mu a$ and $\mu{,}\epsilon\omega\pi\eta$ for $\mu{,}\zeta\omega\pi\eta$, giving

$$\frac{195888}{62351} > \pi > \frac{211875}{67444}$$

or $\qquad 3{\cdot}141697 \ldots > \pi > 3{\cdot}141495 \ldots$

The mean of these figures would give the remarkably close approximation $3{\cdot}141596 \ldots$.

APPROXIMATIONS TO VALUE OF π

Ptolemy gives a value for π expressed by means of sexagesimal fractions (i.e. in units, sixtieths, or *minutes*, and second-sixtieths or *seconds*), namely 3 8' 30", observing that this is very nearly the mean between $3\frac{1}{7}$ and $3\frac{10}{71}$. Ptolemy no doubt deduced this value from his Table of Chords. This gives the lengths of chords of a circle subtended by angles at the centre of $\frac{1}{2}°$, $1°$, $1\frac{1}{2}°$, and so on by half-degrees. The chords are expressed in terms of 120th parts of the diameter. If 1^p is one such part, the Table gives 1^p 2' 50" as the chord subtended by an angle of $1°$. The perimeter of the inscribed regular polygon with 360 sides is 360 times this figure and, dividing by 120 (the number of 'parts' in the diameter), we have, as the value of π, three times 1 2' 50", or 3 8' 30", which is Ptolemy's figure, equivalent to 3·1416.

The Indian mathematician Āryabhatta (born A.D. 476) gives the following calculation for π: 'To 100 add 4; multiply the sum by 8; add 62000 more and thus we have, for a diameter of 2 *ayutas* (=myriads), the approximate length of the circumference of the circle.' That is, $\pi = 62832/20000$ or 3·1416. It has been argued that the use of the myriad suggests a Greek source; but there seems to be no confirmation of this, and Indian authorities dispute it.

Eutocius, in his commentary on Archimedes' *Measurement of a Circle*, says that Apollonius in his Ὠκυτόκιον ('means of quick delivery') used other numbers and found a closer approximation than that of Archimedes.

THE TRISECTION OF ANY ANGLE

It was no doubt when attempting to inscribe in a circle a regular polygon the sides of which are *nine*, or any multiple of nine, in number that the Greeks found themselves confronted with the problem of trisecting an angle other than a right angle. According to Pappus, the ancients first tried 'plane' methods (i.e. those of the straight line and circle only), but failed because the problem is not 'plane' but 'solid' (and requires therefore the use of conics or some equivalent). Not however being, at the time, familiar with conic sections, they first reduced the problem to one of the type known as νεύσεις (*inclinationes* or *vergings*), and afterwards discovered the solution of the latter problem by means of conics.

(a) *Reduction to a νεῦσις solved by conics.*

We need only consider the case of any acute angle, since a right angle can be trisected by drawing an equilateral triangle.

The reduction to a νεῦσις is by analysis.

Let ABC be the given acute angle, and let AC be drawn perpendicular to BC. Complete the parallelogram $ACBF$. Produce FA, and *let E be a point on it such that, if EB be*

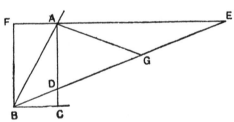

joined meeting AC in D, the intercept DE between AC and AE is equal to $2AB$; or, in the phraseology of νεύσεις, let there be placed between AC and AE a straight line (DE) equal in length to $2AB$ and verging towards B.

Bisect DE at G and join AG.

Then $AG = DG = GE = AB$.

Therefore $\angle ABG = \angle AGB = 2\angle AEG = 2\angle DBC$.

Hence $\angle DBC = \tfrac{1}{3}\angle ABC$.

Thus the problem is reduced to that of drawing BE from B to cut AC and AE in such a way as to make the intercept DE equal to $2AB$.

Pappus shows how to solve this problem in a more general form by means of conics.

Let $ABCD$ be any parallelogram (it need not be a rectangle as Pappus makes it); it is required to draw AEF from A to meet CD and BC produced in points E and F such that EF has a given length.

Suppose the problem solved, EF being of the required length.

Complete the parallelogram $EDGF$.

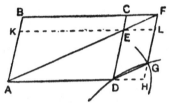

Then, EF being given in length, DG is given in length. Therefore G lies on the circle drawn with D as centre and radius equal to the given length.

Again, if we complete the parallelogram $ABFH$ and draw KEL parallel to AH, we see that, the complements being equal,

$$(BD) = (KH),$$

whence $BC.CD = KL.LH = BF.ED = BF.FG$.

Therefore G lies on a hyperbola with BF, BA as asymptotes and passing through D.

Thus, in order to solve the νεῦσις, we have only to draw (1) the said hyperbola, (2) the circle with centre D and radius equal to the given length.

The intersection of the two curves gives G. Then, to

determine E, F, we draw GF parallel to DC to meet BC produced in F, and join AF meeting CD in E.

(b) *The conchoids of Nicomedes.*

Nicomedes' place is here because he invented a curve for the specific purpose of solving νεύσεις such as the above. We gather that he was intermediate in date between Eratosthenes and Apollonius; he may therefore have been born about 270 B.C.

The curve in question is called by Pappus the *cochloid*, though later, e.g. in Proclus and Eutocius, it is spoken of as the *conchoid* ('shell-shaped'). Pappus mentions four varieties of the cochloidal curves; the 'first', which concerns us here, was used for trisecting any angle and doubling the cube. It was constructed by means of a mechanical device, thus:

AB is a ruler with a slot in it parallel to its length, FE a second ruler fixed at right angles to the first with a peg (C) fixed on it. A third ruler, PC, pointed at P, has a slot in it, parallel to its length, which fits the peg C. D is a fixed peg on PC in a straight line with the slot but on the under-side, and this peg D moves freely along the

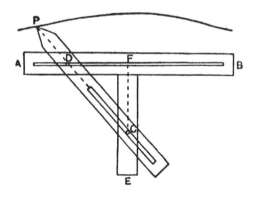

slot in AB. If then the ruler PC moves so that the peg D describes the length of the slot in AB on each side of F, the extremity P of the ruler describes the curve which is called a conchoid or cochloid. Nicomedes called the straight line AB the *ruler*, the fixed point C the *pole*, and the constant length PD the *distance*.

If r be the radius vector CP, and $DP = a$, $CF = b$, the polar equation of the curve is

$$r = a + b \sec \theta.$$

It has AB as an asymptote.

In the application of a suitable conchoid in a particular case we have to place the *pole* C at the point to which the 'inserted' line has to *verge*, and the *ruler* has to coincide with one of the lines between which the 'inserted' line has to be placed. For example, to solve the νεῦσις on p. 148, we must have a conchoid in which the fixed distance DP is equal to $2AB$, the *pole* is at B, and the *ruler* coincides with AC.

Thus we should in general have to construct a fresh appliance for every case. No wonder that, as Pappus says, the conchoid was not always actually drawn, but 'some', for greater convenience, moved a ruler about the fixed point until the intercept was found by trial to be equal to the given length.

(c) *Another reduction to a νεῦσις.*

The collection of lemmas (*Liber Assumptorum*) which has come down to us through the Arabic under the name of Archimedes contains an interesting proposition leading to another reduction to a νεῦσις of the problem of trisecting any angle.

If AB, any chord of a circle with centre O, be produced to C, so that BC is equal to the radius of the circle, and if CO meet the circle in the points D, E, then the arc AE will be three times the arc BD.

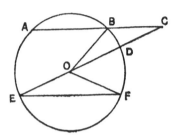

For draw the chord EF parallel to AB, and join OB, OF.

Then, since $BO = BC$,
$$\angle BOC = \angle BCO.$$
But $\angle COF = \angle OEF + \angle OFE = 2\angle OEF$
$$= 2\angle BCO, \text{ by parallels,}$$
$$= 2\angle BOC.$$
Therefore $\angle BOF = 3\angle BOC$,
so that (arc BF) = (arc AE) = 3 (arc BD).

We see from this that, in order to find an arc which is one-third of the arc AE, we have only to draw through A a straight line ABC meeting the circle again in B and EO produced in C, and such that the intercept BC is equal to the radius of the circle.

(d) *Solution by means of conics.*

Pappus gives two direct solutions of the trisection-problem by means of conics, without any preliminary reduction of it to a νεῦσις. The second is especially interesting because it is one of only three known passages in Greek mathematical works where the focus-directrix property of conics appears.

The analysis is as follows.

Let RPS be an arc of a circle which it is required to trisect.

Suppose it done, and let the arc SP be one-third of the arc SPR. Join SP, RP.

Then the angle RSP is equal to twice the angle SRP.

Bisect the angle RSP by the straight line SE meeting RP in E, and draw EX, PN perpendicular to RS.

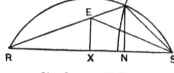

Then $\angle ERS = \angle ESR$, so that $RE = ES$,
whence $RX = XS$, and X is given.
Again, $RS : SP = RE : EP = RX : XN$.
Therefore $RS : RX = SP : NX$.
But $RS = 2RX$; therefore $SP = 2NX$.

It follows that P lies on a hyperbola with S as focus and XE as directrix, and with eccentricity 2.

Accordingly, in order to trisect the arc RPS, we have only to bisect RS in X, draw XE at right angles to RS, and then draw a hyperbola with S as focus, XE as directrix, and 2 as eccentricity.

We may here note that Pappus proves the focus-directrix property of the three conics in a section headed 'Lemmas to the *Surface-Loci* of Euclid', his enunciation being as follows: *If the distance of a point from a fixed point be in a given ratio to its distance from a fixed straight line, the locus of the point is a conic section, which is an ellipse, a parabola, or a hyperbola according as the given ratio is less than, equal to, or greater than, unity.* As this lemma must have been necessary for the understanding of Euclid's treatise, it is a fair inference that Euclid in his *Surface-Loci* assumed the proposition without proof, as being well known. Hence it was probably contained in some treatise current in Euclid's time, perhaps the work of Aristaeus on *Solid Loci*.

THE DUPLICATION OF THE CUBE, OR THE PROBLEM OF THE TWO MEAN PROPORTIONALS

In his commentary on Archimedes' *On the Sphere and Cylinder* Eutocius has preserved for us an invaluable collection of solutions of this famous problem. One of them is that of Eratosthenes, which is introduced by what purports to be a letter from Eratosthenes to Ptolemy. This was Ptolemy III Euergetes I, who, at the beginning of his reign (246 B.C.), persuaded Eratosthenes to come from Athens to Alexandria to be tutor to his son (Philopator). The supposed letter gives the tradition regarding the origin of the problem and the history of its solution down to the time of Eratosthenes. It mentions a 'votive monument' set up by Eratosthenes to Ptolemy, on which was fixed a representation in bronze of Eratosthenes' contrivance for solving the problem. On the pillar was a condensed proof with one figure and, at the end, an epigram. Von Wilamowitz-Möllendorff denies the authenticity of the letter, but thinks the proof and epigram to be the genuine work of Eratosthenes.

Our document says that an ancient tragic poet had represented Minos as dissatisfied with a tomb which he had put up to Glaucus, and which was only 100 feet each way; he therefore ordered it to be made double the size, the poet making him add that each dimension should be doubled for this purpose(!). The poet was, as von Wilamowitz has shown, not Aeschylus or Sophocles or Euripides, but some obscure person who owes the notoriety of his lines to his ignorance of mathematics. Geometers took up the question and made no progress for a long time, until

FINDING OF TWO MEAN PROPORTIONALS

Hippocrates of Chios showed that the problem was reducible to that of finding two mean proportionals in continued proportion between two given straight lines (cf. p. 131 above). Again, after a time, the Delians were told by the oracle that, if they would get rid of a certain plague, they should construct an altar double the size of the existing one, whereupon they went to consult Plato. Plato replied that the oracle meant, not that the god wanted an altar of double the size, but that he intended, in setting them the task, to shame the Greeks for their neglect of mathematics and their contempt for geometry. Plutarch says that Plato referred the Delians to Eudoxus and Helicon of Cyzicus for a solution. However this may be, the problem was studied in the Academy, and solutions are attributed to Eudoxus, Menaechmus, and even (though erroneously) to Plato himself. After Hippocrates' reduction of the problem, it was always attacked in the form of finding two mean proportionals between two given straight lines.

We will now give the recorded solutions in order.

Archytas.

Of all the solutions that of Archytas is the most remarkable, especially in view of its date (first half of fourth century B.C.); for it is a bold construction in three dimensions determining a certain point as the intersection of three surfaces, (1) a right cone, (2) a cylinder, (3) a *tore* or anchor-ring with inner diameter *nil*. The intersection of the two latter surfaces gives (says Archytas) a certain curve (it is in fact a curve of double curvature), and the point required is found as the point in which the cone intersects this curve.

Suppose that we have to find two mean proportionals

between AC and AB, the first being made the diameter of a circle and the second a chord in it.

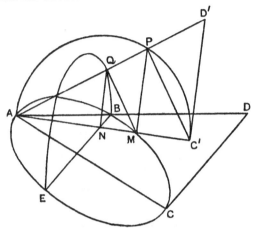

Draw a semicircle on AC as diameter, but in a plane at right angles to that of the circle ABC, and let the semicircle revolve about a straight line through A at right angles to the plane of ABC (thus describing half a *tore* with inner diameter *nil*).

Next draw a right half-cylinder on the semicircle ABC as base. This will cut the half-*tore* in a certain curve.

Lastly, let CD, the tangent to the circle ABC at C, be produced to meet AB produced in D; and suppose the triangle ADC to revolve about AC as axis. This will generate the surface of a right circular cone; the point B will describe a semicircle BQE in a plane at right angles to AC and to the plane of ABC; and the diameter BE will be at right angles to AC. The surface of the cone will meet in some point P the curve which is the intersection of the half-cylinder and the half-*tore*.

Let APC' be the corresponding position of the revolving semicircle, and let AC' meet the circumference ABC in M.

If PM be drawn perpendicular to the plane of the circle ABC, it must meet the circumference of that circle, since P is on the right cylinder standing on ABC as base.

Let AP meet the semicircle BQE in Q, and let AC' meet its diameter BE in N.

Now both semicircles are at right angles to the plane of the circle ABC; therefore so is their line of intersection QN (cf. Eucl. XI. 19).

Therefore QN is at right angles to both BE and AM.

Since QN is at right angles to BE,

$$QN^2 = BN \cdot NE = AN \cdot NM;$$

therefore the angle AQM is right.

But the angle APC' is also right; therefore QM is parallel to PC'.

By similar triangles, therefore,

$$C'A : AP = AP : AM = AM : AQ,$$

that is, $CA : AP = AP : AM = AM : AB$,

and AM, AP are two mean proportionals between AB and AC.

Compounding the ratios, we have

$$AC : AB = (AM : AB)^3;$$

therefore the cube described with AM as side is to the cube of side AB as AC is to AB.

In the particular case where $AC = 2AB$, $AM^3 = 2AB^3$, and the problem is solved.

Eudoxus.

Eudoxus' solution is unfortunately lost. The epigram of Eratosthenes says that Eudoxus used a species of curve (καμπύλον εἶδος). Eutocius had evidently seen some version of the solution which was not correct; for, he says, while in his preface Eudoxus claimed to have discovered a

158 PROBLEM OF TWO MEAN PROPORTIONALS

solution by means of 'curved lines', not only did he make no use of such lines in his proof, but he actually used a certain discrete proportion as if it were continuous. The latter part of this statement compels us to assume that Eutocius' source was in some way defective, for it is inconceivable that a mathematician of the calibre of Eudoxus could have made such a mistake.

Menaechmus.

Menaechmus, a brother of the Dinostratus who used the *quadratrix* to square the circle, was a pupil of Eudoxus. It is said that Alexander asked him to indicate a short cut to geometry, and he replied, 'O King, for travelling over the country there are royal roads and roads for common citizens, but in geometry there is one road for all'. (Euclid is similarly said to have told Ptolemy that there is no royal road to geometry.) Menaechmus is associated by Proclus with Amyclas of Heraclea and Dinostratus as having 'made the whole of geometry more perfect'. He wrote on the technology of mathematics; he discussed the meaning of the word *element*, the distinction between theorems and problems, the convertibility of propositions, and so forth. But his importance for us lies in the fact that the first appearance of conic sections and their properties is in his solutions of the problem of the two mean proportionals; so far as we know, therefore, he was the first to discover them. The properties which he actually uses are the ordinate property of a parabola and the asymptote-property of a rectangular hyperbola.

If x, y be the required two mean proportionals between two straight lines a, b, that is, if $a : x = x : y = y : b$, then clearly
$$x^2 = ay, \ y^2 = bx, \ xy = ab.$$

MENAECHMUS

The properties of the parabola and hyperbola used by Menaechmus are precisely those expressed by these relations when x, y are Cartesian co-ordinates referred to rectangular axes. Menaechmus' first solution used the second and third of the conics so represented, the second solution the first and second.

First solution.

Let AO, OB, placed at right angles, represent the two given straight lines, AO being the greater.

Suppose the problem solved, and let the two mean proportionals be OM measured along BO produced and ON measured along OA produced. Complete the rectangle $OMPN$.

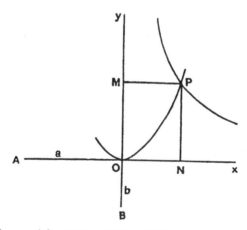

Then, since $AO:OM = OM:ON = ON:OB$,

we have (1) $OB \cdot OM = ON^2 = PM^2$,

so that P lies on a parabola with O for vertex, OM for axis, and OB for *latus rectum*.

Again (2) $AO \cdot OB = OM \cdot ON = PN \cdot PM$,

so that P lies on a hyperbola with O as 'centre' and OM,

ON as asymptotes, and such that the rectangle contained by the straight lines PM, PN drawn from any point P of the curve parallel to one asymptote and meeting the other respectively is equal to the given rectangle $AO \cdot OB$.

If, then, we draw the two curves in accordance with the data, we determine the point P by their intersection, and
$$AO : PN = PN : PM = PM : OB.$$

Second solution.

In this case we draw two parabolas, namely

(1) the parabola with O as vertex, ON as axis, and OA as *latus rectum*,

(2) the parabola with O as vertex, OM as axis, and OB as *latus rectum*.

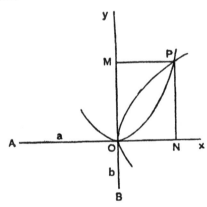

These parabolas determine by their intersection a point P such that

(1) $\qquad OA \cdot ON = PN^2,$

(2) $\qquad OB \cdot OM = PM^2.$

Since $PN = OM$, $PM = ON$, it follows that
$$OA : OM = OM : ON = ON : OB,$$

SOLUTION ATTRIBUTED TO PLATO

and OM, ON are the required mean proportionals between OA, OB.

Solution attributed to Plato.

This solution, which Eutocius alone gives or mentions, can hardly be Plato's, if only because we know that Plato objected on principle to solutions by mechanical means as destroying the good of geometry. It may have been

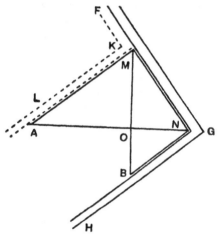

evolved in the Academy by some one contemporary with or junior to Menaechmus, for the arrangement in the figure of the two given straight lines and the two means between them is exactly the same as in Menaechmus' figure; they are arranged in straight lines mutually at right angles and in descending order of magnitude when taken in clockwise order. The difference is that what Menaechmus does by means of conics is here done by means of a mechanical contrivance of a sort akin to that which a shoemaker uses to measure the length of the foot. FGH is a rigid right angle made (say) of wood; KL is a strut which can move along GF, while always remaining parallel to GH, or at right angles to GF.

We have to place the machine so that the inner side of GH always passes through B, and the inner side of the strut (towards G) always passes through A, and then (subject to these conditions) to move the machine, and the strut along it, until (1) the inner angle at G lies on AO produced and (2) the inner angle (towards G) at K lies on BO produced. (This no doubt requires some little manipulation.)

Then the four lines OA, OM, ON, OB take up the same positions as in Menaechmus' figures and, since the angles at M, N are right,

$$MO^2 = AO \cdot ON \text{ and } NO^2 = MO \cdot OB,$$

whence $\quad AO : MO = MO : ON = ON : OB.$

Eratosthenes.

This, too, is a mechanical construction. The device consists of a rectangular frame along which slide three parallelograms (or the triangles which are the halves of them) of height equal to the width of the frame. The parallelograms or triangles move always so that their bases describe one straight line (one edge, say the upper, of the frame), and they can slide over one another.

The original positions of the parallelograms and triangles are shown in the first figure. AX, EY are the sides of the frame; AMF, MNG, NQH (the halves of the parallelograms ME, NF, QG) are the triangles which slide along the frame.

The second figure shows the result of sliding all the triangles except the first (which remains stationary) along from their original positions to positions in which they overlap one another, as AMF, $M'NG$, $N'QH$.

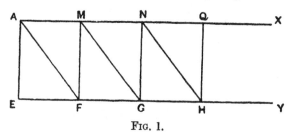

Fig. 1.

Let AE and DH (perpendicular to EY) in the second figure be the two given straight lines.

Let $N'QH$ be the position of the triangle NQH in which

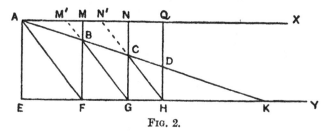

Fig. 2.

QH passes through D, and let the triangle $M'NG$ be such a position of the triangle MNG that the points B, C in which MF, $M'G$, and NG, $N'H$ respectively intersect are in a straight line with A, D.

Produce AD to meet EY in K.

Then $\quad AE:BF = EK:FK = AK:KB$
$$= FK:KG$$
$$= BF:CG.$$
Similarly, $\quad BF:CG = CG:DH.$

Therefore AE, BF, CG, DH are in continued proportion, and BF, CG are the required mean proportionals.

The epigram of Eratosthenes bids the reader 'not to seek to do the difficult business of Archytas' cylinders, or to cut the cone in the triads of Menaechmus, or to compass such a curved form of lines as is described by the god-like Eudoxus'. It adds that the same frame as that used above in Eratosthenes' solution would equally enable any number of means in continued proportion to be interpolated.

Nicomedes.

We gather from Eutocius that Nicomedes was just as proud of his own solution and contemptuous of the others as was Eratosthenes before him.

Nicomedes' solution depends on a certain νεῦσις which he effected by means of his conchoid.

Let AB, BC, placed at right angles, be the two given straight lines. Complete the parallelogram $ABCL$.

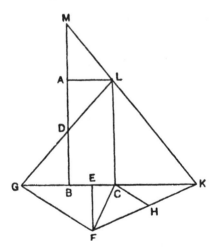

Bisect AB, BC at D, E; join LD and let LD, CB be produced to meet at G. Draw EF at right angles to BC and of length such that $CF = AD$. Join GF, and draw CH parallel to it.

Now from the point F draw FHK cutting CH in H and BC produced in K in such a way that the intercept $HK = CF = AD$. (This is done by means of a conchoid in which F is the pole, CH the 'ruler', and the 'distance' is equal to AD or CF; then, by the property of the conchoid, $HK =$ the 'distance'.)

Join KL, and produce it to meet BA produced in M. Then shall CK, MA be the required mean proportionals. For, since BC is bisected at E and produced to K,
$$BK.KC+CE^2 = EK^2.$$
Adding EF^2 to each, we have
$$BK.KC+CF^2 = KF^2.$$
Now, by parallels, $MA:AB = ML:LK$
$$= BC:CK.$$
But $AB = 2AD$ and $BC = \tfrac{1}{2}GC$;
therefore $\qquad MA:AD = GC:CK$
$$= FH:HK;$$
and, *componendo*, $MD:DA = FK:HK$.

But, by construction, $HK = AD$;
therefore $\qquad MD = FK$, and $MD^2 = FK^2$.
Now $\qquad MD^2 = BM.MA+DA^2$,
and $\qquad FK^2 = BK.KC+CF^2$, from above;
therefore $\quad BM.MA+DA^2 = BK.KC+CF^2$.

But $AD = CF$; therefore $BM.MA = BK.KC$.
Therefore $\qquad CK:MA = BM:BK$
$$= LC:CK,$$
while at the same time $\quad BM:BK = MA:AL$.

Therefore $LC:CK = CK:MA = MA:AL$,

or $AB:CK = CK:MA = MA:BC$.

Apollonius, Heron, Philon of Byzantium.

We take these solutions together because they are really equivalent.

Let AB, AC, placed at right angles, be the two given straight lines. Complete the rectangle $ACDB$, and let E be the point at which the diagonals bisect one another.

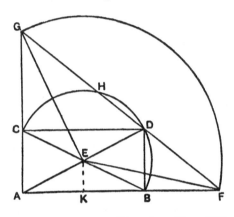

Then a circle with centre E and radius EB will circumscribe the rectangle $ABDC$.

Now (Apollonius) suppose a circle drawn with centre E and cutting AB, AC produced in points F, G such that F, D, G are in one straight line.

Or (Heron) place a ruler so that its edge passes through D and turn it about D until the edge intersects AB, AC produced in points F, G which are equidistant from E.

Or (Philon) turn the ruler about D until it cuts AB, AC produced and the circle about $ABDC$ respectively in points F, G, H such that the intercepts FD, HG are equal.

(All three constructions give the same points, F, G.)

We have now, first, to prove that $AF.FB = AG.GC$.

(a) With Apollonius' and Heron's constructions we have, if K be the middle point of AB,
$$AF.FB + BK^2 = FK^2,$$
and, if we add KE^2 to both,
$$AF.FB + BE^2 = EF^2.$$
Similarly, $\quad AG.GC + CE^2 = EG^2.$

But $BE = CE$ and $EF = EG$;

therefore $\quad AF.FB = AG.GC.$

(b) With Philon's construction, since $GH = FD$,
$$HF.FD = DG.GH.$$
But, since the circle $BDHC$ passes through A,
$$HF.FD = AF.FB \text{ and } DG.GH = AG.GC;$$
therefore $\quad AF.FB = AG.GC$, as before.

It follows that $\quad FA:AG = CG:FB.$

But, by similar triangles,
$$FA:AG = DC:CG = FB:BD;$$
therefore $\quad DC:CG = CG:FB = FB:BD,$

or $\quad\quad AB:CG = CG:FB = FB:AC.$

Diocles and the cissoid.

Diocles was later in date than Apollonius but earlier than Geminus (fl. 70 B.C.), since Geminus described his curve as the *cissoid* ('like an ivy-leaf'). Eutocius quotes two passages from a work of Diocles, *On burning-mirrors* (περὶ πυρείων); one contains a solution by means of conics of the problem of dividing a sphere by a plane in such a way that the volumes of the resulting segments are in

PROBLEM OF TWO MEAN PROPORTIONALS

a given ratio, and the other gives the solution, by means of the cissoid, of the problem of the two mean proportionals.

The cissoid is evolved as follows. AB, DC are diameters of a circle at right angles to one another. Let E, F be points on the quadrants BD, BC respectively such that the arcs BE, BF are equal.

Draw EG, FH perpendicular to DC. Join CE, and let P be the point in which CE, FH intersect.

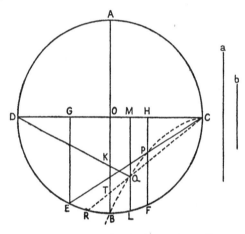

The cissoid is the locus of all the points P corresponding to different positions of E on the quadrant BD and of F at an equal distance from B on the quadrant BC.

If P is any point found by the above construction, it is required to prove that FH, HC are two mean proportionals between DH, HP, or that

$$DH : HF = HF : HC = HC : HP.$$

It is clear from the construction that $EG = FH$ and $DG = HC$,
so that $\qquad CG : GE = DH : HF.$

Now FH is a mean proportional between DH, HC;
therefore $\qquad DH:HF = HF:HC$.

And, by similar triangles, $CG:GE = CH:HP$.

Therefore $DH:HF = HF:CH = CH:HP$.

[Since $DH.HP = HF.CH$, we have, if a is the radius of the circle, and if $OH = x$, $HP = y$, or, in other words, if we use OC, OB as axes of coordinates,

$$(a+x)y = \sqrt{(a^2-x^2)} \cdot (a-x)$$
or $\qquad y^2(a+x) = (a-x)^3$,

which is the Cartesian equation of the cissoid. It has a cusp at C, and the tangent to the circle at D is an asymptote to it.]

Suppose that the cissoid is represented in the figure by the dotted curve. Diocles shows how to find two mean proportionals between two straight lines a, b thus.

Take K on OB such that $DO:OK = a:b$.

Join DK, and produce it to meet the cissoid in Q.

Through Q draw the ordinate LM perpendicular to DC.

Then, by the property of the cissoid, LM, MC are the two mean proportionals in continued proportion between DM, MQ.

And $\qquad DM:MQ = DO:OK = a:b$.

In order to find the two mean proportionals between a, b, take straight lines x, y bearing the same ratio to LM, MC respectively that a bears to DM and b to MQ; then x, y are the required mean proportionals between a and b.

Sporus and Pappus.

These solutions, given separately by Eutocius, are really identical, and also in effect the same as that of Diocles.

The difference is that Sporus and Pappus do not use the cissoid, but a ruler, which they turn about C until, intersecting DK produced in Q, OB in T, and the circle in R, it makes the intercepts QT, TR equal.

Sporus was known to Pappus, and may have been either his master or a fellow student. From Pappus' own account, we gather that Pappus took the credit for the solution.

VIII
FROM PLATO TO EUCLID

THE influence of PLATO on the development of mathematics in Greece can hardly be exaggerated, although (so far as we know) it was due less to any mathematical discoveries of his own than to his enthusiasm for mathematical studies and his exhortations to his pupils and friends to pursue mathematical research. We are told that it was to Plato that the Delians came with the problem of doubling the cube, and that Plato referred them to Eudoxus and Helicon of Cyzicus. And sure enough Eudoxus gave a solution, while another was given by his pupil Menaechmus. Again, in astronomy, Plato is said to have set it as a problem to all earnest students to find 'what are the uniform and ordered movements by the assumption of which the apparent motions of the planets can be accounted for'. One answer to this, representing an advance second to none in the history of astronomy, was given by Heraclides of Pontus, one of Plato's pupils (c. 388–310 B.C.); another was Eudoxus' beautiful hypothesis of concentric spheres, by which he sought to account for the irregularities in the movements of the planets.

Plato regarded mathematics in its four branches, arithmetic, geometry, stereometry, and astronomy, as the first essential in the training of philosophers and of those who should govern his ideal state: 'Let no one destitute of geometry enter', said the inscription over the door of his school. The importance of mathematics, in Plato's view, lay in its value for the training of the mind; its practical utility was of no account in comparison. Arithmetic should be pursued for the sake of knowledge, not for any practical ends such as its use in trade. A very little geometry and arithmetical calculation suffices for the commander of an army; it is the higher and more advanced parts which tend to lift the mind on high, and enable it ultimately to see the final aim of philosophy, the Idea of the Good; these things draw the soul towards truth and create the philosophic attitude of mind.

In the case of geometry in particular, its essence according to Plato is directly opposed even to the language which, for want of better terms, geometers are obliged to use; thus they speak of 'squaring', 'applying' (a rectangle), 'adding', &c., as if the object were to *do* something, whereas the true purpose of geometry is knowledge. Geometry is concerned, not with material things, but with mathematical points, lines, triangles, &c., as objects of pure thought. A diagram in geometry is only an illustration; the triangle which we draw is an imperfect representation of the real triangle of which we think. Constructions, then, or the *processes* of adding, squaring, and the like, are not of the essence of geometry, but are actually antagonistic to it. We can understand, therefore, the story of Plutarch that Plato blamed those who tried to reduce the duplication of the cube to constructions by means of mechanical instruments, 'on the ground that the good of geometry is thereby lost and destroyed, as it is thus made to revert to things of sense instead of being directed upward and grasping at eternal and incorporeal images'.

PLATO AND THE PHILOSOPHY OF MATHEMATICS

In Plato's dialogues we find what appears to be the first serious attempt at a philosophy of mathematics. A passage in one of the Letters (No. 7, to the friends of Dion) is also interesting in this connexion. Speaking of a circle by way of example, Plato observes that there is (1) something called a circle and known by that name; (2) its definition as that in which the distances from its extremities in all directions to the centre are always equal, for this may be said to be the definition of that to which the terms 'round' and 'circle' are applied; again (3) we have the circle which is drawn or turned: this circle is perishable and perishes; not so (4), the essential circle, or the idea of circle (αὐτὸς ὁ κύκλος): it is by reference to this that the other circles exist, and it is different from each of them. Dealing next with the four things thus distinguished, Plato observes that there is nothing essential in (1), the name: it is merely conventional; there is nothing to prevent our assigning the name 'straight line' to what we now call 'round' and vice versa; nor is there any real definiteness about (2), the definition, seeing that it too is made up of parts of speech, nouns, and verbs. The circle (3), the particular circle drawn or turned, is not free from admixture of other things; it is not the mathematical or essential circle.

The hypotheses of mathematics.

In the *Republic* Plato observes that mathematicians take certain things for granted in arithmetic and geometry, without feeling called upon to give any explanation of them, such things namely as odd and even, figures, three kinds of angles, and other things cognate to them in each subject; then, basing themselves on these hypotheses, they proceed at once to go through the rest of the argument until they arrive at the particular conclusion sought. They make use of visible figures and argue about them, but in doing so they are not thinking of these figures but of the things they represent; thus it is the absolute diagonal of the absolute square which is the object of their argument, not the diagonal which they draw; and so on.

Definitions.

Plato paid a good deal of attention to definitions. In some cases his definitions connect themselves with Pythagorean doctrines; in others he seems to have struck out a new line for himself. The division of numbers into odd and even is one of his commonest illustrations; number, he says, is divided equally, i.e. there are as many odd numbers as even. An even number is a number divisible into two equal parts; in one place it is explained as that which is not scalene but isosceles: a curious and apparently unique application of these terms to numbers. The distinctions between 'even-times even', 'odd-times even', 'even-times odd', and 'odd-times odd' occur in Plato; with him 'even-times odd' and 'odd-times even' may describe the same number (e.g. 6), just as they may in Euclid; the terms have not yet acquired the restricted meanings given to them by the neo-Pythagoreans.

In the *Meno* Socrates essays a definition of 'figure'. What is it, he says, that is true of the round, the straight, and the other things which you call figures, and is the same for all? His first suggestion is: 'Let us regard as *figure* that which alone of existing things is associated with colour.' But, says Meno, suppose the interlocutor does not know what colour is. Well, says Socrates, it will be admitted that in geometry there are such things as what we call a surface or a solid, and so on, and from these examples we may learn what we mean by 'figure'; 'figure' is that in which a solid ends, or 'figure' is the limit (or extremity, $\pi\acute{\epsilon}\rho\alpha\varsigma$) of a solid. Thus 'figure' is here regarded as practically equivalent to 'surface', and the association with colour recalls the Pythagorean term for surface, $\chi\rho o\iota\acute{a}$, colour or skin, which Aristotle similarly explains as $\chi\rho\hat{\omega}\mu\alpha$, colour, something inseparable from $\pi\acute{\epsilon}\rho\alpha\varsigma$, an extremity.

There is little doubt that the definition of a line as 'breadthless length' originated in the school of Plato. A straight line is defined by Plato himself as 'that of which the middle covers the ends' (i.e. to an eye at either end looking along the line); and I cannot but think that this is what Euclid tried to express, without any appeal to sight, when he said that a straight line is 'a line which lies evenly with the points on it'. A *point* was defined by the Pythagoreans as a 'unit having position'; Plato apparently objected to this definition, but substituted no other. Aristotle says that he even regarded the genus of points as being a 'geometrical fiction', calling a point the beginning of a line, and often using the term 'indivisible line' in the same sense. Aristotle rejoins that even indivisible lines must have extremities and therefore do not help, while the definition of a point as the beginning of a line is unscientific.

The 'round', or the circle, is defined as 'that in which the farthest points in all directions are at the same distance from the middle (centre)', and Plato gives a similar definition of a sphere.

'If equals be added to unequals', says Plato, 'the sums differ by the same amount as the original magnitudes did'; this is an axiom in a rather more complete form than that subsequently interpolated in Euclid.

SUMMARY OF THE MATHEMATICS IN PLATO

In the *Parmenides* a proposition in proportion is quoted as known, namely that if $a > b$, then $(a+c):(b+c) < a:b$.

The five regular solids.

The so-called 'Platonic figures', namely the five regular solids, are of course not Plato's discovery, for they had been partly investigated by the Pythagoreans and very fully by Theaetetus; they were only called 'Platonic' because of their appearance in the *Timaeus*, where Plato puts together the faces of each solid and appropriates the first four solids to the four elements, the pyramid or tetrahedron to fire, the octahedron to air, the icosahedron to water, and the cube to earth, while the Creator uses the fifth solid, the dodecahedron, for the universe itself.

Archimedes, who described fourteen *semi*-regular solids which are also inscribable in a sphere, is cited in Heron's *Definitions* as having said that 'Plato also knew one of them, the figure with fourteen faces, of which there are two sorts, one made up out of eight triangles and six squares, of earth and air, and already known to the ancients, and the other made up out of eight squares and six triangles, which seems more difficult'. The first of these is easily obtained. Take any cube and bisect the sides of each of its square faces; then, joining the middle 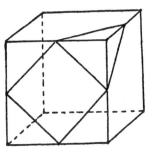 points of each pair of adjacent sides, we obtain, in each face, a square of half the size. The three out of the twenty-four sides of all these squares which are about any one solid angle of the cube form an equilateral triangle; there are eight of these equilateral triangles, and, if we cut off from the corners of the cube the eight pyramids having these triangles as bases, we have the semi-regular solid described. The description of the other semi-regular solid with fourteen faces is incorrect; there are two more such figures, but in one of them the fourteen faces are eight triangles and six *octagons*, while in the other there are six squares and eight *hexagons*. The former is obtained by cutting off from the corners of a cube pyramids on smaller equilateral triangles as bases, the triangles being such that regular octagons, not squares, are left in the six faces; the second is arrived at by cutting off from the corners of an *octahedron* pyramids with square bases such as to leave hexagons in each of the eight faces.

Plato constructs the regular solids by putting together the necessary number of plane faces. For the cube these are squares, for the tetrahedron, octahedron, and icosahedron they are equilateral triangles. But Plato does not take a simple square and a simple equilateral triangle; he makes up his square out of four isosceles right-angled triangles, as shown, and the equilateral triangle out of six equal triangles which are the halves of equilateral triangles formed by drawing a perpendicular from any angular point to the opposite side. This type of triangle, in which the square on the perpendicular is three times the square on the base, was considered by the Pythagoreans to be the most beautiful of all triangles. But the two types of right-angled triangles (the isosceles and the other) used to form the faces of the four solids do not avail in the case of the face of the fifth solid, the dodecahedron; a new kind of triangle had to be used for forming the regular pentagon, namely the isosceles triangle in which each of the base angles is double of the vertical angle (cf. Eucl. IV. 10). Attempts were made, as we learn from Plutarch and Alcinous, to divide up a regular pentagon into elementary scalene 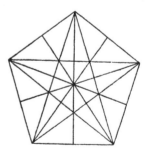 right-angled triangles; the division gave thirty such triangles, but they are different from the elementary triangles which, as above described, make up the faces of the cube, the tetrahedron, the octahedron, and the icosahedron respectively.

Geometric means.

In the *Timaeus* Plato, speaking of numbers 'whether solid or square', observes that between *planes* one mean suffices, but to connect two *solids* two means are necessary. The means are geometric means, and by 'planes' and 'solids' Plato doubtless meant square and cube numbers respectively, so that the theorems quoted are in effect those of Eucl. VIII. 11, 12, namely that between two square numbers there is one mean proportional number, and between two cubes there are two mean proportional numbers. Nicomachus speaks of the two propositions as constituting 'a certain Platonic Theorem'; but it is probably older, and is only called 'Platonic' (like the regular solids) because it appears in the *Timaeus*.

The two geometrical passages in the MENO.

In the first passage Socrates is trying to show that teaching is only reawakening in the mind of the learner the memory of something. He illustrates by putting to the slave a carefully prepared series of questions leading him to recognize that double the square on any straight line is not the square on double the line, but the square on the diagonal of the original square, as shown in the accompanying figure, where the original square is the square on AB, a line two feet in length.

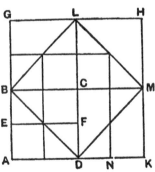

The second passage (86 E–87 C) is much more difficult, and

GEOMETRICAL PASSAGES IN *MENO* 179

there have been many different interpretations of it; but I do not doubt that the following is the correct one. Socrates is seeking a criterion as to whether a given area can be inscribed, in the form of a triangle, in a given circle, that is to say, whether we can inscribe in the circle a triangle equal in area to the given figure, which may be of any shape. (The real condition is clearly that the given area must not be greater than that of the equilateral triangle inscribed in the circle.) Socrates suggests that such a triangle can be inscribed if, and only if, 'the given area is such as when one has applied it [in the form of a rectangle] to the given straight line in the circle, it is deficient by a figure [rectangle] similar to the very figure which is applied'. Here we have practically the phraseology of the 'application of areas'. 'The given straight line in the circle' is an ambiguous expression, but it must mean some definite line in the circle, which would naturally be the *diameter*.

To elucidate this problem, take a circle on AB as diameter, and let AC be the tangent at A. From any

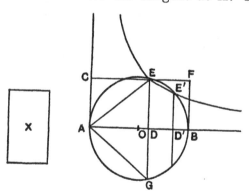

point E on the circle draw ED perpendicular to AB, and complete the parallelograms $ACED$, $DEFB$. Then the rectangle CD is 'applied' to the diameter AB and 'falls short' by the rectangle EB; and EB is similar to CD because $AD:DE = ED:DB$.

If now we produce ED to meet the circle again in G, the isosceles triangle AEG is bisected by AB and is equal to the rectangle CD.

If, then, the rectangle CD is equal to the given area (say X), so is the triangle AEG, and the problem is solved.

What is required therefore is, in other words, to find a point E on the circle such that, if ED be drawn perpendicular to AB, the rectangle $AD.DE$ shall be equal to X. It is clear that E lies on a rectangular hyperbola with AB, AC as asymptotes, and its equation referred to AB, AC as axes of x, y respectively is $xy = b^2$, where b^2 is equal to the given area X. For a real solution it is necessary that b^2 shall not be greater than the area of the equilateral triangle inscribed in the circle, or $\frac{1}{4}(3\sqrt{3}.a^2)$, where a is the radius of the circle. If b^2 is less than this area, there are two solutions; if equal, there is only one solution, for in that case the hyperbola touches the circle in one point.

Solution in integers of $x^2+y^2 = z^2$.

As we have seen (pp. 47–8), Plato is credited with a formula for finding any number of pairs of square numbers the sum of which is a square. The formula is, in our notation,
$$(2n)^2+(n^2-1)^2 = (n^2+1)^2.$$

Incommensurables.

There are several allusions in Plato to the subject of incommensurables. First, we have in the *Republic* the passage distinguishing the 'irrational diameter of 5' from the 'rational diameter of 5'. The reference is to a square with 5 for its *side*; the diagonal is then $\sqrt{50}$, which is Plato's 'irrational diameter of 5'; the 'rational diameter of 5' is the approximation 7, the square root of 49. Secondly, there is the famous passage in the *Theaetetus* about the proof by Theodorus of the incommensurability with unity of the surds $\sqrt{3}, \sqrt{5} \ldots \sqrt{17}$, and the generalization of the theory of such incommensurables by Theaetetus himself. But it would appear that Plato's knowledge of irrationals was not restricted to simple surds; in the *Hippias Major* (303 B, C) it is stated that, just as an even number may be the sum of either two odd or two even numbers, so the sum of two irrationals may be either rational or irrational. An obvious illustration of the former case is afforded by a rational straight line cut 'in extreme and mean ratio'. Euclid proves (XIII. 6) that each of the segments in that case is a certain kind of irrational straight line called an *apotome*. Now Proclus speaks of theorems which Plato 'originated regarding the "section"', and, if this 'section' was what came to be called the 'golden section', namely the division of a straight line 'in extreme and mean ratio', as in Eucl. II. 11 and VI. 30, Plato may well have had this case in mind. The incommensurability of the segments with the whole line and with one another may even have been known to the Pythagoreans.

In addition to theoretical subjects, Plato recognizes mathematical 'arts' or applied mathematics, e.g. the art of measurement and the art of weighing. In the *Protagoras* (356 B) he speaks of the man skilled in weighing 'who puts together first the pleasant and second the painful things, and adjusts the near and the far on the balance': which shows that the principle of the lever was known to Plato, who was no doubt acquainted with the work of Archytas on mechanics (see p. 135 above).

In *optics* Plato has some interesting remarks about the working of mirrors. In the passage of the *Timaeus*

(45 B–46 C) where he gives his theory of vision he observes that, if you look at yourself in a mirror, you do not see yourself but an imaginary person whose right eye is the image of your left eye and vice versa. But, on the other hand, if you look into a hollow cylindrical mirror with its axis vertical, the right portion of the 'visual current' is turned to the left and vice versa, so that the right of the image *is* your right side and vice versa; and if the axis of the mirror is held in a horizontal position, everything appears upside down.

Astronomy.

Plato's *astronomy* was, so far as we can judge from the dialogues, based generally on that of Pythagoras (as distinct from that of the Pythagoreans, who deposed the earth from its place in the centre and made it, like the sun, moon, and planets, revolve round the central fire). The earth for Plato is in the centre; the sun, the moon, and the planets (in the order moon, sun, Venus, Mercury, Mars, Jupiter, Saturn) move in concentric circles about the earth as centre in a plane which cuts at an angle the equator of the heavenly sphere. The following is the order of their speeds: the moon is the quickest; the sun is the next quickest, and Venus and Mercury travel in company with it, each of the three taking about a year to describe its orbit; the next quickest is Mars, the next Jupiter, and the last and slowest is Saturn; the speeds are of course angular speeds, not linear. The vexed question of the interpretation of the famous passage (40 B, C) of the *Timaeus* about the earth has been discussed afresh in the last few years, and opinion seems now to incline more generally towards the view that Plato did there intend to assign some motion to the earth instead of keeping it unmoved in the centre; and this is almost inevitable if the new reading in Burnet's text is accepted. But there is no agreement about the nature of the motion. One alternative was that of Gruppe, that the earth rotates about its axis; another is that of Burnet, that the earth is supposed to describe a *small* orbit about 'the axis which stretches through the universe from pole to pole'. The first of these views is inconsistent, and the second sub-inconsistent, with the rest of the system as described in the *Timaeus* and other dialogues. The latest suggestion (that of A. E. Taylor) is that the motion is an oscillation on the part of the earth about the centre, a backward and forward motion along the axis of the universe, but that Plato is, in the passage in question, explaining what was not his own view but that of Timaeus. This seems highly artificial, and it is unconfirmed by any evidence; in particular, there is nothing in Aristotle or any other ancient author to suggest that Plato was expressing any other view than his own.

There remains the well-known citation by Plutarch of a statement by Theophrastus that 'Plato in his old age regretted that he had given the earth the middle place in the universe, which was not appropriate to it'. It is, of course, permissible to accept this statement on the ground that Theophrastus was in a position to know the facts and would not be likely to misrepresent them. But it is best, I think, to give up the attempt to find in the dialogues a proof of this change of view.

THE SUCCESSORS OF PLATO

In recording the progress of the Elements in the period from Plato to Euclid Proclus gives us little more than a list of names:

'Younger than Leodamas', he says, 'were Neoclides and his

pupil Leon, who added many things to what was known before their time, so that Leon was actually able to make a collection of the Elements more carefully designed in respect both of the number of propositions proved and of their utility, besides which he invented *diorismi*, [the object of which is to determine] when the problem under investigation is possible of solution and when impossible'.

Of Neoclides and Leon nothing more is known, but the formal recognition of the διορισμός, that is, of the necessity of finding, as a preliminary to the solution of a problem, the conditions of possibility, represents a further step towards fixing the terminology of mathematics. The thing itself was not new; the Pythagoreans must have known the limiting condition for a real solution of the equivalent of the quadratic equation $ax - x^2 = B$ connected with the theorem of Eucl. II. 5, and of the more general case corresponding to Eucl. VI. 27, 28, while the condition of the possibility of constructing a triangle out of three given straight lines (= Eucl. I. 22) must have been common knowledge long before Plato's time. The second passage from the *Meno* above referred to also supplies a definite instance of διορισμός.

Proclus' summary continues as follows:

Eudoxus of Cnidos, a little younger than Leon, who had been associated with the school of Plato, was the first to increase the number of the so-called general theorems; he also added three proportions to the three already known, and multiplied the theorems which originated with Plato about the section, applying to them the method of analysis. Amyclas [more correctly Amyntas] of Heraclea, one of the friends of Plato, Menaechmus, a pupil of Eudoxus who had also studied with Plato, and Dinostratus, his brother, made the whole of geometry still more perfect. Theudius of Magnesia had the reputation of excelling in mathematics as well as in the other branches of philosophy; for he put together the Elements admirably and made many partial (or limited) theorems more general. Again, Athenaeus of Cyzicus, who lived about the same time, became famous in other branches of mathematics and most of all in geometry. These men consorted together in the Academy and conducted their investigations in common. Hermotimus of Colophon carried further the investigations already opened up by Eudoxus and Theaetetus, discovered many propositions of the Elements, and compiled some portion of the theory of Loci. Philippus of Medma, who was a pupil of Plato and took up mathematics at his instance, not only carried out his investigations in accordance with Plato's instructions, but also set himself to do whatever in his view contributed to the philosophy of Plato.

Nothing is known of Amyntas, Theudius, Athenaeus, and Hermotimus beyond what is here stated. Of Menaechmus and Dinostratus we have already learnt that the former discovered the conic sections and applied two of them for finding the two mean proportionals, and that the latter used the *quadratrix* of Hippias to square the circle. Philippus of Medma is doubtless the same person as Philippus of Opus, who is said to have copied out for circulation the *Laws* of Plato which had been left unfinished, and to have been the author of the *Epinomis*. He wrote upon astronomy chiefly. Suidas records the titles of the following works by him: *On the Distance of the Sun and Moon*; *On the Eclipse of the Moon*; *On the Size of the Sun, the Moon, and the Earth*; *On the Planets*. From the allusions in Aëtius and Plutarch to *proofs* by Philippus about the shape of the moon we may perhaps infer that he was the first to establish completely the theory of the phases of the moon. In mathematics, according to the same notice by Suidas, he wrote *Arithmetica, Means, On Polygonal Numbers, Cyclica, Optics, Enoptrica* (On mirrors).

Before we pass to Eudoxus, there are three names to be mentioned though they do not appear in Proclus' summary. SPEUSIPPUS, nephew of Plato and the next head of the school, wrote a short treatise *On the Pythagorean Numbers*, a fragment of which, mathematically unimportant, is preserved in the *Theologumena arithmetices*. He had views (so we learn from Proclus) on the relation of theorems and problems. XENOCRATES of Chalcedon (396–314 B.C.), who succeeded Speusippus as the head of the Academy, having been elected by a majority of a few votes over Heraclides, is also said to have written *On Numbers* and a *Theory of Numbers*, besides works on geometry. He was a supporter of 'indivisible lines' (and magnitudes), by which he thought to get over the paradoxical arguments of Zeno. He was mentally slow, so that, comparing him with his other pupil, Aristotle, Plato said 'the one needed the spur, the other a bridle' and 'See what an ass I am training and what a horse he has to run against'. When some one who knew no music or geometry or astronomy wished to attend his lectures, he said, 'Go your ways, for you offer philosophy nothing to lay hold of' (or, perhaps, 'you have not the means of getting a grip of philosophy').

HERACLIDES of Pontus (about 388–310 B.C.), a pupil of Plato, wrote works of the highest class, both in matter and style, on all sorts of subjects, ethical, grammatical, musical, poetical, rhetorical, historical; he wrote geometrical and dialectical treatises also. But his importance for us lies in the fact that he made an epoch-making advance in astronomy. He declared that the apparent daily revolution of the heavenly bodies round the earth was accounted for, not by the circular motion of the stars round the earth, but by the rotation of the earth itself about its own axis.

He also discovered that Venus and Mercury revolve, like satellites, round the sun as centre. There is not sufficient evidence that he came to the same conclusion with regard to the other planets; had he done so, he would have anticipated the hypothesis of Tycho Brahe (with the improvement of the substitution of the earth's rotation about its axis for the daily revolution of the whole system about the earth as centre).

EUDOXUS of Cnidos, an original genius second to none (unless it be Archimedes) in the history of our subject, flourished, according to Apollodorus, in Ol. $103 = 368$–365 B.C., whence we infer that he was born about 408 B.C. and (since he lived 53 years) died about 355 B.C. According to Diogenes Laërtius, he was famous as geometer, astronomer, physician, and legislator. Philosopher and geographer in addition, he commanded and enriched almost the whole field of learning. He went to Italy and Sicily to study geometry with Archytas and medicine with Philistion. In his twenty-third year he went with the physician Theomedon to Athens and there attended, for two months, lectures on philosophy and oratory, and those of Plato in particular; so poor was he that he took up his abode at the Piraeus and trudged to Athens and back on foot each day. Returning to Cnidos, he went thence to Egypt with a letter of introduction to King Nectanebus given him by Agesilaus; he stayed in Egypt sixteen months. While in Egypt he assimilated the astronomical knowledge of the priests of Heliopolis, and himself made observations. The observatory between Heliopolis and Cercesura used by him was still pointed out in Augustus' time; he had another built at Cnidos, where he observed the star Canopus which was not then visible in higher latitudes. He recorded his observations in works attributed to him by Hipparchus, namely the *Mirror* and the *Phaenomena*; it was from the latter that the poem of Aratus was drawn so far as verses 19-732 are concerned. Such was his passion for knowledge of the truth that he would not even indulge in physical speculation about what was inaccessible to observation and experience; instead of guessing at the nature of the sun, he said, he would gladly be burnt up like Phaëthon if at that price he could get to the sun and so ascertain its form, size, and nature.

Hypothesis of concentric spheres.

In astronomy Eudoxus is famous for his elegant hypothesis of concentric spheres put forward to account for the apparent motions of the sun, moon, and planets, and in particular the stations and retrogradations in the case of the planets. The most detailed account of this system is found in Simplicius' commentary on the *De caelo* of Aristotle; Simplicius quotes largely from Sosigenes the Peripatetic (second century A.D.), who drew from Eudemus' *History of Astronomy*. Eudoxus adopted the view which prevailed from the earliest times to the time of Kepler, that circular motion was sufficient to account for the movements of all the heavenly bodies. He represented the motion of each planet as produced by the rotations of four spheres concentric with the earth (conceived as at rest in the centre) and connected in the following way. Each of the inner spheres revolves about a diameter the ends of which (poles) are fixed on the next sphere enclosing it. The outermost sphere represents the daily rotation, the second a motion along the zodiac circle; the poles of the third sphere are fixed on the latter circle; the poles of the fourth sphere (which carries the planet fixed at a point on its equator) are so fixed on the third sphere, and the speeds and directions of rotation so arranged, that the planet describes on the second sphere a curve called the *hippopede* (horse-fetter), like a figure of eight, lying along, and longitudinally bisected by, the zodiac circle. In the case of the sun and moon Eudoxus used three spheres only; the two outer spheres were the same for the sun and moon as for the planets; but (except that the poles of the third sphere were the same for Mercury and Venus) the inner spheres were different for each of the heavenly bodies.

The accompanying figure shows the *hippopede* described by a planet. The circle bisecting it is the zodiac circle.

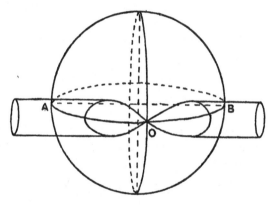

The *hippopede* is really the intersection of the sphere about the earth as centre and containing the zodiac circle with a cylinder touching the sphere internally at the double-point of the *hippopede*. It was an extraordinary stroke of genius to devise four separate revolutions of interconnected spheres the combination of which should make the planet describe this figure-of-eight on the surface of the sphere containing the zodiac circle, and, with the limited means at Eudoxus' disposal, to *see* that this would be the effect.

In geometry Proclus says that Eudoxus increased the

number of the so-called general theorems. We can only conjecture what these are; but it seems possible that the expression may mean those theorems which are true of everything falling under the conception of magnitude, e.g. the definitions and theorems forming part of Eudoxus' own theory of proportion, which applies to numbers, geometrical magnitudes of all sorts, times, &c.

Again, Proclus says that Eudoxus 'extended' or 'increased the number of the propositions about "the section" which originated with Plato, applying to them the method of analysis'. 'The section' is generally understood to mean the division of a straight line in extreme and mean ratio as in Eucl. II. 11 and VI. 30. The propositions of Eucl. XIII. 1–6 all contain properties of a line so divided and may well be due to Eudoxus, though the alternative proofs of them in the form of analysis and synthesis which follow in the MSS. of Euclid are certainly interpolated by some scholiast. Tannery, however, preferred the old view of 'the section' as meaning 'section' of solids, i.e. the investigation of the form of sections of solid figures made by planes, one case of which is that of the conic sections.

But the capital discoveries made by Eudoxus which overshadow everything else are (1) the theory of proportion as expounded in Euclid, Book V, which is applicable to incommensurable as well as to commensurable magnitudes, and (2) the *method of exhaustion*.

Theory of Proportion.

There is no reason to doubt the truth of the dictum of the scholiast to Euclid's Book V, who says that this Book containing the general theory of proportion, which is applicable equally to geometry, arithmetic, music, and all mathematical science, is due to Eudoxus. The new theory appears to have been already familiar to Aristotle. Its discovery was a momentous event, because it enabled geometry to go forward again after it had received the blow which paralysed it for the time. I refer to the discovery of the incommensurable, which made invalid or inconclusive all proofs depending on the theory of proportion then in use, namely the Pythagorean theory, which was arithmetical in character and applied to commensurable magnitudes only.

The intrinsic greatness of the theory needs no further proof when it is remembered that the definition of equal ratios in Euclid, Book V, corresponds exactly to the theory of irrationals due to Dedekind, and that it is word for word the same as Weierstrass' definition of equal numbers.

The Method of Exhaustion.

We have seen that both Anaxagoras and Democritus maintained that mathematical magnitudes are divisible *ad infinitum*. Democritus, too, in more than one special investigation, found himself confronted with the difficulties involved by the assumption of infinitesimals; the question must have come up in his tract on the nature of the contact between a circle and a tangent to it, and also when he came to consider the volume of a pyramid and a cone; the dilemma could hardly be more neatly stated than it is in his remarks about two consecutive parallel sections of a cone (see p. 119 above). The classical attack on infinitesimals is, however, that of Zeno.

ZENO indeed does not seem to have been a mathematician himself; yet the statement of his paradoxes had a profound influence on the course which geometry thereafter took. Antiphon, as we have seen, was really on the right track when he thought to exhaust the area of a circle by continually doubling the number of sides in a series of regular polygons inscribed in the circle, but, in the absence of a sufficient answer to the arguments of Zeno, Antiphon's attempt, in the form which he gave to it, was foredoomed to failure. Zeno's arguments should perhaps be shortly stated here. They are four in number. The first two, the *Dichotomy* and the *Achilles*, deny the possibility of motion on the assumption of the divisibility of magnitudes *ad infinitum*; the last two, the *Arrow* and the *Stadium*, proceed on the opposite hypothesis that space and time are *not* infinitely divisible but are composed of indivisible elements; the four arguments together constitute a complete dilemma. In the *Dichotomy* Zeno argued that there is no motion because that which is moved any distance must arrive at the middle of its course before it arrives at the end; but, before it has traversed the half of the distance, it must have traversed the half of that half, and so on *ad infinitum*; hence the motion can never even *begin*. The *Achilles* asserts that Achilles can never catch up the tortoise because he must first reach the point at which the tortoise started; when he has done so, the tortoise has gone a little farther and the same argument applies, and so on; hence the tortoise must always be some distance ahead. The moving *Arrow*, at any indivisible instant, must either be at rest or in motion. Now the arrow cannot move in the instant, supposed indivisible, for, if it changed its position, the instant would at once be divided. If, however, it is not in motion in the instant, it must be at rest, and, as time is made up of such instants, it must remain always at rest. The *Stadium* is more complicated, but comes to the same thing; on the hypothesis of relative motion, it proves that an instant, supposed to be indivisible, must actually be divided.

EUDOXUS

The *Method of Exhaustion* due to Eudoxus was the answer to Zeno from the mathematical point of view. It showed that it was not necessary to assume the existence of the infinitely small, but that it was sufficient for the mathematician's purpose to assume that, by continual division of a magnitude, we can ultimately arrive at a magnitude *as small as we please*. Thus geometry was extricated from the apparent *impasse*. We have the authority of Archimedes for assigning the credit of this to Eudoxus, for he says that Eudoxus was the first to give the scientific proof of the two theorems discovered by Democritus about the volume of a pyramid and a cone, and he adds that these propositions and two others (that circles are to one another in the duplicate ratio of their diameters, and that spheres are to one another in the triplicate ratio of their diameters) were proved by means of the same lemma which he himself used for finding the area of a segment of a parabola, namely that 'of unequal lines, unequal surfaces, or unequal solids, the greater exceeds the less by a magnitude such as is capable, if added (continually) to itself, of exceeding any magnitude of those which are comparable to one another', i.e. of magnitudes of the same kind as the original magnitudes. This lemma, now known as the Axiom of Archimedes, is really equivalent to the definition of magnitudes having a ratio to one another (Eucl. V, Def. 4). 'Magnitudes are said to have a ratio to one another which are capable, when multiplied, of exceeding one another', a definition forming part of Eudoxus' theory of proportion. From all this we may infer that Eudoxus invented the *method of exhaustion* which we find used in Euclid for proving the very propositions referred to about the areas of circles and the volumes of spheres, pyramids, and cones. The lemma actually used by Euclid is not in form precisely the same as that of Archimedes; Euclid uses his own proposition, X. 1, to the effect that, if there be two unequal magnitudes and from the greater there be subtracted more than its half (or the half itself), from the remainder more than its half (or the half), and so on continually, there will at length be left some magnitude which will be less than the lesser of the given magnitudes. But the difference between the two lemmas constitutes no difficulty, because (1) Archimedes himself uses the Euclidean lemma in one proof of his proposition about the area of a parabolic segment, and (2) Archimedes' lemma in the form of Eucl. V, Def. 4, is in effect used by Euclid to prove his proposition X. 1.

We are told, on the authority of Eudemus, that Hippocrates of Chios was the first to prove that circles are to one another as the squares on their diameters. We are constrained to suppose that Hippocrates' method fell short in some way of the full *method of exhaustion*. Perhaps Hippocrates proceeded on the lines of Antiphon's attempted quadrature, gradually exhausting the circles, and then in effect 'took the limit' without clinching the proof by *reductio ad absurdum*.

ARISTOTLE was clearly not a professional mathematician, and he does not in his works show any acquaintance with the higher branches—he makes no allusion to conic sections, for example—but he was fond of mathematical illustrations, and he throws a flood of light on the first principles of mathematics as accepted in his time. He gives the clearest distinctions between *axioms* (which are common to all sciences), *definitions, hypotheses*, and *postulates* (which are different for different sciences, since they relate to the particular subject-matter of each science). Axioms with him are 'common (things)', 'common axioms', or 'common opinions' (cf. the term 'common notions', κοιναὶ ἔννοιαι, by which they are described in the text of Euclid); these are universal self-evident truths which every one must know and admit if he is to learn anything at all; Aristotle's favourite example is 'If equals be subtracted from equals, the remainders are equal'. A *postulate* is an assumption which (besides being restricted to the subject-matter of the particular science) differs from an axiom in that it is not self-evident. It is something which, e.g., the geometer assumes (for reasons known to himself) without demonstration (though properly a subject for demonstration) and without any assent on the part of the learner, or even against his opinion rather than otherwise. As regards *definitions*, Aristotle is clear that they say nothing as to whether the thing defined exists or not; they only require to be understood. Existence is only postulated in the case of a few fundamental things, e.g. the *unit* and *magnitude*; in geometry we assume points and lines only, and the existence of everything else has to be proved. Aristotle also has interesting remarks about the geometer's hypotheses. It is untrue, he says, to assert that a geometer's hypotheses are false because he assumes that a line which he has drawn is a foot long when it is not, or straight when it is not straight. The geometer bases no conclusion on the particular line being what he has assumed it to be; he argues about what it *represents*, and the figure itself is a mere illustration.

Aristotle's citations of definitions and propositions in geometry, when compared with the corresponding things in Euclid, enable us to judge to some extent what changes Euclid himself made in the exposition of the subject by earlier writers on the Elements. The text-book in use in the Academy would no doubt be that of Theudius, and this may have been Aristotle's source. Aristotle has the equivalent of Euclid's Defs. 1-3, 5, 6 of Book I, but for a straight line he gives Plato's definition only: whence we conclude that Euclid's definition was his own, as also was his definition of a plane, which is adapted from that of a straight line. Some terms seem to have been defined in Aristotle's time which Euclid does not define, e.g. κεκλάσθαι, 'to be inflected', νεύειν, 'to verge' or 'tend' (to a certain point). Aristotle was apparently acquainted with Eudoxus' theory of proportion; he frequently uses the terminology of proportions, and he defines similar figures as Euclid does.

There are in Aristotle indications of proofs differing from Euclid's. The most remarkable case is a proof of the theorem of Eucl. I. 5 given in the *Posterior Analytics*. This proof uses 'mixed' angles ('angles' formed by straight lines with circular arcs), and makes assumptions about them which seem not more obvious than the proposition to be proved. The triangle taken is OAB, where OA, OB

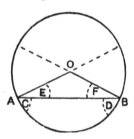

are two radii of a circle with centre O. The angles are denoted by single letters as shown, C, D being the 'angles' formed by the straight line AB and the arc AB. There is the further peculiarity of notation that Aristotle also uses A and B to denote the same angles as E and F respectively, since by AC he means the sum of E and C, and by BD the sum of F and D. Here, says Aristotle, we must base our assertion of the equality of the angles AC (i.e. $E+C$) and BD (i.e. $F+D$) upon the truth of the general proposition that *the angles of semicircles are equal*, and, similarly, we can only say that the angles C, D are equal because of the truth of the general theorem that *the two angles of all segments are equal*; lastly, the equality of the angles E, F, which are arrived at by subtracting equal angles from equal angles, can only be inferred if we first assume generally that, if equals be subtracted from equals, the remainders are equal. It is clear from this proof that mixed angles, and assumptions or propositions about them, played a larger part in earlier text-books than they do in Euclid, where they appear only once or twice as a survival (in Book III).

Two passages in Aristotle bearing on the theory of *parallels* seem to show that the theorems of Eucl. I. 27, 28 are pre-Euclidean; but another passage suggests that there was some flaw amounting to a *petitio principii* in the current theory of parallels, and it seems probable that this was due to the absence of some Postulate, such as Euclid's Post. 5; whence we infer that it was Euclid himself who first, by this expedient, got rid of the *petitio principii* in question.

A difference of method is also indicated in regard to the theorem of Eucl. III. 31 that the angle in a semicircle is a right angle. Two passages of Aristotle show that before Euclid's time the proposition was first proved to be true of one particular angle at the circumference, namely that formed by joining the extremities of the diameter to the middle point of the circumference. The radius drawn to the said middle point is at right angles to the diameter, and we have two isosceles right-angled triangles. Two angles of these 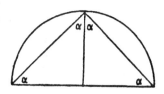 triangles, one in each triangle, which are at the middle point of the circumference are halves of right angles, so that their sum is a right angle. Thus *one* angle in a semicircle is right, and it remains to complete the proof of the general theorem by means of the proposition that angles in the same segment are equal to one another. Euclid, by his more general proof, avoids the necessity for this.

Aristotle is already aware of the principle on which Eudoxus based the method of exhaustion: 'If I continually add to a finite magnitude, I shall exceed every assigned magnitude, and similarly, if I continually subtract, I shall fall short of any assigned magnitude.'

Aristotle has some propositions which are not found in Euclid. (1) The exterior angles of any polygon are together equal to four right angles: this is evidently a Pythagorean proposition, as is (2) the proposition that only two solids can fill up space, namely the pyramid and the cube. (3) The locus of a point such that its distances from two given points are in a given ratio (not being a ratio of equality) is a circle: this is a proposition quoted by Eutocius from Apollonius' *Plane Loci*, but Aristotle gives a proof of it (in *Meteor*. III. 5) not differing substantially from that of Apollonius, which shows that there was a standard proof of it current before Euclid's time. (4) Of all closed lines starting from a point, returning to it again and including a given area, the circumference of a circle is the shortest: this shows that isoperimetry (comparison of the content of different figures having their contours equal) was in vogue long before the time of Zenodorus.

Aristotle, again, has some illuminating remarks on the continuous and the infinite. The continuous cannot be made up of indivisible parts; in the continuous the boundary or limit between two consecutive parts is one and the same. The infinite exists only potentially, and not in actuality; nor does it exist even potentially except in the sense of infinite subdivision. It cannot exist even potentially in the sense of exceeding every finite magnitude as the result of successive additions. The correct view of the infinite, says Aristotle, is the opposite of that commonly held; it is not that which has nothing outside it, but that which always has something outside it.

'My argument, however', he adds, 'does not anyhow rob mathematicians of their study, although it denies the existence of the infinite in the sense of actual existence as something increased to such an extent that it cannot be gone through; for, as it is, they do not even need the infinite or use it, but only require that the finite (straight line) shall be as long *as they please* . . . Hence it will make no difference to them for the purpose of demonstration.'

This is a fair description of the line taken by Eudoxus in the method of exhaustion which was to become classical.

The *Mechanica* included in the Aristotelian *corpus* is not by Aristotle, but is very close in date, as we infer from the terminology. The Aristotelian origin of many of the ideas in the *Mechanica* is proved by their occurrence in Aristotle's genuine writings. For example, the principle of the lever is, in the *Mechanica*, explained as due to the fact that the extremity of the longer arm describes the greater circle, while in the *De caelo* we are told that it is natural that the speeds of circles should be in the proportion of their sizes. In the *De caelo* and the *Physics* we have a principle enunciated which has been regarded as containing the germ of the principle of 'virtual velocities': 'A smaller and lighter weight will be given more movement if the force acting on it is the same. . . . The speed of the lesser body will be to that of the greater as the greater body is to the lesser.' We find also in the *Mechanica* the 'parallelogram of velocities'.

The Aristotelian tract *On indivisible Lines* is not Aristotle's, but is, with great probability, attributed to his pupil Theophrastus. The tract is directed against Xenocrates, the definite supporter of the doctrine of indivisible lines. It is not mathematically important, but it cites the equivalent of certain definitions and propositions in Euclid, especially Book X (on irrationals); the writer may, however, have drawn on Theaetetus rather than Euclid.

SPHAERIC

AUTOLYCUS of Pitane was the teacher of Arcesilaus (about 315–241/0 B.C.), also of Pitane, the founder of the Middle Academy. He flourished therefore about 310 B.C. or a little earlier, so that he was an elder contemporary of Euclid. Two works by him have come down to us, one *On the Moving Sphere*, the other *On Risings and Settings*. The former at any rate was included in the list of works forming the 'Little Astronomy', as it was afterwards called to distinguish it from the 'Great Collection' ($\mu\epsilon\gamma\acute{a}\lambda\eta$ $\sigma\acute{v}\nu\tau a\xi\iota s$). The subject is the geometry of the sphere in its application to astronomy. The treatises are of special interest because Autolycus is the earliest Greek mathematician from whom original treatises have come down entire. That he wrote earlier than Euclid is clear from the fact that in his work of like content, the *Phaenomena*, Euclid makes use of propositions appearing in Autolycus, though, as usual, without indicating their source. The form of Autolycus' propositions is exactly the same as that which Euclid gives to his propositions; we have first the enunciation in general terms, then the closer description with reference to a diagram with letters marking various points on it, then the demonstration, and lastly, in some cases but not in all, the conclusion stated in terms similar to those of the enunciation. Thus it is clear that Greek geometrical propositions had already before Euclid's time taken the form which we recognize as classical.

But, further, it is important to note that even Autolycus makes use of a number of propositions relating to the sphere without giving any proof of them or quoting any authority. This indicates that there was already in existence in his time a text-book on the elementary geometry of the sphere, the propositions in which were generally known to mathematicians. As many of these propositions are proved in the *Sphaerica* of Theodosius, a work compiled two or three centuries later, we may assume that the lost text-book proceeded on much the same lines as that of Theodosius, with much the same order of propositions. Who was the author of this pre-Euclidean text-book it is impossible to say. Tannery thought it could hardly be attributed to any one but Eudoxus; but it is not safe, in such a case, to attribute a work to a particular person merely because we know of no one else of whom we can say with certainty that he was capable of writing it.

IX
EUCLID

WE have little or no detailed information about the life and personality of any of the great mathematicians of Greece. Euclid himself is no exception. Practically all that we know of him comes from Proclus' summary, but it is evident that Proclus based himself upon inference more than anything else. He speaks of him as 'Euclid who put together the Elements, collecting many of Eudoxus' theorems, perfecting many of Theaetetus', and also bringing to irrefragable demonstration the things which were only somewhat loosely proved by his predecessors'; and he infers that he lived in the time of the first Ptolemy (306–283 B.C.) because Archimedes mentions him, and Archimedes came immediately after the first Ptolemy; Proclus evidently knew nothing of his birthplace or of the dates of his birth and death. He only adds the story of Euclid's reply to Ptolemy that there is no royal road to geometry.

The Arabs, who eagerly assimilated Greek geometry, so far adopted Euclid that they wished to connect him with the East (the same predilection made them describe Pythagoras as a pupil of the wise Salomo, Hipparchus as the exponent of Chaldaean philosophy or as 'the Chaldaean', Archimedes as an Egyptian, and so on); according to the Arabs, Euclid was the son of Naucrates and grandson of Zenarchus, a Greek born at Tyre and domiciled at Damascus. They identified him so completely with geometry that they interpreted his name (which they pronounced variously as Uclides or Icludes) as meaning the 'key of geometry' (from *Ucli*, a key, and *Dis*, a measure or, as some say, geometry); and again they said that the Greek philosophers used to put up on the doors of their schools the well-known notice 'Let no one come to our school who has not first learnt the Elements of Euclid' (an obvious adaptation of Plato's ἀγεωμέτρητος μηδεὶς εἰσίτω, 'let no one enter who has no geometry').

Euclid probably received his mathematical training from the pupils of Plato in Athens; but he himself founded a school at Alexandria, where, according to Pappus, Apollonius of Perga afterwards 'spent a long time with the pupils of Euclid'.

There is another story of him which one would like to believe true. According to Stobaeus, some one who had begun to read geometry with Euclid had no sooner learnt the first theorem than he asked, 'What shall I get by learning these things?' whereupon Euclid called the slave and said, 'Give him threepence, since he must needs make gain out of what he learns.'

Pappus praises Euclid for his modesty and his consideration for others. Thus Euclid, he says, regarded Aristaeus as deserving credit for the discoveries he had made in conics (Aristaeus had written a treatise on *Solid Loci*, that is to say, on conics, no doubt regarded as loci), and made no attempt to anticipate him or to construct afresh the same system, such was his scrupulous fairness and his exemplary kindliness to all who could advance mathematical science to however small an extent. Pappus is here comparing Euclid with Apollonius, to the disadvantage of Apollonius, whom he evidently regarded as too self-assertive, probably because in the prefaces to the Books of his *Conics* Apollonius seemed to Pappus to give too little credit to Euclid for his earlier work on the same subject.

But we can believe Pappus' testimony to Euclid's character, for his extant works betray no sign of any claim to be original. In the *Elements*, for instance, there is not a word of preamble; he plunges at once into his subject: 'A point is that which has no part'! And although he made great changes in the exposition, altering the arrangement of whole Books, redistributing propositions between them, and inventing new proofs where the new order made the earlier proofs inapplicable, it is safe to say that Euclid made no more alterations than his own acumen and the latest special researches (such as Eudoxus' theory of proportion) showed to be necessary to make the treatment more scientific. He even showed unnecessary respect for tradition, as when he retained certain definitions never afterwards used, and when in solitary passages in Book III he permits himself to speak of the 'angle *of* a semicircle' and the 'angle *of* a segment'.

THE *ELEMENTS*

Euclid has always been known almost exclusively as the author of the *Elements*, ὁ στοιχειωτής, as the Greeks from Archimedes onwards called him instead of using his name. This wonderful book, notwithstanding its imperfections, remains the greatest elementary text-book in mathematics that the world is privileged to possess. Scarcely any other book except the Bible can have circulated more widely the world over or been more edited and studied. Immediately on its appearance it superseded all other Elements, and that so completely that no others have survived. Archimedes already cites propositions by the Book and number; so do all later Greek mathematicians. Even in Greek times the most accomplished mathematicians, e.g. Heron and Pappus (to say nothing of Porphyry, Simplicius, and Proclus) wrote commentaries. The great Apollonius of Perga was moved by Euclid's work to discuss the first principles of geometry. His *General Treatise* (ἡ καθόλου πραγματεία), as it was called, seems to have contained suggestions for improvement, e.g. a new definition of an angle and alternative constructions for the problems of I. 10, 11, 23. These last seem to show that Apollonius wished to give a more *practical* turn to the beginnings of the subject, herein anticipating the tendency which has of late been so much in evidence. So true is it that 'there is nothing new under the sun' or, as Aristotle forcibly puts it more than once, 'it is not once nor twice but times without number that the same ideas make their appearance in the world'. Apollonius' attempt to prove the Axioms (if we may judge by what Proclus gives as his attempted proof of Ax. 1) was thoroughly misconceived.

Even in ancient times certain things in Euclid were the subject of acute controversy. We know from Aristotle that in his time the theory of parallels had not yet been put on a scientific basis. Euclid seems to have been the first to see the necessity of some postulate and to formulate one. But the fifth Postulate was found a great stumbling-block. Why should a fact which is the converse of an ordinary proposition proved by Euclid himself not be proved? Some tried to prove it, like Ptolemy, Proclus, and (according to Simplicius) one Diodorus, as well as 'Aganis'; we have three of these attempted proofs, which of course all, tacitly or otherwise, make some equivalent assumption. Posidonius and Geminus substituted an *equidistance* theory of parallels.

Zeno of Sidon objected even to Eucl. I. 1 as not being conclusive unless we assume that neither two straight lines nor two circumferences can have a common part, and Posidonius wrote a whole book to controvert Zeno. It was the habit of the Epicureans, says Proclus, to ridicule Eucl. I. 20 (proving that two sides of a triangle are together greater than the third) as being evident even to an ass and requiring no proof. But the Epicureans and Sceptics objected to the whole of mathematics, mainly on the ground of the opposition between the fundamental hypotheses of mathematics and the data of sense. There are no such things, they said, as mathematical points, lines, &c. Even if points exist, you cannot make up a line out of points. It is absurd to define a line as that which, if it be turned about one of its extremities, will always touch a plane; a line being, say, length without breadth, and therefore an unsubstantial thing, cannot be turned round at all; and so on. I mention these things in order to show that mathematicians had to contend with Philistines and others even in Greece.

Cicero is the first Latin author to mention Euclid; but it is not likely that in his time Euclid had been translated into Latin. As Cicero says elsewhere, the Romans did not care for geometry except so far as it was useful for measurements and calculations. Philosophers studied Euclid, but no doubt in the original Greek. Martianus Capella speaks of the effect of the mention of the problem 'how to construct an equilateral triangle on a given straight line' among a company of philosophers, who, recognizing the first proposition of the *Elements*, straightway break out into encomiums on Euclid. Beyond a fragment in a Verona palimpsest of a free reproduction of some propositions from Books XII and XIII dating apparently from the fourth century, there is no trace of any Latin version before Boëtius (born about A.D. 480), to whom Magnus

Aurelius Cassiodorus and Theodoric attribute a translation of Euclid. The so-called geometry of Boëtius (which is, however, only a compilation from various sources, put together in the eleventh century) is anything but such a translation, for it contains only the definitions of Book I, the five Postulates, three Axioms, certain definitions from Books II, III, IV, the enunciations (without proofs) of Book I, ten propositions of Book II and a few from Books III, IV, and, lastly, a passage indicating that the editor will now give something of his own, which proves to be a literal translation of the proofs of Eucl. I. 1–3. This shows that the pseudo-Boëtius had a Latin translation of Euclid from which he extracted these proofs. Moreover, the text of the definitions from Book I shows traces of correct readings which are not found even in the Greek MSS. of the tenth century, but which appear in Proclus and other ancient sources.

All the Greek texts of the *Elements* before Peyrard's (1814–18) were based on MSS. containing Theon's recension (fourth century A.D.). They purport in their titles to be 'from the edition of Theon' or 'from the lectures of Theon', and they contain, in the proposition VI. 33, an application to *sectors* of circles which Theon in his commentary on Ptolemy's *Syntaxis* claims as having been proved by himself in his edition of the *Elements*. When, therefore, Peyrard found in the Vatican the great MS. gr. 190 (now known as P) which contained neither the words from the titles of other MSS. quoted above nor the addition to VI. 33, it was clear that here was what, on the face of it, represented a more ancient edition than Theon's. This is confirmed by the fact that the copyist of P (or rather its archetype) had the two recensions before him and systematically gave the preference to the earlier one; thus the first hand in P has on XIII. 6 a marginal note that 'this theorem is not given in most copies of the new edition, but is found in those of the old'. The *editio princeps* of the Greek text by Simon Grynaeus (Basel 1533) was based on two manuscripts of the sixteenth century which are among the worst. Gregory in his great edition (Oxford 1703) followed the *editio princeps* in the main, only consulting the manuscripts bequeathed by Savile to the University where the Basel text differed from the Latin translation by Commandinus. Even Peyrard only corrected the Basel text by means of P, instead of rejecting it altogether and starting afresh. E. F. August's edition (1826-9) followed P more closely and used the Viennese MS. gr. 103 as well. It was reserved for Heiberg to bring out the new and authoritative text based on P and the best of the Theonine MSS., and taking account of external sources such as Heron and Proclus. Authors earlier than Theon, e.g. Heron, generally agree with our best manuscripts, and Heiberg concludes that the *Elements* were most spoiled by interpolations about the third century, since Sextus Empiricus (about A.D. 200) had a correct text, while Iamblichus had an interpolated one.

A remarkable difference between the best and the inferior manuscripts is in the number and arrangement of the Postulates and Axioms. Our ordinary editions based on Simson had three Postulates and twelve Axioms. Of the twelve Axioms, the eleventh (that all right angles are equal) is, in the genuine text, the Fourth Postulate, and the twelfth is the Fifth Postulate (the Parallel-Postulate) The genuine Postulates are thus five in number. Of the Axioms or Common Notions Heron recognized only three, and Proclus only these and two others (that things which coincide are equal, and that the whole is greater than the part); all the rest, therefore, are probably interpolated, including the assumption that 'two straight lines cannot enclose a space'.

The first Latin translations which we possess in a complete form were made not from the Greek but from the Arabic. The Caliphs al-Manṣūr (754–75) and al-Ma'mūn (813–33) obtained from the Byzantines manuscripts of Euclid among other authors. The *Elements* were translated in the reign of ar-Rashīd (786–809) by al-Ḥajjāj b. Yūsuf b. Maṭar, who also made a second version abridged from the other, but with corrections and explanations, for al-Ma'mūn; six Books of the latter version survive in the Codex Leidensis 399. 1, which has been edited (as to four Books) with Latin translation by Besthorn and Heiberg. The next translation was by Isḥāq b. Ḥunain b. Isḥāq al-'Ibādī (died 910); this translation, as revised by Thābit b. Qurra (died 901), exists in two manuscripts in the Bodleian. Isḥāq's version seems to be a model of good translation; while attempting to get rid of difficulties and unevennesses in the Greek text, the translator gave a faithful reproduction of it. The third extant Arabic version, by Naṣīraddīn aṭ-Ṭūsī (born at Ṭūs in Khurāsān in 1201), is not a translation but a rewritten version based on the older Arabic translations.

The known Latin translations begin with that of Athelhard, an Englishman, of Bath, made about A.D. 1120. It was made from the Arabic, as is proved by the occurrence in it of Arabic words, but Athelhard must also have had before him a translation of (at least) the enunciations based ultimately on the Greek text, and going back to the old Latin version used by the Pseudo-Boëtius and the *Gromatici*. But some sort of translation, or fragments of one, must have reached England earlier still, namely about 924–30, if we may judge by the old English verses:

> The clerk Euclide on this wyse hit fonde
> Thys craft of gemetry yn Egypte londe
> Yn Egypte he tawghte hyt ful wyde,
> In dyvers londe on every syde.
> Mony erys afterwarde y understonde
> Yer that the craft com ynto thys londe.
> Thys craft com into England, as y yow say,
> Yn tyme of good Kyng Adelstone's day.

Next Gherard of Cremona (1114–87) is said to have translated the '15 Books of Euclid' from the Arabic, as he undoubtedly translated an-Nairīzī's Commentary on Books I–X. This translation of the *Elements* was formerly supposed to be lost, but Björnbo claims (1904) to have discovered it in manuscripts at Paris, Boulogne-sur-Mer, Bruges, and (as regards Books X–'XV') at Rome. Gherard's translation was independent of Athelhard's, and gives a word for word rendering of an Arabic manuscript containing a revised and critical edition of Thābit's version.

The third translation from the Arabic was made about 150 years after that of Athelhard by Johannes Campanus. It was not independent of Athelhard's, as is clear from the fact that, in all manuscripts and editions, the definitions, postulates, and axioms and the 364 enunciations are word for word identical in Athelhard and Campanus. But Campanus' translation is the clearer and more complete of the two; the arrangement is also different in that Athelhard regularly puts the proofs before the enunciations, instead of following the usual order. It may be that Campanus used Athelhard's translation, but altered and improved it by means of other Arabic originals.

Gherard of Cremona, in addition to translating the

Elements and an-Nairīzī's commentary thereon, made a whole series of translations from the Arabic of Greek treatises, including the *Data* of Euclid, the *Sphaerica* of Menelaus and Theodosius, and the *Syntaxis* of Ptolemy; he also translated Arabian geometrical works such as the *Liber trium fratrum* and, in addition, the algebra of Muḥammad b. Mūsā. The interest in Greek and Arabian mathematicians thus aroused quickly led to fruitful results in the brilliant works of Leonardo of Pisa (Fibonacci). Leonardo first published in 1202, and then later (1228) brought out an improved edition of, his *Liber Abaci* giving the whole of arithmetic and algebra as known to the Arabs, but in a free and independent style of his own. In like manner, in his *Practica geometriae* (1220), he collected (1) all that the *Elements* of Euclid and Archimedes' works *On the Measurement of a Circle* and *On the Sphere and Cylinder* had taught him of the measurement of plane figures bounded by straight lines, solids bounded by planes, the circle, and the sphere respectively, (2) divisions of figures after the manner of Euclid's book *On Divisions* (of figures), but carried further, (3) some trigonometry. Leonardo is, however, a solitary figure in a waste which extended over the next three centuries; it is as if the talent he had left to the Latin world had lain hidden in a napkin and earned no interest. Roger Bacon (1214–94), though no doubt he exaggerated a little, is witness to the neglect of geometry in education. The philosophers of his day, he says, despised geometry, languages, &c., declaring that they were useless; and people in general, finding no utility in any science such as geometry, could hardly (unless they were boys forced to it by the rod) be induced to study so much as three or four propositions of Euclid, while the fifth proposition was called *Elefuga* or *fuga miserorum*, a punning identification of 'escape from the Elements' with 'escape from troubling' (ἔλεος).

In the Universities in this country and abroad during the fourteenth and fifteenth centuries little geometry was required from candidates for degrees. To have attended lectures on a few books (not more than six) of the *Elements* was a usual qualification. At Oxford in the middle of the fifteenth century two Books of Euclid were read, and no doubt the Cambridge course was similar.

With the issue, however, of the first printed editions of the *Elements* the study of Euclid received a great impetus.

The first printed edition was published at Venice by Erhard Ratdolt in 1482. It contained Campanus' translation from the Arabic already mentioned. This beautiful and very rare book was not only the first printed edition of Euclid, but also the first printed mathematical book of any importance. In the margins of $2\frac{1}{2}$ inches were printed the figures of the propositions. Ratdolt says in his dedication that at that time, although books by ancient and modern authors were being printed every day in Venice, little or nothing mathematical had appeared; this fact he puts down to the difficulty caused by the diagrams, which no one up to that time had succeeded in printing; he adds that after much labour he had discovered a method by which figures could be produced as easily as letters. How eagerly the opportunity of spreading geometrical knowledge was seized is proved by the number of editions which followed in the next few years. Even 1482 saw two forms of the book, though they only differ in the first sheet. Another edition appeared at Ulm in 1486, and another at Vicenza in 1491.

Bartolomeo Zamberti (Zambertus) was the first to bring out a translation of the whole of the *Elements* from the Greek; this appeared at Venice in 1505. The most important Latin translation is, however, that of Commandinus (1509–75), who followed the Greek text more closely than his predecessors and added some ancient scholia, as well as good notes of his own; this translation, which appeared in 1572, was the basis of most translations down to that of Peyrard, including Simson's, and therefore of all those editions, numerous in England, which gave Euclid 'chiefly after the text of Dr. Simson'.

The first complete English translation (1570) is that of Henry Billingsley, who was Sheriff of London in 1584 and was elected Lord Mayor, on a death vacancy, on December 31, 1596. This is a monumental work of 928 folio pages, with a preface by John Dee and notes extracted from all the most important commentaries from Proclus down to Dee himself, a magnificent tribute to Euclid. About the same time Henry Savile began to give unpaid lectures on the Greek geometers; those on Euclid (of 1620) do not extend beyond I. 8, but they are valuable because they grapple with the difficulties connected with the preliminary matter, the definitions, &c., and the tacit assumptions made in the first propositions. It was, however, in the period from about 1660 to 1730, during which Wallis and Halley were Professors at Oxford, and Barrow and Newton at Cambridge, that the study of Greek mathematics was at its height in England. Barrow's admiration for Euclid was unbounded. His Latin version (*Euclidis Elementorum Libri XV breviter demonstrati*) appeared in 1655, and several more editions followed down to 1732; the first English edition appeared in 1660 and was followed by others in 1705, 1722, 1732, and 1751. We are told that Newton, when he first bought a Euclid in 1662 or 1663, thought it a 'trifling book', as the propositions seemed to him obvious; afterwards, however, on Barrow's advice, he studied the *Elements* carefully and derived, as he himself stated, great benefit therefrom. The unique status of Euclid as a text-book in England, which lasted until recently, may perhaps be said to date from the publication of Robert Simson's *Elements of Euclid*, which first appeared both in Latin and in English in 1756. It was a full translation, mainly based on Commandinus, of Books I–VI, XI, and XII, but enriched by valuable notes and suggestions for improvements; it was the basis of most editions down to Todhunter's.

As Euclid's propositions are in the form which is recognized as classical, though that form was not invented by him, it will be useful to give here some account of certain technical terms used by the Greeks in connexion with their form of exposition.

In its completest form a proposition contained six parts: (1) the πρότασις, or *enunciation*, in general terms; (2) the ἔκθεσις or *setting-out*, which states the data, e.g. the particular lines or figures given and denoted by letters, on which the argument is to be illustrated; (3) the διορισμός, *definition* or *specification*, which is a restatement of what it is required to do or prove, but in terms of the particular data, the object being to fix our ideas; (4) the κατασκευή, *construction* or *machinery*, which includes any additions to the original figure by way of construction in order to facilitate the proof; (5) the ἀπόδειξις, the *proof* itself; (6) the συμπέρασμα, or *conclusion*, which reverts to the enunciation and states (often in the same general terms) what has been proved or done. A particular proposition may be without some of these parts (e.g. no *construction* beyond the data may be necessary), but three parts are indispensable to all propositions, the enunciation, proof, and conclusion.

A special term was used in connexion with problems,

namely διορισμός, the same word as we have seen above in (3). There it means a restatement, in terms of particular data, of what it is required to prove or do; here it means the statement of the conditions under which the solution of the problem is possible, or, in its most complete form, a criterion as to 'whether what is sought is impossible or possible, and how far it is practicable and in how many ways' (Proclus). It is introduced, like the other kind of διορισμός, by the phrase δεῖ δή, 'it is thus necessary' (or 'required'). Cf. I. 22, 'Out of three straight lines which are equal to three given straight lines to construct a triangle: thus it is necessary that two of the given straight lines taken together in any manner should be greater than the remaining straight line'.

The *Elements* is a *synthetic* treatise, proceeding straight forward from the known and simple to the unknown and more complex; *analysis*, therefore, which reduces the unknown, or less known, and more complex to the known, has no place in the exposition, though no doubt it played its part in the discovery of the proofs. But *reductio ad absurdum*, a method of proof to which Euclid often has to resort, is a variety of analysis; for analysis begins with the *reduction* (ἀπαγωγή) of the proposition required to be proved, which we hypothetically assume to be true, to something simpler which we can immediately recognize as true or false; the case where the reduction leads to a conclusion obviously false is the *reductio ad absurdum* (ἡ εἰς τὸ ἀδύνατον ἀπαγωγή).

What we call a *corollary* was for the Greeks a *porism* (πόρισμα), something provided or ready-made, i.e. some result incidentally revealed in the course of the demonstration of the main proposition. The name *porism* was also applied to a special kind of substantive proposition, as in

Euclid's three Books of *Porisms*, now lost (see pp. 262-5 below).

The word *lemma* means simply something *assumed*. Archimedes uses it of what is now known as the Axiom of Archimedes, the principle used by Eudoxus and others in the method of exhaustion. More commonly it is an auxiliary proposition requiring proof, but assumed in the place where it is wanted in the main proof for convenience' sake only, in order that the argument may not be interrupted or unduly lengthened. It may be proved in advance, but is often left over to be proved afterwards (ὡς ἑξῆς δειχθήσεται, 'as will be proved immediately').

The content of the 'Elements'.

Book I of the *Elements* necessarily begins with the essential preliminary matter, the *Definitions* (ὅροι), *Postulates* (αἰτήματα), and *Common Notions* (κοιναὶ ἔννοιαι). The *Common Notions* are what we know as *Axioms*, for which Aristotle has the alternative names 'common (things)', 'common opinions'.

Many of the *Definitions* are open to criticism. Two at least seem to be original, namely those of (1) a straight line and (2) a plane. Unsatisfactory as these are, they seem to be capable of a simple explanation. Plato had defined a straight line as 'that of which the middle covers the ends' (i.e. to an eye placed at one end and looking along the line), and Euclid's 'line which lies evenly with the points on itself' may well be an attempt to express Plato's idea in terms excluding any appeal to sight; so also with the definition of a plane. But most of the definitions were probably taken from earlier text-books; this is no doubt the reason why some were included which are never used in the *Elements*, e.g. the definitions of *oblong*, *rhombus*, and *rhomboid*. A square and different kinds of triangles are defined, but there is no definition of a parallelogram. After the existence of a parallelogram is proved (in I. 33), it is first called a *parallelogrammic area* (i.e. an area contained by parallel straight lines), and then (I. 35) the name is shortened to *parallelogram*. To the definition of the diameter of a circle is added the statement that 'such a straight line also bisects the circle' (a discovery attributed to Thales); no doubt this is really a theorem, but the addition was necessary in order to justify the definition of a *semi*-circle immediately following, namely 'the figure contained by the diameter and the circumference cut off by it.'

The five *Postulates* are more important, for they embody the distinctive principles of Euclidean geometry. The first three are commonly regarded as the postulates of *construction*, since they assert the possibility (1) of drawing a straight line joining two given points, (2) of producing a straight line continuously in either direction, (3) of describing a circle with a given centre and 'distance'. Euclid here postulates the existence of real (mathematical) straight lines and circles, of which the straight lines and circles that can in practice be drawn are only imperfect illustrations. Postulates 1 and 2 also imply that the straight line in the first case and the produced portion in the second case are *unique*; in other words, they imply that 'two straight lines cannot enclose a space' and that 'two straight lines cannot have a common segment', and they obviate the necessity for a separate statement of these facts (the reference to the former 'axiom' in I. 4 is in fact interpolated).

Postulates 4 (that all right angles are equal) and 5 (the Parallel-Postulate) might seem to be of an entirely different character. Euclid, however, having to lay down *some* postulate as a basis for a theory of parallels, actually formulated a postulate which also supplies a criterion indispensable in constructions, namely a means of knowing whether two straight lines drawn in a figure will or will not meet if produced. This is one of the advantages of Euclid's Postulate as compared with other equivalents such as Playfair's; and Euclid actually employs it for this purpose as early as I. 44. No doubt Postulate 4 about right angles is often classed as a theorem. But, if we are to prove it, we can only do so by applying one pair of adjacent right angles to another pair; and this method would not be valid except on the assumption of the *invariability of figures*, which would, therefore, strictly speaking, have to be asserted as an antecedent postulate. Euclid had, in any case, to place Postulate 4 before Postulate 5, because Postulate 5 would be no criterion at all unless right angles were determinate magnitudes.

Of the *Common Notions* it is probable that only five (at most) are genuine, the first three and two others, namely 'Things which coincide with [lit. 'fit on'] one another are equal to one another' (4), and 'The whole is greater than the part' (5). The objection to (4) is that its subject-matter belongs to a special science, namely geometry, whereas, according to Aristotle, 'axioms' or 'common notions' are general truths common to all sciences (e.g., if equals be subtracted from equals, the remainders are equal). Neither of the two supposed Axioms 4 and 5 seems to be quoted in terms by Euclid himself; thus in I. 4, where he might have quoted the former, he says simply, 'The base BC will coincide with the base EF and will be equal to it', without referring to any axiom. It seems probable, therefore, that these two Common Notions, though recognized by Proclus, were generalizations from particular inferences found in Euclid, and were inserted after his time.

The propositions of Book I fall into three distinct groups. The first, consisting of Propositions 1–26, deals mainly with triangles, their construction, and their properties in the sense of the relation of their parts, the sides and angles, to one another, including the three congruence-theorems; it also treats of two intersecting straight lines making 'vertically opposite' angles, and one straight line standing on another and making 'adjacent' angles; and it contains a few simple problems of construction, the drawing of perpendiculars, and the bisection of a given angle and of a given straight line. The second group, beginning with I. 27, establishes the theory of parallels, and leads up to the proposition that the sum of the three angles of any triangle is equal to two right angles (32). The third group, beginning with I. 33, 34, which introduce the parallelogram for the first time, deals generally with parallelograms, triangles, and squares with reference to their areas. Propositions 44, 45 are of the greatest importance, being the first cases of the Pythagorean method of 'application of areas': to apply to a given straight line, in a given rectilineal angle, a parallelogram equal to a given triangle (or rectilineal figure); the solution depends on I. 43 proving that the 'complements' of the parallelograms about the diameter of a parallelogram are equal in area, and is equivalent to the algebraical operation of dividing the product of two quantities by a third. I. 46 shows how to construct a square on any given straight line as side, and I. 47 is the great Pythagorean theorem of the square on the hypotenuse of a right-angled triangle. With a converse to the latter theorem the Book ends (I. 48).

Book II is a continuation of the third section of Book I relating to the transformation of areas. It deals mainly with *rectangles* (which appear for the first time) and squares, and not with parallelograms in general, and it shows the equality of sums of rectangles and squares to other such sums. Much use is made of the *gnomon*; this is defined (Def. 2) with reference to any parallelogram, but the gnomons actually used are those belonging to squares. The whole Book is part of the *geometrical algebra* which, with the Greeks, had to take the place of our algebra. The first ten propositions of Book II give the equivalent of the following algebraical identities:

1. $a(b+c+d+\ldots) = ab+ac+ad+\ldots$
2. $(a+b)a+(a+b)b = (a+b)^2$,
3. $(a+b)a = ab+a^2$,
4. $(a+b)^2 = a^2+b^2+2ab$,
5. $ab+\{\frac{1}{2}(a+b)-b\}^2 = \{\frac{1}{2}(a+b)\}^2$,
   or $(\alpha+\beta)(\alpha-\beta)+\beta^2 = \alpha^2$,
6. $(2a+b)b+a^2 = (a+b)^2$,
   or $(\alpha+\beta)(\beta-\alpha)+\alpha^2 = \beta^2$,
7. $(a+b)^2+a^2 = 2(a+b)a+b^2$,
   or $\alpha^2+\beta^2 = 2\alpha\beta+(\alpha-\beta)^2$,
8. $4(a+b)a+b^2 = \{(a+b)+a\}^2$,
   or $4\alpha\beta+(\alpha-\beta)^2 = (\alpha+\beta)^2$,
9. $a^2+b^2 = 2[\{\frac{1}{2}(a+b)\}^2+\{\frac{1}{2}(a+b)-b\}^2]$,
   or $(\alpha+\beta)^2+(\alpha-\beta)^2 = 2(\alpha^2+\beta^2)$,
10. $(2a+b)^2+b^2 = 2\{a^2+(a+b)^2\}$,
    or $(\alpha+\beta)^2+(\beta-\alpha)^2 = 2(\alpha^2+\beta^2)$.

As we have seen (pp. 103-4), Propositions 5 and 6 enable us to solve geometrically the equivalent of the quadratic equations

(1) $ax-x^2 = b^2$, or $\begin{cases} x+y=a, \\ xy = b^2, \end{cases}$

and   (2) $ax+x^2 = b^2$, or $\begin{cases} y-x = a, \\ xy = b^2. \end{cases}$

The procedure is geometrical throughout, and the various areas in all the Propositions 1-8 are actually shown in the figures.

Propositions 9 and 10 were (as we have seen, p. 57 above) used to solve a problem in numbers, namely that of finding any number of successive pairs of integral numbers ('side-' and 'diameter-' numbers) satisfying the equations

$$2x^2 - y^2 = \pm 1.$$

Of the remaining propositions, II. 11 and II. 14 give the geometrical equivalent of the solution of the quadratics

$$x^2 + ax = a^2$$
and $x^2 = ab,$

while Propositions 12 and 13 prove for any triangle with sides a, b, c, the equivalent of the formula

$$a^2 = b^2 + c^2 - 2bc \cos A.$$

It is worth noting that, just as Book I seems designed to lead up to the Pythagorean proposition I. 47 and its converse, so Book II gives, in its last two propositions but one, a generalization of that theorem with any triangle taking the place of the right-angled triangle.

Book III is on the geometry of the circle, including the relations between circles cutting or touching one another. It begins with definitions. 'Equal circles' are defined as circles with equal diameters or radii (the Greeks had no single word for 'radius'; they called it 'the (straight line) from the centre', ἡ ἐκ τοῦ κέντρου); if this 'definition' were proved, it could only be by superposition. A 'tangent' is defined (Def. 2) as 'a straight line which meets a circle but, if produced, does not cut it'. A chord is simply 'a straight line in a circle', and chords are equally distant, or more or less distant, from the centre, according as the perpendiculars on them from the centre are equal, greater, or less (Defs. 4, 5). Euclid defines not only a segment of a circle and the 'angle in a segment' (Defs. 6, 8), but also the 'angle *of* a segment' (Def. 7). The last-named definition, as well as the part of Proposition 16 about the 'angle *of* a semicircle', are the last survivals in Greek geometry of the 'angle *of* a segment' (the 'mixed' angle made by the curve with the base of the segment at either end); these survivals show Euclid's almost excessive respect for tradition, the 'angle' in question being of no practical use in demonstrations. The last definitions are those of a *sector* of a circle and of 'similar segments'; the word for *sector*, τομεύς, is said to have been suggested by the shape of the 'shoemaker's knife' (σκυτοτομικὸς τομεύς). The definition of 'similar segments' assumes provisionally (pending the proof in III. 21) that the angle in a segment is one and the same at whatever point on the circumference it is formed.

The propositions of Book III may be roughly classified thus. Central and chord properties account for six propositions (1, 3, 4, 9, 14, 15). Three propositions throw light on the form of the circle (2, showing that it is everywhere concave towards the centre, and 7, 8, comparing the respective lengths of all straight lines drawn to the circumference from a single point (other than the centre), internal or external. Propositions 5, 6, 10, 11, 13 and the interpolated Proposition 12 deal with two circles cutting or touching one another. Tangent properties, including the drawing of a tangent, occupy Propositions 16–19; it is in 16 that we have the survival of the 'angle *of* a semicircle' and of its complement, the 'angle' between the curve and the tangent at the extremity of the diameter, the latter angle (afterwards called the κερατοειδής or 'hornlike' angle) being proved to be less than any rectilineal angle. These 'mixed' angles, occurring here and in Proposition 31 only, appear no more in serious Greek geometry, though controversy about the nature of the 'hornlike angle' went on in the works of commentators down to Clavius, Peletarius, Vieta, Galilei, and Wallis. Propositions 20–34 are concerned with segments, angles in segments and at the centre, &c. The Book ends with three important propositions (35–7) to the effect that, 'given a circle and any point O internal or external to it, if any straight line through O meet the circle in P, Q, the rectangle $PO . OQ$ is constant and, in the case where O is external to the circle, is equal to the square on the tangent to the circle from O.

Book IV continues the geometry of the circle, with special reference to the problems of inscribing and circumscribing to the circle certain rectilineal figures which can be so inscribed and circumscribed by means of the geometry of the straight line and circle only, namely, a triangle equiangular with a given triangle (2, 3), a square (6, 7), a regular pentagon (11, 12), a regular hexagon (15), and a regular polygon with fifteen sides (16), and the corresponding problems of inscribing or circumscribing a circle to a triangle (4, 5), a square (8, 9), and the other figures mentioned (13, 14, 15 Por.). IV. 10 is the important Pythagorean proposition, used in the construction of a regular pentagon, 'To construct an isosceles triangle having each of the angles at the base double of the remaining angle', which again depends on the Pythagorean proposition (II. 11) showing how to divide a given straight line in extreme and mean ratio. The regular fifteen-sided figure (Prop. 16) was found useful in astronomy, the obliquity of the ecliptic being taken to be about 24° or one-fifteenth of 360°. The whole of the Book seems to be unquestionably Pythagorean.

Book V expounds the new theory of proportion applicable to incommensurable as well as commensurable magnitudes, and to magnitudes of every kind (straight lines, angles, areas, volumes, numbers, times, &c.). Greek mathematics can boast of no finer discovery than this theory, due to Eudoxus, which first put on a sound footing so much of geometry as depended on the use of proportions. The scholiast who attributes the discovery of the theory to Eudoxus is equally clear that the actual arrangement and sequence of Book V is due to Euclid himself. The ordering of the propositions and the development of the proofs are indeed masterly and worthy of Euclid; as Barrow said, 'there is nothing in the whole body of the Elements of a more subtle invention, nothing more solidly established and more accurately handled, than the doctrine of proportionals'.

The Definitions of Book V are naturally of supreme importance. The definition (3) of ratio as 'a sort of relation (ποιὰ σχέσις) in respect of size (πηλικότης) between two magnitudes of the same kind' tells us little, certainly; but Definition 4 ('Magnitudes are said to have a ratio to one another which are capable when multiplied of exceeding one another') makes amends, for not only does it show that the magnitudes must be of the same kind, but, while including incommensurable as well as commensurable magnitudes, it *excludes* the relation between a finite magnitude and a magnitude of the same kind which is infinitely great or infinitely small; it is also practically equivalent to the 'Axiom of Archimedes' (so-called), which lies at the root of the method of exhaustion. Most important of all is the fundamental definition (5) of magnitudes which are in the same ratio: 'Magnitudes are said to be in the same ratio, the first to the second and the third to the fourth, when, if any equimultiples whatever be taken of the first and third, and any equimultiples whatever of the second and fourth, the former equimultiples alike exceed, are alike equal to, or are alike less than, the latter equimultiples taken in corresponding order.' Perhaps the greatest tribute to this wonderful definition is its adoption by Weierstrass as a definition of equal numbers. For a most attractive explanation showing its exact significance and its absolute sufficiency the reader should refer to De Morgan's articles on Ratio and Proportion in the *Penny Cyclopaedia* (vol. xix, 1841), largely reproduced in *The Thirteen Books of Euclid's Elements* (vol. ii, pp. 116-24). Euclid adds (7) a definition of 'greater ratio': 'When of the equimultiples the multiple of the first magnitude exceeds the multiple of the second, but the multiple of the third does not exceed the multiple of the fourth, then the first is said to have a *greater ratio* to the second than the third has to the fourth'; here Euclid takes (possibly for brevity) only one criterion for greater ratio, the other possible criterion being that, while the multiple of the first is *equal* to that of the second, the multiple of the third is *less* than that of the fourth. A proportion may be in three or four terms (Defs. 8, 9, 10); 'corresponding' or 'homologous' terms mean antecedents in relation to antecedents and consequents in relation to consequents (11). Definitions 12–16 explain the terms used for the transformation of ratios: (α) ἐναλλάξ, *alternando*, transforms the proportion $a:b = c:d$ into $a:c = b:d$; (β) *inversion* (ἀνάπαλιν, *inversely*) turns the ratio $a:b$ into $b:a$; (γ) *composition*, σύνθεσις (συνθέντι, lit. 'to one who has compounded', = componendo) turns $a:b$ into $(a+b):b$; (δ) *separation*, διαίρεσις (διελόντι, lit. 'to one who has separated', = *separando*) turns $a:b$ into $(a-b):b$; (ε) *conversion*, ἀναστροφή (ἀναστρέψαντι = *convertendo*) turns $a:b$ into $a:(a-b)$. Lastly, we have definitions (17, 18) of *ex aequali* (sc. *distantia*), δι' ἴσου, and *ex aequali* 'in disturbed proportion' (δι' ἴσου ἐν τεταραγμένῃ ἀναλογίᾳ); the first infers from $a:b = A:B$ and $b:c = B:C$ that $a:c = A:C$, and the second infers from $a:b = B:C$ and $b:c = A:B$ that $a:c = A:C$.

As the content of the wonderful Book V is too little known, it is worth while to summarize it with the aid of modern notation. In the summary the letters $a, b, c \ldots$ will mean *magnitudes* in general and the letters $m, n, p \ldots$ integral *numbers*; thus ma, mb are equimultiples of a, b.

The first six propositions are arithmetical theorems about multiples and equimultiples.

$\begin{cases} 1. & ma+mb+mc+\ldots = m(a+b+c+\ldots). \\ 5. & ma-mb = m(a-b). \end{cases}$

$\begin{cases} 2. & ma+na+pa+\ldots = (m+n+p+\ldots)a. \\ 6. & ma-na = (m-n)a. \end{cases}$

3. Equimultiples of equimultiples are themselves equimultiples.

4. If $a:b = c:d$, then $ma:nb = mc:nd$; or the equimultiples in Def. 5 are themselves proportionals.

All these propositions except (4) are proved by separating the multiples used into their units. (4) is proved by taking equimultiples of the equimultiples, namely pma and pmc of ma, mc, and qnb, qnd of nb, nd. Then, by 3, the new equimultiples are equimultiples of a, c and b, d respectively. Since $a:b = c:d$, the new equimultiples

satisfy the criterion of Def. 5, whence conversely
$$ma : nb = mc : nd.$$

7, 9. If $a = b$, then $\left.\begin{array}{l} a:c = b:c \\ \text{and} \quad c:a = c:b \end{array}\right\}$; and conversely.

8, 10. If $a > b$, then $\left.\begin{array}{l} a:c > b:c \\ \text{and} \quad c:b > c:a \end{array}\right\}$; and conversely.

7, 8 are proved by using Defs. 5 and 7, and the converses are proved by *reductio ad absurdum*.

11. If $\quad a:b = c:d,$
and $\quad c:d = e:f,$
then $\quad a:b = e:f.$

12. If $\quad a:b = c:d = e:f\ldots$
then $\quad a:b = (a+c+e+\ldots) : (b+d+f+\ldots).$

13. If $\quad a:b = c:d,$
but $\quad c:d > e:f,$
then $\quad a:b > e:f.$

14. If $\quad a:b = c:d,$
then, according as $a > = < c$, $b > = < d$;
that is, the criterion of Def. 5 is true if, instead of equimultiples, we take *once* the magnitudes respectively.

15. $a:b = ma:mb.$ This follows from 12.

16–18 prove the legitimacy of transforming a proportion *alternando, separando, componendo* respectively; that is, they prove that, if the original proportion is true, the transformed proportion is also true.

19. If $\quad a:b = c:d,$
then $\quad (a-c) : (b-d) = a:b.$

The transformation of a proportion by *inversion* is not given, probably because it is obvious from Def. 5; transformation by *conversion* is not given either, but it follows, as 19 does, by using 17 combined with 16.

20-3 establish the truth of inferences from two proportions *ex aequali* and *ex aequali* 'in disturbed proportion' respectively, 20 being preliminary to 22 and 21 to 23; i.e. it is proved

(22) that, if $\quad a:b = d:e,$
and $\quad b:c = e:f,$
then, *ex aequali*, $\quad a:c = d:f,$
and (23) that, if $\quad a:b = e:f,$
and $\quad b:c = d:e,$

then, *ex aequali* 'in disturbed proportion', $a:c = d:f$.

The Book concludes with

24. If $\quad a:c = d:f,$
and $\quad b:c = e:f,$
then $\quad (a+b):c = (d+e):f.$

25. If $a:b = c:d$, and of the four terms a is the greatest (so that d is the least), then

$$a+d > b+c.$$

Some slight defects are found in the Book as it has reached us, and perhaps, therefore, it never received the final touches from Euclid's hand; but these defects can all be corrected without much difficulty, as Simson showed in his admirable edition. M. J. M. Hill has gone further and, after long and unremitting labour, has recently contributed valuable papers to the *Mathematical Gazette* supplementing Euclid and making of his system a consistent and well-rounded whole.

Book VI applies the general theory of proportion set out in Book V to plane geometry. The first proposition, proving that triangles and parallelograms of the same height are respectively as their bases, and the last (33), to the effect that in equal circles angles at the centre or at the circumference respectively are as the arcs on which they stand, both use the method of equimultiples and apply the test of proportion laid down in V. Def. 5. The fundamental proposition (2) that two sides of a triangle cut by any parallel to the third side are divided proportionally, and the converse, gives the means of solving the problems of cutting off from a straight line a prescribed part (9), of cutting a given straight line proportionally to a given divided straight line (10), of finding a third proportional to two straight lines (11), and a fourth proportional to three (12). Proposition 3 proves that the internal bisector of an angle of a triangle cuts the opposite side into parts which have the same ratio as the sides containing the angle, and the converse. Next come propositions showing the alternative conditions for the similarity of two triangles, namely equality of all the angles respectively (4), proportionality of pairs of sides in order (5), equality of one angle in each with proportionality of the sides containing the equal angles (6), and (the 'ambiguous case') equality of one angle in each and proportionality of the sides containing other angles (7). Proposition 8 proves that the perpendicular from the right angle in a right-angled triangle to the opposite side divides the triangle into two triangles which are similar to the original triangle and to one another, a proposition used in solving the problem of finding a mean proportional between two given straight lines (Prop. 13, the Book VI version of II. 14). In Propositions 14, 15 Euclid proves that, in parallelograms or triangles of equal area which have one angle equal to one angle, the sides about the equal angles are reciprocally proportional, and the converse. It is then proved (16), by means of 14, that, if four straight lines are proportional, the rectangle contained by the extremes is equal to that contained by the means, and conversely; Proposition 17 contains the particular case of three proportional straight lines, where the rectangle contained by the extremes is equal to the square on the mean. Propositions 18–22 deal with similar rectilineal figures; 19 (with Porism) and 20 are specially important, proving that similar triangles, and similar polygons generally, are to one another in the duplicate ratio of corresponding sides, and that, if three straight lines are proportional, then, as the first is to the third, so is the figure described on the first to the similar figure similarly described on the second.

Proposition 23 (equiangular parallelograms have to one another the ratio compounded of the ratios of their sides) is highly important in itself, and also because it introduces us to the method of compounding, i.e. multiplying, ratios, a practical method of very wide application in Greek geometry. Euclid has never defined 'compound ratio' or the 'compounding' of ratios; the meaning of 'compound ratio' and the method of compounding are made clear by this proposition. The equiangular parallelograms are placed so that two equal angles as BCD, GCE are vertically opposite at C, or BCG, ECD are straight lines. Complete the parallelogram $DCGH$. The ratio 'compounded of the ratios of the sides' of the parallelograms AC, CF is the ratio compounded of the ratios $BC:CG$ and $DC:CE$, and is obtained

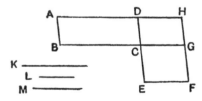

thus. Take any straight line K and find another, L, such that
$$BC:CG = K:L.$$

Again, find a straight line, M, such that
$$DC : CE = L : M.$$
Then the ratio 'compounded of the ratios of the sides' is equal to the ratio compounded of the ratios $K : L$ and $L : M$, that is, the ratio $K : M$.

But (VI. 1) $\quad (ABCD) : (DCGH) = BC : CG,$
$$= K : L;$$
and $\quad\quad\quad (DCGH) : (CEFG) = DC : CE,$
$$= L : M.$$
Therefore, *ex aequali* (V. 22)
$$(ABCD) : (CEFG) = K : M.$$

The important Proposition 25 (to construct a rectilineal figure similar to one, and equal to another, rectilineal figure) is one of the famous propositions attributed to Pythagoras; it is doubtless Pythagorean, since it employs precisely the problems in 'application of areas' contained in Eucl. I. 44, 45. The given figure (P, say) to which the required figure is to be similar is transformed (by I. 44) into a parallelogram on the same base BC. Then the figure (Q) to which the required figure is to be *equal* is (by I. 45) transformed into a parallelogram on the base CF (in a straight line with BC) and of equal height with the other parallelogram. Then $(P) : (Q) = BC : CF$. It is now only necessary to take a straight line GH a mean proportional between BC and CF, and to describe on GH as base a rectilineal figure similar and similarly situated to P in which BC is the base (VI. 18). The correctness of the construction is proved by VI. 19 Por.

In the vital Propositions 27, 28, 29 the Pythagorean *application of areas* appears in its most general form, equivalent to the geometrical solution of the most general form of quadratic equation where that equation has a real and positive root. The method is fundamental in Greek geometry; it is, for instance, the foundation of Euclid's Book X (on irrationals) and of the whole treatment of conic sections by Apollonius of Perga. The problems of Propositions 28, 29 are thus enunciated: 'To a given straight line to apply a parallelogram equal to a given rectilineal figure and *deficient* (or *exceeding*) by a parallelogrammic figure similar to a given parallelogram'; and Proposition 27 proves the διορισμός, or determination of the condition of possibility of solution, in the case of *deficiency* (28): 'The given rectilineal figure must not be greater than the parallelogram described on the half of the straight line and similar to the defect.'

We will first examine the problem of Proposition 28. We are already familiar with the notion of applying a parallelogram to a straight line AB so that it *falls short* by a certain other parallelogram. Suppose that D is the given parallelogram to which the *defect* has to be similar. Bisect AB at E, and on the half EB describe the parallelo-

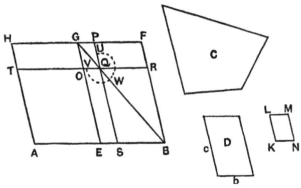

gram $GEBF$ similar and similarly situated to D. Draw the diagonal GB and complete the parallelogram $HABF$.

Through any point T on HA draw $TOQR$ parallel to AB meeting GE, GB and FB in O, Q, R respectively, and through Q draw PQS parallel to HA or GE. Then $TASQ$ is a parallelogram applied to AB but *falling short* by a parallelogram ($QSBR$) which is similar and similarly situated to EF (24) and therefore to D (21). (In the same way, if T had been on HA *produced* and TR had met GB *produced*, we should have had a parallelogram applied to AB but *exceeding* by a parallelogram similar and similarly situated to D.)

Now consider the parallelogram AQ applied to AB but falling short by SR which is similar and similarly situated to D. Since $(AO) = (ER)$ and $(OS) = (QF)$, the parallelogram AQ is equal to the gnomon UWV, and the problem is therefore to find the gnomon UWV such that its area is equal to that of the given rectilineal figure C. And clearly the gnomon cannot be greater than the parallelogram EF. Therefore C must not be greater than EF; and this is the διορισμός proved in 27.

Since now the gnomon UWV has to be equal to C, the parallelogram $GOPQ$ must be equal to the difference between (EF) and C. Hence we have merely to draw in the angle FGE the parallelogram $GOQP$ equal to $(EF) - C$ and similar and similarly situated to D. Euclid does in fact describe the parallelogram $LKNM$ equal to $(EF) - C$ and similar and similarly situated to D, and then draws $GOQP$ equal to it. The problem is then solved, $TASQ$ being the required parallelogram applied to AB and falling short in the manner required.

We will now show that Euclid's geometrical solution corresponds exactly to our algebraical solution of a quadratic equation. Let $AB = a$, and let $b : c$ be the ratios of the sides of D as shown in the figure. We may take for x, the unknown quantity, either of the sides of the *defect SR*. If $SB = x$, $QS = \frac{c}{b}x$. Then, if m is a certain constant (actually the sine of an angle of one of the parallelograms), the area (AQ) is $m.AS.SQ$ or $m(a-x)\frac{c}{b}x$. This is equal to C, and the equation to be solved is

$$ax - x^2 = \frac{b}{c} \cdot \frac{C}{m}.$$

To solve this, we change the sign throughout and complete the square on the left-hand side; thus

$$(\tfrac{1}{2}a - x)^2 = \tfrac{1}{4}a^2 - \frac{b}{c} \cdot \frac{C}{m},$$

and $\qquad x = \tfrac{1}{2}a \mp \sqrt{\left(\tfrac{1}{4}a^2 - \frac{b}{c} \cdot \frac{C}{m}\right)}.$

Now Euclid actually constructs the parallelogram $GOPQ$. The area of this is

$$m.GO.OQ = m.\frac{c}{b}.OQ^2 = m.\frac{c}{b}(\tfrac{1}{2}a - x)^2;$$

that is, Euclid in effect finds $(\tfrac{1}{2}a-x)^2$ just as we do. The solution in the figure corresponds to the *negative* sign before the radical

$$x = \tfrac{1}{2}a - \sqrt{\left(\tfrac{1}{4}a^2 - \frac{b}{c} \cdot \frac{C}{m}\right)};$$

but Euclid was, of course, aware that there are two solutions, and how he could exhibit the second in the figure.

For a real solution C must not be greater than $m(\tfrac{1}{2}a)^2 \cdot \frac{c}{b}$, which is the area of (EF). This is what is proved in Euclid's Proposition 27.

The solution of Proposition 29 is similar *mutatis mutandis*,

THE ELEMENTS. BOOKS VI, VII

but no διορισμός is necessary, a solution being always possible.

Proposition 30 uses 29 for the purpose of dividing a straight line 'in extreme and mean ratio'. Proposition 31 extends the theorem of I. 47, showing that that theorem is true not only of squares but of three similar plane figures (of whatever shape) described upon the three sides of the right-angled triangle and similarly situated with reference to the sides.

Except in the respect that it is based on the new theory of proportion, Book VI does not appear to contain any matter that was not known before Euclid's time. The extension of I. 47 is assumed by Hippocrates of Chios for semicircles described on the sides of a right-angled triangle as diameters.

Books VII, VIII, IX are arithmetical in the Greek sense, that is to say, they deal mainly with the nature and properties of (integral) numbers. Book VII begins with definitions, including those of a *unit*, a *number*, and the varieties of numbers, *even, odd, even-times even, even-times odd, odd-times odd, prime, prime to one another, composite, composite to one another, plane, solid, square, cube, similar plane* and *similar solid* numbers, and a *perfect* number. There are also definitions of terms employed in the numerical theory of proportion, namely *a part* (= an aliquot part or submultiple), *parts* (= a certain number of such parts, equivalent to a proper fraction), *multiply*; and we have finally the definition of four proportional numbers, stating that 'numbers are proportional when the first is the same multiple, the same part, or the same parts, of the second that the third is of the fourth', i.e. numbers a, b, c, d are proportional if, when $a = \frac{m}{n} b$, $c = \frac{m}{n} d$, where m, n are any integers (though the definition does not in terms cover the case where $m > n$). The mode of presentation is geometrical in the sense that numbers are throughout represented by straight lines and not by numerical signs.

The propositions of Book VII fall into four main groups. Propositions 1-3 give the method of finding the greatest common measure of two or three numbers in essentially the same form as that in which our text-books have it; the test for two numbers being prime to one another (namely, that no remainder measures the preceding divisor till 1 is reached) comes in the first proposition. Propositions 4-19 set out the numerical theory of proportion; 4-10 are preliminary, dealing with numbers which are 'a part' or 'parts' of other numbers, or 'the same part' or 'parts' of other numbers respectively, and so connecting the theory with the definition of proportionals. 11 and 13 are transformations of proportions corresponding to V. 19 and V. 16, while 12 corresponds to V. 12 and 14 to V. 22 (the *ex aequali* proposition).

Proposition 15 proves that, if $1 : m = a : ma$ ('if the third number measures the fourth the same number of times that the unit measures the second'), then alternately

$$1 : a = m : ma.$$

This result is used (16) to prove that $ab = ba$, or, in other words, that the order of multiplication is indifferent. Two simple propositions (17, 18), based on 16, namely that $b : c = ab : ac$ and $a : b = ac : bc$, lead to the important Proposition 19 that, if $a : b = c : d$, then $ad = bc$, and conversely, which corresponds to VI. 16 for straight lines.

Propositions 20, 21 about 'the least numbers of those which have the same ratio with them' prove that, if m, n are such numbers, and a, b any others in the same ratio,

m measures a the same number of times that n measures b, and that numbers prime to one another are the least of those which have the same ratio with them. These propositions lead up to propositions (22–32) about numbers prime to one another, prime numbers, and composite numbers; of which we may mention: (24) if two numbers be prime to any number, their product will also be prime to the same number; (28) if two numbers be prime to one another, their sum will be prime to each of them, and, if the sum be prime to either, the original numbers will be prime to one another; (30) if any prime number measures the product of two numbers, it will measure one of the two; (32) any number either is prime or is measured by some prime number.

Propositions 33–9 are directed to finding the least common multiple of two or three numbers; 33 is preliminary, using the G. C. M. for the purpose of solving the problem, 'Given as many numbers as we please, to find the least of those which have the same ratio with them'.

In Book VII Euclid was probably following earlier models, while making improvements of his own. Propositions corresponding to VII. 20, 22, 33 are presupposed in the fragment of Archytas already referred to (pp. 136–7).

Book VIII deals largely with numbers in continued 'proportion', i.e. in geometrical progression. If we denote the terms by $a^n, a^{n-1}b, a^{n-2}b^2 \ldots ab^{n-1}, b^n$, we learn that, if a^n, b^n are prime to one another, the terms of the series are the least of those which have the same ratio with them, and vice versa (1, 3); 2 shows how to find the series when the ratio $a:b$ is given in its lowest terms. If a^n does not measure $a^{n-1}b$, no term measures any other, but if a^n measures b^n, it measures $a^{n-1}b$ (6, 7). According as a^2 does or does not measure b^2, and according as a^3 does or does not measure b^3, a does or does not measure b, and vice versa (14–17). If a, b, c are in geometrical progression, so are a^2, b^2, c^2 ... and a^3, b^3, c^3 ... respectively (13).

Given any number of ratios between numbers as $a:b$, $c:d$ ..., Proposition 4 shows how to find a series p, q, r ... in the least possible terms such that

$$p:q = a:b, \ q:r = c:d, \ \ldots.$$

This is done by finding the least common measure first of b, c and then of other pairs of numbers as required. This proposition enables us to compound any number of ratios between numbers in the same way as ratios between straight lines are compounded in VI. 23; the proposition (5) corresponding to VI. 23 then follows to the effect that plane numbers have to one another the ratio compounded of the ratios of their sides.

Propositions 8–12 and 18–21 deal with the interpolation of geometric means between numbers. If $a:b = e:f$, and there are n geometric means between a and b, there are n geometric means between e and f (8). If a^n, $a^{n-1}b$, $a^{n-2}b^2$ ... ab^{n-1}, b^n is a G. P. of $n+1$ terms, so that there are $n-1$ geometric means between a^n and b^n, there are the same number of geometric means between 1, a^n and between 1, b^n respectively (9); and conversely, if 1, a, a^2 ... a^n, and 1, b, b^2 ... b^n are series in geometrical progression, there are the same number $(n-1)$ of geometric means between a^n, b^n as there are between 1, a^n and between 1, b^n respectively (10). There is one mean proportional number between two square numbers (11) and between two similar plane numbers (18), and conversely, if there is one mean proportional number between two numbers, the numbers are similar plane numbers (20); there are two geometric means between two cube numbers (12)

THE ELEMENTS. BOOKS VIII, IX

and between two similar solid numbers (19), and conversely, if there are two geometric means between two numbers, the numbers are similar solid numbers (21). These propositions are stated for square and cube numbers by Plato in the *Timaeus*, and Nicomachus calls them 'Platonic' accordingly. Lastly, similar plane numbers have the same ratio as a square has to a square (26), and similar solid numbers have the same ratio as a cube has to a cube (27).

Book IX begins with some simple propositions such as the following: the product of two similar plane numbers is a square (1), and, if the product of two numbers is a square number, the numbers are similar plane numbers (2); the product of two equal or unequal cubes is a cube (3, 4); if a^3B is a cube, B is a cube (5), and if A^2 is a cube, A is a cube (6). Propositions 8–13 prove certain relations between the terms of a series in geometrical progression in which 1 is the first term. If the series is 1, $a, b, c, \ldots k$, then (9), if a is a square (or a cube), all the succeeding terms are squares (or cubes); if a is not a square, the only squares in the series are the term following a, namely b, and all alternate terms after b; if a is not a cube, the only cubes in the series are the fourth term (c), the seventh, tenth, &c., terms (leaving out two terms throughout); the seventh, thirteenth, &c., terms (leaving out five terms each time) are both square and cube (8, 10). The interesting theorem follows (11 and Porism) that, if 1, $a_1, a_2 \ldots a_n$ are terms in geometrical progression, and a_r, a_n are any two terms, a_r being less than a_n, then a_r will measure a_n, and $a_n = a_r \cdot a_{n-r}$; this is of course equivalent to the formula $a^{m+n} = a^m \cdot a^n$. It is next proved (12, 13) that, if 1, $a, b, c \ldots k$ are numbers in geometrical progression, and k is measured by any primes, a is measured by the same; and if a is prime, k will not be measured by any numbers except those which occur in the series. Proposition 14 is the equivalent of the important theorem that *a number can only be resolved into prime factors in one way*. Propositions 16–19 deal with the conditions under which it is possible, or impossible, that there should be an integral third proportional to two, or an integral fourth proportional to three, given numbers. Next, by a proof which is the same as that usually given in our algebras, Euclid proves (20) that *the number of prime numbers is infinite*. After some easy propositions (21–34) about odd, even, even-times odd, and even-times even numbers respectively, Euclid gives in the last two propositions of the Book an elegant summation of a G. P. of n terms (35), and a proof of the criterion for the formation of 'perfect' numbers (36).

The summation of the G. P. amounts to the following. Suppose $a_1, a_2, a_3 \ldots a_{n+1}$ to be $n+1$ terms in G. P.

Then
$$\frac{a_{n+1}}{a_n} = \frac{a_n}{a_{n-1}} = \ldots = \frac{a_2}{a_1},$$

and, *separando*, $\dfrac{a_{n+1}-a_n}{a_n} = \dfrac{a_n-a_{n-1}}{a_{n-1}} = \ldots = \dfrac{a_2-a_1}{a_1}.$

Adding the antecedents and the consequents, we have (VII. 12)
$$\frac{a_{n+1}-a_1}{a_n+a_{n-1}+\ldots+a_2+a_1} = \frac{a_2-a_1}{a_1}$$
which gives $a_n+a_{n-1}+ \ldots +a_2+a_1$ or $\Sigma_1^n a$.

In Proposition 36 Euclid proves that, if the sum of any number of terms of the series $1, 2, 2^2 \ldots 2^n$ is prime, the product of the said sum and of the last term, or
$$(1+2+2^2+ \ldots +2^n)2^n$$
is a *perfect* number, i.e. is equal to the sum of all its factors.

In the arithmetical Books all numbers are, as we said,

represented by straight lines. This applies throughout, whether the numbers are linear, plane, or solid, or any other kinds of numbers: thus a product of two or three numbers is represented, not by a rectangle or a solid, but by a straight line.

Book X is perhaps the most remarkable, as it is certainly the most finished, of all the Books in the *Elements*. It deals with irrationals, by which must be understood, in general, *straight lines* which are irrational in relation to any particular straight line assumed as rational; and it investigates every possible variety of straight line corresponding to what we should express in algebra by $\sqrt{(\sqrt{a} \pm \sqrt{b})}$, where $\sqrt{a}$ and $\sqrt{b}$ are surds and incommensurable with one another. The subject did not originate with Euclid. We know that not only the fundamental proposition X. 9 (proving that squares which have not to one another the ratio of a square number to a square number have their sides incommensurable in length, and vice versa), but also a large part of the further development of the subject, was due to Theaetetus. But, as Pappus says, in a commentary partly extant in Arabic, Euclid systematized the theory, making precise the definitions of rational and irrational magnitudes, setting out a number of orders of irrational magnitudes and exhibiting their whole extent.

To begin with the definitions. 'Commensurable' magnitudes can be measured by one and the same measure; 'incommensurable' magnitudes have no common measure (1). Straight lines incommensurable in length may be 'commensurable in square' or 'incommensurable in square' according as the squares on them can or cannot be measured by one and the same area (2). Given a straight line which we agree to call rational, Euclid regards as rational not only any straight line commensurable in length with the given straight line, but also any straight line commensurable with it in square though not in length; if, however, a straight line is commensurable neither in length nor in square with the given rational straight line, it is irrational (3). On the other hand, while the square on the straight line assumed as rational is rational, any area incommensurable with it is irrational (4). Thus, if ρ is a straight line assumed as rational, not only is $k\rho$ rational, but also $\sqrt{k} \cdot \rho$, where k is a non-square number or a fraction m/n which, when reduced to its lowest terms, is not square. In regard, therefore, to rational *straight lines* (only) Euclid takes a somewhat broader view than we have met before. On the other hand, the straight lines $(1 \pm \sqrt{k})\rho$ and $(\sqrt{k} \pm \sqrt{\lambda})\rho$ corresponding to $\sqrt{a} \pm \sqrt{b}$ in algebra (when $\sqrt{a}$, $\sqrt{b}$ are not commensurable) are irrational.

The area $\sqrt{k} \cdot \rho^2$ which may be regarded as a rectangle with sides ρ and $\sqrt{k} \cdot \rho$ is a *medial* rectangle or area, and the side of a square equal to it, or $k^{\frac{1}{4}}\rho$, is a *medial straight line*, the first in Euclid's classification of irrational straight lines (it is, of course, the mean proportional between ρ and $\sqrt{k} \cdot \rho$). The medial straight line may take any equivalent forms, e.g. $\sqrt{(b\sqrt{A})}$ or $\sqrt[4]{(AB)}$.

The Book opens with the famous proposition (X. 1) which is the basis of the method of exhaustion used in Book XII, namely that, if from any magnitude there be subtracted more than its half (or its half), from the remainder again more than its half (or its half), and so on continually, there will at length remain a magnitude less than any assigned magnitude of the same kind. Proposition 2 uses the operation for finding the greatest common measure of two magnitudes as a test of their commensurability or otherwise; Propositions 3, 4 find the greatest common measure, where there is one, of two or three magnitudes, just as VII. 2, 3 do for numbers. Easy propositions (5–8) lead up to the fundamental theorem of Theaetetus (9). Propositions 17, 18 prove the equivalent of the fact that the roots of the quadratic equation $ax - x^2 = \frac{1}{4}b^2$ are commensurable or incommensurable with a according as $\sqrt{(a^2 - b^2)}$ is commensurable or incommensurable with a. Propositions 19–21 deal with rational and irrational rectangles, and Propositions 23–8 with *medial* rectangles and straight lines. The difference between two *medial* areas, e.g. $\sqrt{k} \cdot \rho^2$ and $\sqrt{\lambda} \cdot \rho^2$ cannot be rational (26); this is equivalent to proving, as we do in algebra, that $\sqrt{k} - \sqrt{\lambda}$ cannot be equal to k'. Next Euclid finds (27, 28) medial straight lines commensurable in square only, (1) containing a *rational* rectangle, e.g. $k^{\frac{1}{4}}\rho$ and $k^{\frac{3}{4}}\rho$, and (2) containing a *medial* rectangle, as $k^{\frac{1}{4}}\rho$, $\lambda^{\frac{1}{2}}\rho/k^{\frac{1}{4}}$.

Two lemmas follow, the object of which is to find (1) two square numbers the sum of which is a square number, (2) two square numbers the sum of which is not square. Euclid's solution in the first case has been given above (p. 48); the numbers found in the second case are $mp^2 \cdot mq^2$ and $\{\frac{1}{2}(mp^2 - mq^2) - 1\}^2$.

Propositions 29, 30 find two *rational* straight lines x, y commensurable in square only such that $\sqrt{(x^2 - y^2)}$ is (1) commensurable, (2) incommensurable, with x, and Propositions 31, 32 four pairs of *medial* straight lines x, y commensurable in square only satisfying the four possible combinations of the conditions of xy being rational or medial and $\sqrt{(x^2 - y^2)}$ commensurable or incommensurable with x. Euclid then finds (33–5) *three* pairs of lines x, y *incommensurable in square* satisfying the respective sets of conditions (1) $x^2 + y^2$ rational, xy medial, (2) $x^2 + y^2$ medial, xy rational, (3) $x^2 + y^2$, xy both medial and incommensurable with one another.

With Proposition 36 begins Euclid's exposition of compound irrational straight lines, each of which is the sum or difference of two straight lines incommensurable in length. The first set contains six with the positive sign (Props. 36–41) and six with the negative sign (Props. 73–8). The first pair is the sum and difference of two rational straight lines commensurable in square only, e.g. $\rho \pm \sqrt{k} \cdot \rho$ (where, of course, ρ may be of the form a or $\sqrt{A}$). The second, third, fourth, fifth, and sixth pairs are the sums and differences of the pairs of lines x, y found in Props 27, 28, 33, 34, 35 respectively. The names of the first pair are *binomial* and *apotome* respectively; those of the other five pairs are more complicated. As a matter of fact, these six pairs of compound irrationals are the positive roots of different equations of the form

$$x^4 \pm 2\alpha x^2 \cdot \rho^2 \pm \beta \rho^4 = 0,$$

where ρ is a rational straight line and α, β have different characters and value (α, but not β, may contain a surd, as $\sqrt{m}$ or $\sqrt{(m/n)}$, as well as rational numbers).

Take the equation $x^4 - 2\alpha x^2 \cdot \rho^2 + \beta \rho^4 = 0$; then, solving for x^2, we have
$$x^2 = \rho^2 \{\alpha \pm \sqrt{(\alpha^2 - \beta)}\}.$$

Now x is a compound irrational which has to be expressed as the sum or difference of two terms. Therefore we have to express $\sqrt{\{\alpha \pm \sqrt{(\alpha^2 - \beta)}\}}$ as the sum or difference of two terms. We should find these terms (u, v, say) thus.

Suppose that $u^2 + v^2 = \alpha$,
and $2uv = \sqrt{(\alpha^2 - \beta)}$, or $4u^2v^2 = \alpha^2 - \beta$.

By subtraction,
$$(u^2 - v^2)^2 = \beta, \text{ or } u^2 - v^2 = \sqrt{\beta}.$$

Therefore $u^2 = \frac{1}{2}(\alpha + \sqrt{\beta})$, and $v^2 = \frac{1}{2}(\alpha - \sqrt{\beta})$, and the required compound irrational straight lines are

$$\sqrt{\{\tfrac{1}{2}(\alpha+\sqrt{\beta})\}} \pm \sqrt{\{\tfrac{1}{2}(\alpha-\sqrt{\beta})\}}.$$

Euclid does the exact geometrical equivalent of this working in Propositions 54-9 and 91-6.

Propositions 42-7 and 79-84 prove that each of the twelve compound irrational straight lines forming the first set is divisible into its terms in only one way. In particular, Proposition 42 is equivalent to the well-known theorem in algebra that,

if $\quad a+\sqrt{b}=x+\sqrt{y}, \quad$ then $\quad a=x, b=y,$
and,

if $\sqrt{a}+\sqrt{b}=\sqrt{x}+\sqrt{y},$ then $a=x, b=y,$ or $a=y, b=x.$

In Propositions 48-53 and 85-90 Euclid sets out the second set of six pairs of compound irrationals which are called the *first, second, third, fourth, fifth* and *sixth binomials*, and the *first, second, third, fourth, fifth* and *sixth apotomes* respectively, according as the terms are connected by the positive or negative sign. These irrationals are the positive roots of quadratic equations of the form

$$x^2 \pm 2\alpha x \cdot \rho \pm \beta\rho^2 = 0,$$

where α, β have different values and character, as before.

Take the equation $x^2 - 2\alpha x \cdot \rho + \beta\rho^2 = 0$; this gives

$$x = \rho\{\alpha \pm \sqrt{(\alpha^2-\beta)}\}.$$

It remains to prove the reciprocal connexion between the two sets of compound irrationals in pairs. We should express it by saying that one of the second set is the *square* of its analogue in the first set, and that one of the first set is the *square root* of its analogue in the second set. Euclid states the facts in a geometrical form, but his geometrical proofs correspond to what we should do in algebra

(Propositions 54–65 and 91–102). For example, Proposition 54 proves that the side of a square equal to the rectangle contained by ρ and the '*first binomial*' is a '*binomial*', and Proposition 60 proves that the square on a '*binomial*' if applied to a rational straight line (σ, say) has for its breadth a '*first binomial*', and so on.

Straight lines commensurable in length with any of the twelve compound irrationals are irrationals of the same type and order respectively (66–70 and 103–7). Finally, it is proved at the end of Proposition 72 and in Proposition 111 and the explanation following it that the medial and the twelve other irrationals are all different from one another.

Propositions 112–14 are the equivalent of rationalizing the denominators of the fractions $c^2/(\sqrt{A}\pm\sqrt{B})$ or $c^2/(a\pm\sqrt{B})$ by multiplying numerator and denominator by $(\sqrt{A}\mp\sqrt{B})$ or $(a\mp\sqrt{B})$ respectively.

Fuller details will be found in *The Thirteen Books of Euclid's Elements*, vol. iii.

What, it may be asked, is the specific object of the elaborate classification in Book X? The most probable explanation seems to be this. In algebra we can express any root of an equation such as those mentioned above in symbolic form by means of surds. The Greeks had no such symbols; the roots of the equivalent equations found by geometry are always straight lines, any one of which looks like any other. The Greeks, therefore, seem to have thought it necessary to compile a sort of repertory of results, described in definitions instead of by symbols, so that if, for instance, a certain straight line which has to be found in a particular case is proved to be a particular irrational, a 'binomial' or 'apotome', a 'major' or a 'minor' irrational, and so on, this would be accepted as

THE ELEMENTS. BOOKS X-XI 247

a sufficient solution; cf. the straight line proved in XIII. 17 to be an 'apotome', that in XIII. 6 proved to be a 'first apotome', and other similar cases occurring in Pappus.

Books XI-XIII are almost entirely concerned with geometry in three dimensions. The definitions are in Book XI, and include those of a straight line, and a plane, at right angles to a plane, the inclination of a plane to a plane (dihedral angle), parallel planes, equal and similar solid figures, solid angle, pyramid, prism, sphere, cone, cylinder, and parts of them, cube, octahedron, icosahedron, and dodecahedron. The sphere is defined, not as having all the points on its surface equidistant from the centre, but as the figure comprehended by the revolution of a semicircle about its diameter; this is clearly with an eye to the propositions in Book XIII where the regular solids have to be 'comprehended' in a sphere respectively.

The order of propositions in Book XI is fairly analogous to the order followed in Books I and VI. A straight line is wholly in a plane if a portion of it is in the plane (Prop. 1), and two intersecting straight lines are in one plane, as is a triangle also (2). Straight lines perpendicular to planes are next dealt with (4-6, 8, 11-14), then parallel straight lines not all in the same plane (9, 10, 15), parallel planes (14, 16), planes at right angles to one another (18, 19), solid angles contained by three plane angles (20, 22, 23, 26) or by more plane angles (21). The rest of the Book is mainly on parallelepipedal solids. Thus parallelepipedal solids on the same or equal bases and between the same parallel planes (i.e. having equal heights) are equal (29-31). Parallelepipedal solids of equal height are to one another as their bases (32). Similar parallelepipedal solids are in the triplicate ratio of corresponding sides (33). In equal parallelepipedal solids the bases are reciprocally proportional to the heights, and conversely (34). If four straight lines be proportional, so are similar parallelepipedal solids similarly described upon them, and conversely (37).

In Book XII the *method of exhaustion* plays the leading part, being used to prove successively that circles are to one another as the squares on their diameters (Props. 1, 2), that pyramids of the same height and with triangular bases are to one another as their bases (3-5), that any cone is, in content, equal to one third part of the cylinder which has the same base with it and equal height (10), that cones and cylinders of the same height are to one another as their bases (11), that similar cones and cylinders are to one another in the triplicate ratio of the diameters of their bases (12), and finally that spheres are to one another in the triplicate ratio of their diameters (16-18).

Proposition 5 is extended to pyramids with polygonal bases by Proposition 6; and Proposition 7 proves that any prism with triangular bases is divided into three pyramids with triangular bases and equal in content, whence it follows that any pyramid with triangular base (and therefore also any pyramid with polygonal base) is equal to one third part of the prism having the same base and equal height. Lastly, we have propositions about pyramids, cones, and cylinders similar to those in Book XI about parallelepipeds and in Book VI about parallelograms; similar pyramids are in the triplicate ratio of corresponding sides (8), and in equal pyramids, cones, and cylinders the bases are reciprocally proportional to the heights, and conversely (9, 15).

The method of exhaustion, as applied in Euclid, rests of course on X. 1 as lemma. The case of the pyramid (pyramids with triangular bases and of the same height are to one another as their bases) may be given as an illustration.

It is first proved (Proposition 3) that, given any pyramid, as *ABCD*, on the base *BCD*, if we bisect the six edges at the points *E, F, G, H, K, L*, and draw the straight lines shown in the figure, we divide the pyramid into two equal prisms and two equal pyramids *AFGE*, *FBHK* similar to the original pyramid (the equality of the prisms is proved in XI. 39), and that the sum of the two prisms is greater than half the original pyramid. Proposition 4 proves that, if each of two given pyramids of the same height be so divided, and if the small pyramids in each be similarly divided, then the smaller pyramids left over from that division similarly divided, and so on to any extent, the sums of all the pairs of prisms in the two given pyramids respectively will be to one another as the respective bases. Let the two pyramids and their volumes be denoted by *P, P′* respectively, and their bases by *B, B′* respectively. Then, if *B : B′* is not equal to *P : P′*, it must be equal to *P : W*, where *W* is some volume either less or greater than *P′*.

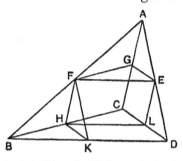

I. Suppose $W < P'$.

By X. 1 we can divide *P′* and the successive pyramids in it into prisms and pyramids until the sum of the small pyramids left over in it is less than *P′ − W*, so that

$$P' > \text{(prisms in } P') > W.$$

Suppose this done, and *P* divided similarly.
Then (XII. 4)
(sum of prisms in *P*) : (sum of prisms in *P′*) = *B : B′*
$= P : W$, by hypothesis.

But $P >$ (sum of prisms in P);
therefore $W >$ (sum of prisms in P').
But W is also less than the sum of the prisms in P': which is impossible.

Therefore W is not less than P'.

II. Suppose $W > P'$.

We have, inversely,
$$B' : B = W : P$$
$$= P' : V, \text{ where } V \text{ is some solid less than } P.$$

But this can be proved impossible, exactly as in Part I. Therefore W is neither greater nor less than P', so that
$$B : B' = P : P'.$$

Book XIII crowns the work by showing how to construct and to 'comprehend in a sphere' each of the five regular solids, the pyramid or tetrahedron (Prop. 13), the octahedron (14), the cube (15), the icosahedron (16), and the dodecahedron (17): 'comprehending in a sphere' means of course finding the sphere which circumscribes each solid, and this involves the determination of the relation of a 'side' (i.e. edge) of the solid to the radius of the sphere: in the case of the first three solids the relation is actually determined, while in the case of the icosahedron the side of the figure is shown to be the irrational straight line called 'minor' (cf. X. 76), and in the case of the dodecahedron an 'apotome'. Preliminary propositions relate to straight lines cut in extreme and mean ratio (1–6) and to pentagons (7, 8), and it is proved that if, in a regular pentagon, two diagonals (straight lines joining angular points next but one to each other) be drawn meeting in a point, each of them is divided at the point in extreme and mean ratio, and the greater segment is equal to the side of the pentagon. Propositions 9, 10 relate to the sides of a regular pentagon, decagon, and hexagon all inscribed in one circle. If p, d, h be the sides of the figures respectively, $h+d$ is cut in extreme and mean ratio, h being the greater segment (9); this is equivalent (since $h = r$, the radius of the circle) to saying that $(r+d)d = r^2$, whence $d = \frac{1}{2}r(\sqrt{5}-1)$. Proposition 10 proves that $p^2 = h^2+d^2$ or r^2+d^2, whence we can deduce $p = \frac{1}{2}r\sqrt{(10-2\sqrt{5})}$. Euclid does not find p, the side of the pentagon, in this way; but he proves that it is the irrational straight line called 'minor' (Prop. 11); we can in fact separate the above expression into two terms and deduce $p = \frac{1}{2}r\sqrt{(5+2\sqrt{5})} - \frac{1}{2}r\sqrt{(5-2\sqrt{5})}$. XIII. 12 proves that, if a is the side of an equilateral triangle inscribed in a circle with radius r, $a^2 = 3r^2$.

The constructions (only) for the several solids are as follows:

1. The regular pyramid or *tetrahedron*.

Given D, the diameter of the sphere which is to circumscribe the tetrahedron, Euclid draws a circle with radius r such that $r^2 = \frac{1}{3}D \cdot \frac{2}{3}D$, or $r = \frac{1}{3}\sqrt{2} \cdot D$, inscribes an equilateral triangle in the circle, and then erects from the centre of it a straight line perpendicular to its plane and of length $\frac{2}{3}D$. The lines joining the extremity of this perpendicular to the angular points of the equilateral triangle form, with the triangle itself, the required tetrahedron.

2. The *octahedron*.

If D be the diameter of the circumscribing sphere, a square is inscribed in a circle with D as diameter, and from its centre straight lines are drawn in both directions perpendicular to its plane and of length equal to the radius of the circle. Joining the extremities of the perpendiculars to the four angular points of the square, we have the required octahedron.

3. The *cube*.

D being the diameter of the circumscribing sphere, draw a square with side a such that $a^2 = D \cdot \tfrac{1}{3}D$, and describe a cube on this square as base.

4. The *icosahedron*.

Given D, the diameter of the circumscribing sphere, describe a circle with radius r such that $r^2 = D \cdot \tfrac{1}{5}D$. Inscribe a regular decagon in the circle. From the angular points draw straight lines at right angles to the plane of the circle (on one side of it) and of length r. This determines the angular points of another regular decagon inscribed in an equal parallel circle. Join alternate angular points in one decagon, making a regular pentagon inscribed in the same circle; do the same in the other circle, but so that the angular points of the second pentagon are not opposite those of the first pentagon. Join the angular points of one pentagon to the nearest angular points of the other; this gives ten equilateral triangles forming part of the surface of the required solid. To find the remaining faces, draw from the centre of each circle (outwards, i.e. in the direction away from the other circle in each case) perpendiculars of such length that the lines joining the extremity of each perpendicular to the five angular points of the nearer of the pentagons are all equal to the side of the pentagon. This gives the ten equilateral triangles which complete the required icosahedron. (The length of each perpendicular is actually equal to the side of the regular decagon inscribed in the circles.)

5. The *dodecahedron*.

Given a sphere with diameter D, Euclid first inscribes in it a *cube*. He then draws regular pentagons which have the sides of the cube for 'diagonals' in the manner shown in the annexed figure, thus. In one face *BF* let *HM, NO* be straight lines joining the middle points of opposite sides

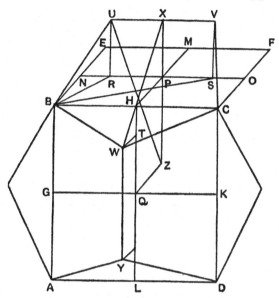

and intersecting at right angles at *P*; and in the face *BD* let *HL, GK* be similarly drawn, meeting in *Q*. Divide *PN, PO, QH* in extreme and mean ratio at *R, S, T*, and let *PR, PS, QT* be the greater segments. Draw (outwards) *RU, PX, SV* at right angles to the face *BF*, and *TW* at right angles to the face *BD*, such that each of these perpendiculars is equal to *PR* or *PS*. Join *UV, VC, CW, WB, BU*.

It is then shown that *UVCWB* is one of the required pentagonal faces by proving that the figure is (1) equilateral, (2) in one and the same plane, and (3) equiangular. The other pentagons are similarly drawn in relation to the other edges of the cube.

Book XIII ends with Proposition 18, which arranges in

order of magnitude the edges of the five regular solids inscribed in one and the same sphere; and an addendum proves that no other regular solids exist except the five.

The *editio princeps* of the Greek text of the *Elements* and most of the early editions contain two Books, XIV and XV, purporting to belong to Euclid's work, but these are not by Euclid. 'Book XIV' is by Hypsicles, the reputed author of an astronomical tract Ἀναφορικός (*De ascensionibus*) still extant (the earliest surviving Greek work in which we find the division of a circle into 360 degrees), and of works on the harmony of the spheres and on polygonal numbers, which are lost. 'Book XIV' contains some interesting propositions about the relative dimensions of the dodecahedron and icosahedron inscribed in one and the same sphere; it is valuable also for its statement that Apollonius wrote a tract on the comparison of the two figures, in which he proved that, as the surface of the dodecahedron is to the surface of the icosahedron, so is the solid content of the dodecahedron to that of the icosahedron, 'because the perpendicular from the centre of the sphere to the pentagon of the dodecahedron and to the triangle of the icosahedron is the same'. We also learn from Hypsicles that Aristaeus, in a work entitled *Comparison of the five figures*, proved that the same circle circumscribes the pentagon of the dodecahedron and the triangle of the icosahedron inscribed in the same sphere.

'Book XV' is also concerned with the regular solids, but is badly arranged and is of no particular interest, except for the fact that in the third section of it we are given rules for determining the dihedral angles between the faces meeting in any edge of any one of the regular solids, and the rules are attributed to 'Isidorus, our great teacher', who is doubtless Isidorus of Miletus, the architect of the church of St. Sophia at Constantinople (about A.D. 532).

EUCLID'S OTHER WORKS

Euclid wrote a number of treatises besides the *Elements*. We will begin with those which have survived, and first with the *Data*, because it belongs to plane geometry, the subject-matter of Books I–VI of the *Elements*. There are several senses in which, in Greek geometry, things are said to be 'given'; Euclid begins by defining them. Areas, straight lines, angles, ratios, and the like are said to be given *in magnitude* 'when we can find others equal to them' (in other words, when we can determine them). Rectilineal figures are given *in species* when their angles are severally given and also the ratios of the sides to one another (cf. the definition of similar rectilineal figures in VI, Def. 1). Points, lines, and angles are given *in position* 'when they always occupy the same place', by which we are no doubt to understand that, by whatever method you find them, you always find them in the same place. A circle is given *in position and in magnitude* when the centre is given in position and the radius in magnitude.

The object of the type of proposition formulated in the *Data* is to prove that, if in a given figure certain parts or relations are given, other parts or relations are also given, in one or other of the senses defined. It is manifest that a collection of propositions of this form is calculated to shorten the procedure in the analysis preliminary to a problem or proof; this is no doubt the reason why Pappus included the *Data* of Euclid in the *Treasury of Analysis*. Provided that we know that a certain thing is given, it is often unnecessary to carry out the actual operation of determining it.

As we should expect, much of the subject-matter of the *Data* is the same as that of the *Elements*, but in a different form. Thus the solution of the quadratic equations $ax \pm x^2 = b^2$ associated with Eucl. II. 5, 6 is equivalent to the solution of the simultaneous equations

$$y \mp x = a, \qquad xy = b^2,$$

and this is the form the question takes in *Data* 84, 85: 'If two straight lines contain a given area in a given angle, and if the difference (or sum) of them be given, then shall each of them be given'; the proofs depend directly on those of Props. 58, 59, 'If a given area be applied to a straight line falling short (or exceeding) by a figure given in species, the breadths of the defect (or excess) are given'. The areas are of course parallelograms.

We will give the proof of Prop. 59 (the case of 'excess').

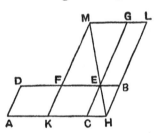

Let the given area AB be applied to AC, exceeding by the figure CB given in species. I say that each of the sides HC, CE is given.

Bisect DE in F, and construct on EF the figure FG similar and similarly situated to CB (VI. 18). Therefore FG, CB are about the same diagonal (VI. 26). Complete the figure.

Then FG, being similar to CB, is given in species, and, since FE is given, FG is given in magnitude (Prop. 52).

But AB is given; therefore $AB+FG$, that is to say, KL, is given in magnitude. But it is also given in species, being similar to CB. Therefore the sides of KL are given.

Therefore KH is given, and, since KC ($=FE$) is also given, the difference CH is given.

And CH has a given ratio to HB; therefore HB is also given (Prop. 2).

A few more enunciations may be given: 'If a (rectilineal) figure given in species be described on a straight line given in magnitude, the figure is given in magnitude' (Prop. 52); 'If a triangle have one angle given, the rectangle contained by the sides including the angle has to the area of the triangle a given ratio' (Prop. 66).

Proposition 93 is interesting. In a circle ABC let the chord BC cut off a segment containing a given angle BAC, and let the angle be bisected by AE meeting BC in D and the circumference again in E.

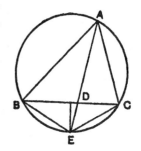

It is required to prove that the ratio $(BA+AC):AE$ is given, and that the rectangle $(BA+AC).DE$ is also given.

Join BE. Then, since the circle is given in magnitude, and BC cuts off a segment containing a given angle, BC is given.

Similarly BE is given (since E bisects the arc BEC).

Therefore the ratio $BC:BE$ is given (it is easy to see that it is equal to $2 \cos \frac{1}{2} A$).

Now, since AD bisects the angle BAC,
$$BD:DC = BA:AC\;;$$
therefore $\quad AC:DC = (BA+AC):(BD+DC)$
$$= (BA+AC):BC.$$

But, since the triangles ABE, ADC are similar,
$$AC:DC = AE:BE\;;$$
therefore $\quad AE:BE = (BA+AC):BC,$
or $\quad (BA+AC):AE = BC:BE$, which is a given ratio.

Again, since the triangles ADC, BDE are similar,

$$BE:ED = AC:CD = (BA+AC):BC, \text{ from above.}$$

Therefore $(BA+AC).ED = BC.BE$, which is a given area.

One other work in pure geometry by Euclid has reached us, and that through the Arabs only. This is the book *On divisions (of figures)*, περὶ διαιρέσεων βιβλίον. Two books of similar content have come down to us, both from the Arabic. The first, by one Muḥammad Bagdadinus (died 1141) was discovered in a Latin translation by John Dee, handed over by him to Commandinus, and published in the joint names of Commandinus and Dee in 1570. This work, however, can hardly have been a reproduction, still less a translation, of Euclid's book, since it does not contain certain propositions quoted from it by Proclus in his Commentary on Book I of the *Elements*. On the other hand, the second book, an Arabic treatise on the division of figures, discovered by Woepcke in a Paris manuscript and translated and published by him in 1851, does correspond to Proclus' description. It contains the propositions cited by him, is a well-ordered and compact treatise ending with the words 'end of the treatise', and (except that most of the proofs are omitted, presumably as being too easy) clearly represents the whole of Euclid's book. The four proofs of propositions which are given are elegant and genuinely Euclidean, while a lemma 'To apply to a straight line a rectangle equal to the rectangle contained by AB, AC, and deficient by a square' has a true Greek ring. The treatise has now been well edited by R. C. Archibald (Cambridge 1915) on the basis of Woepcke's text and the portion of Leonardo of Pisa's *Practica geometriae* dealing with the division of figures, which may itself be a restoration and extension of some version from the Arabic.

The general idea of the propositions is the division of figures (triangles, parallelograms, trapezia, quadrilaterals, circles, &c.) by straight lines into two or more parts having areas either equal or in prescribed ratios. The dividing straight lines may be transversals drawn through some given point, which may be exterior or interior to the figure, or a vertex, or a point on a side, and so on; or again they may be merely lines parallel to one another or to a side.

A characteristic problem is the following (cf. Woepcke 19, 20). From a triangle ABC to cut off, by a straight line GH passing through a given point D within the triangle, a triangle BGH equal in area to a certain fraction (m/n) of the original triangle. We proceed in this way.

Draw DE parallel to BC meeting AB in E. Then DE, EB are given in magnitude. So are AB, BC.

Let $AB = a$, $BC = b$, $DE = h$, $EB = k$.

Suppose that the problem is solved by the straight line GH drawn through D.

We choose some one length as our unknown (x), say, GB. Then, since the triangles GBH, ABC have the common angle B,

$$GB \cdot BH = \frac{m}{n} \cdot AB \cdot BC.$$

Therefore $\qquad x \cdot BH = \dfrac{m}{n} ab.$

Again, since DE is parallel to BC,
$$BH : ED = GB : GE,$$
or
$$BH = hx/(x-k).$$

Eliminating BH, we have
$$\frac{hx^2}{x-k} = \frac{m}{n}ab,$$
or
$$x^2 = \alpha(x-k), \quad \text{where} \quad \alpha = \frac{m}{n}ab/h.$$

We have, therefore, to solve the equation
$$\alpha x - x^2 = \alpha k.$$

This is exactly what Euclid does. He first finds F on BA such that $BF = \alpha$ (this he does by applying to the straight line DE, or h, a rectangle equal to $\frac{m}{n}ab$), and then 'applies to BF (α) a rectangle equal to $BF \cdot BE$ (αk) and deficient by a square'.

This gives BG or x, and we have then only to draw GH through D in order to cut off the required area.

Two other problems are interesting.

Proposition 28: To divide into two equal parts, by a straight line, the area formed by an arc of a circle BEC and two straight lines AB, AC forming an angle.

Bisect BC at D, and draw DE at right angles to BC meet-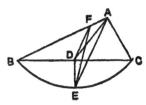ing the arc at E. Then clearly the broken line ADE divides the area $ABEC$ into two parts equal in area.

Join AE, and draw DF parallel to it meeting AB in F. Join FE. Then FE is the straight line required.

For the triangles ADE, AFE, in the same parallels, are equal.

ON DIVISIONS (OF FIGURES)

Add to each the area AEC.

Therefore the area $AFEC$ is equal to the area $ADEC$, and therefore to half the area of the figure $ABEC$.

Proposition 29: To draw in a given circle two parallel chords cutting off, between them, a given fraction m/n of the circle.

(The fraction must be such that we can by 'plane' methods cut off, by a chord, an arc equal to m/nths of the whole circumference. Euclid takes the case $m/n = \frac{1}{3}$.)

Let the arc ADB be m/n of the circumference, and let D be its middle point. Join A, B to the centre O. Draw OC parallel to AB. Join CB, and draw DE parallel to BC meeting the circle again in E.

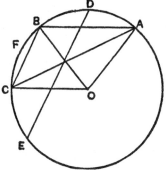

Then shall BC, DE intercept between them an area equal to m/n of that of the circle.

Join AC. Then the triangles AOB, ABC, in the same parallels AB, OC, are equal.

Add to each the segment ADB.

Therefore

(sector $AOBD$) = (segmt. ADB) + ($\triangle ABC$)

= (segmt. ABC) − (segmt. BFC).

But, since BC, DE are parallel, the arcs BD, CE are equal; therefore the sum of the arcs BD, CE is equal to the arc ADB.

Add to each the arc BC;

thus the arcs ABC, DBE, and therefore the segments ABC, DBE, are equal.

Therefore, from above,

(sector $AODB$) = (segmt. DBE) − (segmt. BFC).

The difference between the segments DBE, BFC is the part of the circle intercepted between BC, DE, which is therefore equal to the sector $AODB$, that is, by hypothesis, to m/n of the area of the circle.

We pass to the geometrical works which, so far as we know at present, are lost. To the domain of elementary geometry belongs the *Pseudaria*, as it is called by Proclus, or *Pseudographemata*, as a commentator on Aristotle has it. This was a 'Book of Fallacies', designed to put beginners in geometry on their guard against arguments which, not being based on the true principles of the subject, might lead them to wrong conclusions. The plan seems to have been to show, by illustrative examples, the true and false methods of deduction side by side.

One of the most grievous losses that fate has inflicted upon us is that of Euclid's *Porisms* in three Books. Pappus writes somewhat fully about it in his general preface describing the books constituting the *Treasury of Analysis*, and it is clear, not only that it belonged to higher geometry, but that it was a comprehensive and difficult work. Unfortunately Pappus' indications are not clear enough to give any certain view of the contents of the book. Hence, not unnaturally, the problem of trying to restore it has had a great fascination for distinguished mathematicians ever since the revival of learning. The great Fermat (1601–65) gave his idea of a 'porism' and illustrated it by five examples, but did not succeed in connecting them with Pappus' description or in producing (as he had hoped) a restoration of the treatise. Robert Simson (1687–1768) made the first decisive step towards the solution of the problem. In his tract on Porisms he proves the first porism given by Pappus in each of the ten cases which, according to Pappus (or an interpolator), Euclid distinguished; then, besides other propositions from Pappus, he gives 29 'porisms', some of which are meant to illustrate various classes distinguished by Pappus. After further speculations by Lawson (1777), Playfair (1794), W. Wallace (1798), and Lord Brougham (1798), three Frenchmen, A. J. H. Vincent, P. Breton (de Champ) and Chasles entered the field, and Chasles produced 'Les trois livres des Porismes d'Euclide rétablis . . .' (Paris, 1860). This work is historically important because it was in the course of his researches on this subject that Chasles was led to the idea of anharmonic ratios; and he was probably right in thinking that the *Porisms* were propositions belonging to the modern theory of transversals and to projective geometry. But Chasles, who makes 'Porisms' out of Pappus' various *lemmas* to Euclid and comparatively easy deductions from them, can hardly be regarded as having succeeded in making a satisfactory restoration, and the problem still remains for the most part unsolved.

A 'Porism' was regarded as a type of proposition intermediate between a theorem and a problem. Thus Proclus says that, whereas in a theorem we have to *prove* something, and in a problem to *make* or *construct* something, a *porism* deals with something which is already there but has to be found, brought to light or 'produced' ($\pi o \rho i \zeta \epsilon \iota \nu$); to find, e.g. the centre of a circle, or the greatest common measure of two commensurable magnitudes, is a 'porism'. If we may trust the text of Pappus, a whole class of porisms were of the nature of *loci*, a porism only falling short of a locus-theorem in respect of its hypothesis. Pappus observes that it is characteristic of these porisms that it is possible to comprehend a number of them in one enunciation, which he states as follows:

If, in a system of four straight lines which cut one another two and two, three points on one straight line be given, while the rest (of the points of intersection) except one lie on straight lines given in position, the remaining point will also lie on a straight line given in position.

We gather that Euclid gave ten cases of this. A still more general enunciation by Pappus had better, for perspicuity, be given in modern language, as it is by Gino Loria, who points out that Simson was the first to interpret it properly:

If a complete n-lateral be deformed so that its sides respectively turn about n points on a straight line, and $n-1$ of its $\frac{1}{2}n(n-1)$ vertices move on as many straight lines, the remaining $\frac{1}{2}(n-1)(n-2)$ of its vertices likewise move on as many straight lines; but it is necessary that it should be impossible to form with the $n-1$ vertices any triangle having for sides the sides of the polygon.

Pappus says that the three Books of *Porisms* contained 171 propositions, besides 38 lemmas, and that at the beginning of the work was this proposition:

I. If from two given points straight lines be drawn meeting on a straight line given in position, and one cut off from a straight line given in position (a segment measured) to a given point on it, the other will also cut off from another (straight line a segment) having to the first a given ratio.

Pappus observes further that the sort of thing to be proved took various forms, of which he gives twenty-nine, e.g. 'that such and such a point lies on a straight line given in position', 'that the ratio of such and such a pair of straight lines is given', &c., &c.

Enough has been said to show that the *Porisms* was a treatise of an advanced kind, worthy of the reputation of the author of the *Elements*.

Of Euclid's *Conics* Pappus says that it consisted of four Books, and that these four Books were completed by Apollonius, who added four more. Euclid's work seems to have been lost by Pappus' time, for he goes on to speak of 'Aristaeus who wrote the *still extant* five Books of *Solid Loci* connected with the conics'. The latter work was presumably a treatise on conics regarded as loci, for 'solid loci' in Greek geometry meant loci which were conics, as distinct from 'plane loci', which were straight lines and circles.

Euclid still used the old names for the three conics (sections of a right-angled, acute-angled, and obtuse-angled, cone respectively), as is evident from a sentence in his *Phaenomena* to the effect that, 'if a cone or cylinder be cut by a plane not parallel to the base, this section is a section of an acute-angled cone, which is like a shield (θυρεός)'; and it is probable that his methods were less general than those of Apollonius. Subject to this qualification, Euclid no doubt gave the essential properties proved by Apollonius up to the point in Apollonius' Book III where he takes up the question of the conic as a 'three- or four-line locus', since it is only for some propositions bearing on this question that Apollonius claims originality. When Archimedes alludes to propositions proved in the 'Elements of Conics', he must refer to the works of Euclid and Aristaeus.

Another work of Euclid's, the *Surface-Loci* (τόποι πρὸς ἐπιφανείᾳ), is also mentioned by Pappus as belonging to the *Treasury of Analysis*. The two lemmas which Pappus adds unfortunately do not enable us to form any definite idea of the nature of the contents of this lost work. The

Greek name suggests that 'surface-loci' might be loci *traced on* surfaces, and some colour is lent to this by a passage in Pappus where, after saying that the equivalent of the *quadratrix* may be obtained geometrically 'by means of loci on surfaces as follows', he proceeds to use a spiral traced on a cylinder (the cylindrical helix). But it is possible that 'surface-loci' may also have included loci which *are* surfaces, e.g. cones and cylinders. In any case what interests us is the second lemma to the treatise given by Pappus, in which he states, and gives a complete proof of, the focus-directrix property of the three conics, namely, that *the locus of a point the distance of which from a given point is in a given ratio to its distance from a fixed straight line is a conic section, which is an ellipse, a parabola or a hyperbola according as the given ratio is less than, equal to, or greater than, unity*.

The other works of Euclid belong to applied mathematics. Two have survived in Greek, the *Phaenomena* and the *Optics*, and are included in the Teubner text by Heiberg and Menge.

The *Phaenomena* deals with the geometry of the sphere so far as required for observational astronomy. Like Autolycus' work *On the Moving Sphere*, it considers the special circles on the heavenly sphere, the equator and parallel circles, the zodiac or ecliptic, and the *horizon* which here appears for the first time as a single word in its technical sense: 'let the name *horizon* be given to the plane through us (as observers) passing through the universe and separating off the hemisphere which is visible above the earth'.

The *Optics* exists in two versions, Euclid's own, and a recension by Theon of Alexandria with a preface which seems to be a reproduction by a pupil of explanations given by Theon in lectures. The book is a kind of elementary treatise on perspective, explaining how things *look* from different points of view or at different distances, as compared with what they *are*; it may have been intended as a corrective of heterodox ideas such as that of the Epicureans who maintained that the heavenly bodies (e.g. the sun) *are* of the size they *look*. In the Definitions the process of vision is regarded as it was by Plato. Rays are supposed to proceed from the eye and to impinge on the object seen, instead of the other way about. These rays are straight and 'traverse the distances (or dimensions) of great magnitudes'; the figure formed by the visual rays is a cone having its vertex in the eye and its base at the extremities of the object; things seen under a greater angle appear greater, those seen under a lesser angle smaller, and those seen under equal angles equal.

A few propositions of interest may be mentioned. The apparent sizes of two equal and parallel objects are *not* inversely proportional to their distances from the eye: to establish this (Prop. 8), Euclid proves the equivalent of the fact that, if α, β be two angles such that $\alpha < \beta < \frac{1}{2}\pi$, then

$$\tan \alpha / \tan \beta < \alpha/\beta.$$

Proposition 19 assumes that the angles of incidence on a mirror and reflection therefrom are equal, 'as is explained in the Catoptrica (theory of mirrors)'. If an eye sees a sphere, it sees less than half of it, and the contour of what is seen appears as a circle; if the eye approaches nearer to the sphere, the portion seen becomes less, though it appears greater (23-7). The diameters of a circle will all appear equal, or the circle will really look like a circle, if (and only if) the line joining the eye to the centre of the circle is (1) perpendicular to the plane of the circle, or (2), not being perpendicular to that plane, is equal to the radius of the circle. How can we move a straight line about in a plane so that it may always appear of the same length to an eye placed at a fixed point ? By keeping it as a chord in a circle in which the eye is either the centre or a point on the circumference (provided, in the latter case, that the extremities of the chord are not on different sides of the eye and neither extremity coincides with it), Props. 37, 38. To find a point at which two unequal straight lines will appear equal in length, place the two unequal straight lines BC, CD so as to form one straight line BCD, and describe on BC as base a segment of a circle greater than a semicircle, and on CD as base (and on the same side) a similar segment; the segments will then intersect at F (say), and F will be the required point (45).

The *Catoptrica* (theory of mirrors) included by Heiberg in the same volume with the *Optics* is not by Euclid, but is a compilation made much later (possibly by Theon) from ancient works on the subject. It states as an axiom the fact that, if an object be placed just out of sight at the bottom of an empty vessel, it will become visible over the edge if water is poured in: a fact which Archimedes is said to have proved in a *Catoptrica*.

Proclus and Marinus attribute to Euclid a work on the *Elements of Music*. Two musical treatises attributed to Euclid are extant, the *Sectio Canonis* (κατατομὴ κανόνος) and the *Introductio harmonica*, which were edited by Jan in the *Musici Graeci*. The latter tract is, however, by Cleonides, a pupil of Aristoxenus; the *Sectio*, which gives the Pythagorean theory of music, was thought by Jan to be genuine, but the latest editor, Menge, thinks that it may only have been extracted from the genuine work of Euclid by some editor of less ability.

Euclid is also supposed, though on doubtful authority, to have written on mechanics. The Arabian list of his works includes among those held to be genuine 'the book of the Heavy and Light', which is apparently the *De levi et ponderoso* included in the Basel Latin translation of 1537 and in Gregory's Euclid. The lettering of the figures shows that it comes from the Greek. The tract consists of nine definitions or axioms and five propositions, and represents pretty closely the same point of view as we find in Aristotle's dynamics, which persisted until its falsity was proved by Benedetti (1530–90) and Galilei (1564–1642).

A further fragment was translated from the Arabic in 1851 by Woepcke under the title 'Le livre d'Euclide sur la balance'. Although spoiled by some commentator, this fragment seems to go back to a Greek original, and to have been an attempt to prove the principle of the lever by means of common notions. A third fragment unearthed by Duhem from manuscripts in the Bibliothèque Nationale contains four propositions purporting to be 'Liber Euclidis de ponderibus secundum terminorum circumferentiam'. The first proposition, like the Aristotelian *Mechanica*, connects the law of the lever with the size of the circles described by its extremities; the others attempt to give a theory of the balance, taking account of the lever itself and assuming that a portion of it may be supposed to be detached and replaced by an equal weight suspended from the middle point of the said portion. It is possible that the fragments may have come from some work or works written before Archimedes' time, perhaps by a contemporary or contemporaries of Euclid.

X
ARISTARCHUS OF SAMOS

INTERMEDIATE between Euclid and Archimedes comes ARISTARCHUS of Samos, famous as the Copernicus of antiquity. Perhaps about twenty-five years earlier in date than Archimedes, he made an observation of the summer solstice in 281/0 B.C. He was a pupil of Strato of Lampsacus and, like him, wrote on physical subjects, vision, light, and colours. But the Greeks knew him as 'Aristarchus the mathematician', a title which he justified by his brilliant application of mathematics to astronomy in the one work of his which is extant, *On the sizes and distances of the Sun and Moon*. We gather that his principal interest was in astronomy. He is credited with having invented an improved sun-dial, the so-called σκάφη (boat), in which the pointer was erected vertically, not on a plane, but in the middle of a concave hemispherical surface, and enabled the direction and height of the sun to be read off by means of lines marked on the surface of the hemisphere; it is not, however, clear in what this sun-dial differed from the so-called πόλος which, along with the *gnomon*, the Greeks are supposed to have introduced from Babylon.

Aristarchus' crowning achievement was the hypothesis he put forward regarding the solar system, wherein he anticipated Copernicus. No trace of this hypothesis appears in the tract *On the sizes and distances of the Sun and Moon*, whence it is reasonable to infer that that treatise was antecedent to its formulation; but Archimedes, who is our authority, leaves us in no doubt about the nature of the hypothesis.

'Aristarchus', says Archimedes, 'brought out a book consisting of some hypotheses, wherein it appears, from the assumptions made, that the universe is many times greater than the universe as commonly understood by astronomers, his hypotheses being "that the fixed stars and the sun remain unmoved, that the earth revolves about the sun in the circumference of a circle, the sun lying in the middle of the orbit, and that the sphere of the fixed stars, situated about the same centre as the sun, is so great that the circle in which he supposes the earth to revolve bears such a proportion to the distance of the fixed stars as the centre of the sphere bears to its surface"'.

(In other words, the size of the earth's orbit about the sun is negligible in comparison with the size of the whole universe.)

As we have seen, Heraclides of Pontus (along with one Ecphantus, a Pythagorean) is credited with the first declaration that the earth rotates about its own axis once in twenty-four hours; and Aristarchus clearly followed him in this. Heraclides further held that Venus and Mercury move in circles round the sun, like satellites, and the next step (whoever took it) was to extend this theory to the superior planets Mars, Jupiter, and Saturn. We may, therefore, safely assume that Aristarchus would hold that all the planets, no less than the earth, revolved about the sun.

The heliocentric hypothesis found few adherents. Seleucus, of Seleucia on the Tigris, was a convinced supporter of it, but it was speedily abandoned altogether, mainly owing to the great authority of Hipparchus.

Archimedes further tells us that Aristarchus discovered that the apparent angular diameter of the sun is about 1/720th part of the zodiac circle; in other words, half a degree. This discovery again may have been made later than the date of the treatise on sizes and distances, in which the said angle is assumed to be 2°.

The treatise in question is a fine piece of geometrical work, thoroughly classical in form and style; it is especially interesting because it is the first extant specimen of pure geometry used with a *trigonometrical* object, in which respect it is a sort of forerunner of Archimedes' *Measurement of a Circle*.

Aristarchus gives first the assumptions on which his work is based; then follow a well-ordered series of eighteen propositions in the course of which he obtains his results. One of the assumptions is that mentioned above, that the angular diameter of the sun is 2°. The others are as follows:

(1) The moon receives its light from the sun.

(2) The earth is in the relation of a point and centre to the sphere in which the moon moves.

(3) When the moon appears to us halved, the great circle which divides the dark and the bright portions of the moon is in the direction of our eye (the effect of which is that at the time of dichotomy the centres of the sun, moon, and earth, say S, M, E, form a triangle right-angled at the centre of the moon).

(4) When the moon appears to us halved, its distance from the sun is then less than a quadrant by one-thirtieth of a quadrant (that is, the angle SEM in the right-angled triangle is 87°).

(5) The breadth of the earth's shadow is (that) of two moons.

The second assumption is made only for the purpose of simplifying the problem by avoiding any question of parallax. *Some* assumption of the nature of (4) is necessary in order to work out the figures, since everything depends on the shape of the right-angled triangle SEM. One of its acute angles has therefore to be given some definite value.

Aristarchus assumes the angle SEM to be $87°$, whereas it is in fact $89° 50'$. On the assumptions made, Aristarchus declares that he is in a position to prove

(1) that the distance of the sun from the earth is greater than eighteen and less than twenty times the distance of the moon from the earth;

(2) that the diameter of the sun has the same ratio as aforesaid to the diameter of the moon;

(3) that the diameter of the sun has to the diameter of the earth a ratio greater than $19:3$, but less than $43:6$.

Aristarchus is as good as his word, for these results are rigorously established in Propositions 7, 9 and 15 of the treatise.

The essence of the matter is that, in order to prove his theorems, Aristarchus requires values for certain ratios which are the equivalent of the trigonometrical sines and cosines of certain small angles, and in particular those of the angles $3°$ and $1°$. He does not find their actual values, but shows that they lie between certain limits. He uses for this purpose certain propositions which he assumes without proof, thereby implying that they were generally known; these propositions are equivalent to the facts that

(1) if α be what we call the 'circular measure' of an angle, and α be less than $\frac{1}{2}\pi$, then the ratio $\sin\alpha/\alpha$ decreases, and the ratio $\tan\alpha/\alpha$ increases, as α increases from 0 to $\frac{1}{2}\pi$;

(2) if β be the circular measure of another angle less than $\frac{1}{2}\pi$, and $\alpha > \beta$, then

$$\frac{\sin\alpha}{\sin\beta} < \frac{\alpha}{\beta} < \frac{\tan\alpha}{\tan\beta}.$$

Aristarchus deals, of course, not with circular measures, sines, and tangents, but with angles (expressed, not in degrees, but as fractions of right angles), arcs of circles, and chords, but he obtains results equivalent to the following:

$$\tfrac{1}{18} > \sin 3° > \tfrac{1}{20}, \quad \text{(Prop. 7)}$$
$$\tfrac{1}{45} > \sin 1° > \tfrac{1}{60}, \quad \text{(Prop. 11)}$$
$$1 > \cos 1° > \tfrac{89}{90}, \quad \text{(Prop. 12)}$$
$$1 > \cos^2 1° > \tfrac{44}{45}. \quad \text{(Prop. 13)}$$

It is the first of these results which he requires for his main proposition (7) that the distances of the sun and moon respectively from the earth are in a ratio greater than 18 : 1 but less than 20 : 1.

Space does not allow of our reproducing more than one of Aristarchus' propositions; we select, therefore, the proposition just cited, as an example.

In the accompanying figure A is the centre of the sun, B that of the earth, and C that of the moon at the moment of dichotomy, so that the angle ACB is right.

$ABEF$ is a square, and AE a quadrant of the sun's circular orbit.

Join BF, and bisect the angle FBE by the straight line BG, so that $\angle GBE = \tfrac{1}{4}R$ or $22\tfrac{1}{2}°$.

I. Now, by Hypothesis 4, $\angle ABC = 87°$,
so that $\angle HBE = \angle BAC = 3°$.
Therefore $\angle GBE : \angle HBE = \tfrac{1}{4}R : \tfrac{1}{30}R$
$= 15 : 2$,
so that $GE : HE\ [=\tan GBE : \tan HBE]$
$> \angle GBE : \angle HBE$
$> 15 : 2. \quad \ldots \ldots \ldots \quad (1)$

The ratio which has to be proved greater than 18 : 1 is the ratio $AB : BC$ or $BH : HE$, which is $> FE : HE$.

Now $FG : GE = FB : BE$,

whence $\quad FG^2 : GE^2 = FB^2 : BE^2 = 2 : 1$,
and $\quad\quad\quad FG : GE = \sqrt{2} : 1$
$\quad\quad\quad\quad\quad\quad > 7 : 5$.

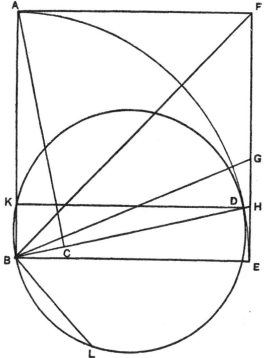

Therefore $\quad FE : EG > 12 : 5$, or $36 : 15$.
Compounding with (1) above, we have
$$FE : HE > 36 : 2 \text{ or } 18 : 1.$$
Therefore, *a fortiori*, $AB : BC > 18 : 1$.

II. To prove that $BA < 20\, BC$.

Let BH meet the quadrant AE in D. Draw DK parallel to EB, and about the triangle BKD circumscribe the circle BKD. Let BL be a chord in this circle equal to its radius (r, say).

Now $\angle BDK = \angle DBE = \frac{1}{30}R$,

so that (arc BK) $= \frac{1}{60}$ (circumference of circle).

Therefore (arc BK) : (arc BL) $= \frac{1}{60} : \frac{1}{6}$
$$= 1 : 10.$$

But (arc BK) : (arc BL) $< BK : r$.

[This is equivalent to $\dfrac{\alpha}{\beta} < \dfrac{\sin\alpha}{\sin\beta}$, where $\alpha < \beta < \tfrac{1}{2}\pi$.]

Therefore $r < 10 BK$,

and $BD < 20 BK$.

But $BD : BK = AB : BC$;

therefore $AB < 20 BC$.

The further results obtained by Aristarchus are arrived at in a series of propositions, the diagrams for which represent positions of the moon as it passes through the shadow of the earth during an eclipse. The propositions are admirably worked out, but are rather more complicated than that just given. In Proposition 11 Aristarchus proves that the diameter of the moon is less than 2/45ths, but greater than 1/30th, of the distance of the centre of the moon from our eye.

XI
ARCHIMEDES

ARCHIMEDES, a native of Syracuse, was killed in the sack of that city by the Romans under Marcellus in the Second Punic War, 212 B.C.; he must therefore have lived from about 287 to 212 B.C. He was the son of Phidias, an astronomer, and was intimate with, if not related to, Hieron, king of Syracuse, and his son Gelon. He spent some time in Egypt, and it was there that he is said to have discovered the so-called 'Archimedean' screw for pumping water. He studied, no doubt, at Alexandria with the successors of Euclid, and it was probably there that he made the acquaintance of Conon of Samos, the courtier-astronomer, for whom, both as a mathematician and as a friend, he expresses the highest regard, and of Eratosthenes of Cyrene. To the former he commonly communicated his discoveries before publishing them; while it was to Eratosthenes that he addressed the *Method*, as well as (if we may judge by its heading) the famous *Cattle-Problem*.

The historians Polybius and Livy, as well as Plutarch, tell picturesque stories about the machines used by Archimedes against the Romans in the siege of Syracuse. He contrived (so we are told) catapults so ingeniously constructed as to be equally serviceable at long or short range, machines for discharging showers of missiles through holes in the walls, and others consisting of long movable poles projecting from the walls which either dropped heavy weights on the enemy's ships, or grappled their prows by means of an iron hand or a beak like that of a crane, then lifted them into the air and let them fall again. Marcellus is said to have derided his own engineers thus:

Shall we not make an end of fighting against this geometrical Briareus who uses our ships as cups to ladle water from the sea, drives off our *sambuca* ignominiously with cudgel-blows and, by the multitude of missiles that he hurls at us all at once, outdoes the hundred-handed giants of mythology?

The Romans, however, were in such abject terror that, 'if they did but see a piece of rope or wood projecting from the wall, they would say, "There it is!" and, declaring that Archimedes was setting some engine in motion, would turn tail and run away'. These inventions were, as Plutarch says, only the 'diversions of geometry at play', and Archimedes himself thought nothing of them, insomuch that,

although these inventions had obtained for him the reputation of more than human sagacity, he did not deign to leave behind him any written work on such subjects, but, regarding as ignoble and sordid the business of mechanics and every sort of art which is directed to use and profit, he placed his whole ambition in those speculations the beauty and subtlety of which are untainted by any admixture of the common needs of life.

Archimedes did indeed write one mechanical book, *On Sphere-making*, which is lost; this described the construction of a sphere to imitate the motions of the sun, moon, and planets, a contrivance which, according to Cicero who saw it, actually represented the periods of the moon and the apparent motion of the sun with such accuracy that it would even (over a short period) show the eclipses of the sun and moon. In any case Archimedes was much occupied with astronomy. Livy calls him 'unicus spectator caeli siderumque'; Hipparchus speaks of observations by him of the solstices made with a view to determining the length of the year. Macrobius says he discovered the distances of the planets; and he himself describes in his *Sand-reckoner* a rough apparatus by which he measured the apparent angular diameter of the sun.

Archimedes wrote on theoretical mechanics, laying down the principles and proving the fundamental propositions. It was by theory that he solved the problem *To move a given weight by a given force*; it was 'in reliance on the irresistible cogency of his proof' that he declared to Hieron that any given weight could be moved by any given force (however small): 'Give me a place to stand on and I will move the earth' (πᾶ βῶ καὶ κινῶ τὰν γᾶν, as he said in his Doric dialect). Plutarch relates that,

when Hieron was struck with amazement and asked Archimedes to reduce the problem to practice and to give an illustration of some great weight moved by a small force, he fixed upon a ship of burden from the king's arsenal which had only been drawn up with great labour by many men, and, loading her with many passengers and a full freight, himself the while sitting far off, with no great effort, but only holding the end of a compound pulley (πολύσπαστος) quietly in his hand and pulling at it, he drew the ship along smoothly and safely as if she were moving through the water.

The story that Archimedes set the Roman ships on fire by means of burning-glasses or concave mirrors is not found in any authority earlier than Lucian; but it is quite likely that he discovered some form of burning-mirror, e.g. a paraboloid of revolution, which would reflect to one point all rays falling on its concave surface in a direction parallel to its axis.

Archimedes' own view of the relative significance of his many discoveries is vividly shown by his request to his friends and relatives that they should place on his tomb a representation of a cylinder circumscribing a sphere, with an inscription stating the ratio which the cylinder bears to the sphere; he evidently regarded the discovery of this ratio as his greatest achievement. Cicero, when quaestor in Sicily, found the tomb in a neglected state and repaired it; it has now vanished and no one knows where he was buried.

Archimedes' preoccupation by his abstract studies is illustrated by a number of stories. He would, we are told, forget all about his food and such necessities of life, and would be drawing geometrical diagrams in the ashes of the fire or, when anointing himself, in the oil on his body. When he discovered in the bath the solution of the question referred to him by Hieron whether a certain crown supposed to have been made of gold did not in fact contain a certain amount of silver, he ran home naked, shouting 'Εὕρηκα, εὕρηκα'. The story of his death at the hands of a Roman soldier in the sack of Syracuse is told in various forms; according to Tsetzes, he said to a Roman soldier who found him absorbed in some diagrams which he had drawn on the dust and came too close, 'Stand away, fellow, from my diagram', and the soldier, incensed, killed him.

It may be useful to summarize here Archimedes' main contributions to mathematics. In geometry his work consists chiefly of original researches into the quadrature of curvilinear plane figures and the quadrature and cubature of curved surfaces. These investigations, beginning where Euclid's Book XII left off, actually (in the words of Chasles) 'gave birth to the calculus of the infinite conceived and brought to perfection successively by Kepler, Cavalieri, Fermat, Leibniz, and Newton'. Archimedes in fact found, by what is equivalent to *integration*, the area of a parabolic segment, the area of a spiral, the surface and volume of a sphere and a segment of a sphere, and the volumes of any segments of the solids of revolution of the second degree. In arithmetic he calculated approximations to the value of π, and in the course of these calculations showed that he could find approximate values for the square roots of large or small non-square numbers; he further invented a system of arithmetical nomenclature by which he could express in language enormously large numbers, in fact all numbers up to that which we should write as 1 followed by 80,000 million million ciphers. In mechanics he laid down certain postulates and, on the basis of these postulates, established certain fundamental theorems on magnitudes balancing about a point and on centres of gravity, going so far as to find the centre of gravity of any segment of a parabola, a semicircle, a cone, a hemisphere, a segment of a sphere, and a right segment of a paraboloid of revolution. As we shall see, Archimedes made most ingenious use of mechanics in aid of geometry, namely for the purpose of discovering the areas and volumes of certain figures. Lastly, he invented the whole science of hydrostatics, which again he carried so far as to give a complete investigation of the positions of rest and stability of a right segment of a paraboloid of revolution floating in a fluid with its base either upwards or downwards, but so that the base is entirely above or entirely below the surface of the fluid.

The treatises themselves are, without exception, models of mathematical exposition; the gradual unfolding of the plan of attack, the masterly ordering of the propositions, the stern elimination of everything not immediately relevant, the perfect finish of the whole, combine to produce a deep impression, almost a feeling of awe, in the mind of the reader. There is here, as in all the great Greek mathematical masterpieces, no hint as to the kind of analysis by which the results were first arrived at; for it is clear that they were not *discovered* by the steps which lead up to them in the finished treatise. If the geometrical treatises had stood alone, Archimedes might seem, as Wallis said, 'as it were of set purpose to have covered up the traces of his investigations, as if he had grudged posterity the secret of his method of inquiry, while he wished to extort from them assent to his results'. And indeed (again in the words of Wallis) 'not only Archimedes, but nearly all the ancients, so hid from posterity their method of Analysis (though it is clear that they had one) that more modern mathematicians found it easier to invent a new Analysis than to seek out the old'. But the romantic discovery (so recently as 1906) of the *Method* of Archimedes in a Constantinople palimpsest has changed the position so far as Archimedes is concerned; for here he does tell us how he discovered certain theorems in quadrature and cubature, namely, by the mechanical method of (theoretically) weighing elements of one figure against elements of another simpler figure the mensuration of which was already known. Archimedes at the same time makes it clear that, while the mechanical method is useful for suggesting the truth of theorems, he does not regard it as furnishing a scientific proof of them; this can only be effected by rigorous and orthodox geometrical methods.

'Certain things', he says, 'first became clear to me by the mechanical method, although they had to be demonstrated by geometry afterwards, because their investigation by the said method did not furnish an actual demonstration. But it is of course easier to supply the proof when we have previously acquired, by the method, some knowledge of the questions than it is to find it without any previous knowledge.' 'This', he adds, 'is a reason why, in the case of the theorems that the volumes of a cone and a pyramid are one-third of the volumes of the cylinder and prism respectively having the same base and equal height, the proofs of which Eudoxus was the first to discover, no small share of the credit should be given to Democritus, who was the first to state the fact, though without proof.'

Finally, he tells us that the very first theorem which he discovered by means of mechanics was that of the separate treatise on the 'Quadrature of the parabola', namely, that the area of any segment of a parabola cut off by a chord is four-thirds of that of the triangle which has the same base and height.

The following is the order in which the treatises appear in Heiberg's edition:

(5) *On the Sphere and Cylinder*, I, II.
(9) *Measurement of a Circle*.
(7) *On Conoids and Spheroids*.
(6) *On Spirals*.
(1) *On Plane Equilibriums*, I.
(3) *On Plane Equilibriums*, II.
(10) *The Sand-reckoner (Psammites)*.
(2) *Quadrature of the Parabola*.
(8) *On Floating Bodies*, I, II.
(?) *Stomachion* (a fragment).
(4) *The Method*.

The above was not, however, the order of composition, which, to judge by statements in Archimedes' own prefaces, and by the use in certain treatises of results obtained in others, was rather that indicated by the numbers in brackets on the left in the above list. The treatise *On Floating Bodies* was formerly only known in the Latin translation by William of Moerbeke, but the Greek text has now been in great part restored by Heiberg from the Constantinople MS. containing the *Method* and the fragment of the *Stomachion*.

A *Liber Assumptorum* which has reached us through an Arabic translation by Thābit b. Qurra under the name of Archimedes is not by Archimedes in its present form, since the propositions mention his name several times; but some of these propositions may easily be of Archimedean origin, notably those about the geometrical figures called ἄρβηλος ('shoemaker's knife') and σάλινον (perhaps 'salt-cellar') respectively, and the proposition (8) bearing on the trisection of any angle.

The *Cattle-Problem* in epigrammatic form purports by its heading to have been communicated by Archimedes to the mathematicians at Alexandria in a letter to Eratosthenes. Whether the epigrammatic form is due to Archimedes or not, there is no sufficient reason for doubting the possibility that the substance of the epigram was set as a problem by Archimedes.

There are various traces of lost works as follows:

1. Pappus refers to investigations relating to polyhedra, and, in connexion with the five regular solids, describes thirteen other solids discovered by Archimedes which are *semi*-regular, being contained by polygons equilateral and equiangular but not all similar.

2. Archimedes himself refers to a tract, which he dedicated to Zeuxippus, dealing with the *naming of numbers* (κατονόμαξις τῶν ἀριθμῶν) and expounding the system of nomenclature for very large numbers used in the *Sand-reckoner*, as already mentioned (pp. 19–20).

3. The mechanical treatises evidently contained much more than is preserved in the two Books *On Plane Equilibriums*. Archimedes himself quotes propositions determining the centre of gravity (1) of a cone, (2) of any segment of a right-angled conoid (i.e. a paraboloid of revolution), the latter of which, he says, was proved in

the *Equilibriums*. In the *Quadrature of a Parabola* he assumes that, if a body hangs at rest from a point, the centre of gravity of the body and the point of suspension are in the same vertical line. Pappus states the same property with reference to a point of *support*, and Heron says that Archimedes lays down a certain procedure in a 'Book on Supports'. Simplicius refers to 'problems on the centre of gravity', κεντροβαρικά, 'such as those of Archimedes', showing how to find the centre of gravity, and adds the explanation that the centre of gravity is 'the point in a body such that, if the body is hung from it, the body will remain at rest in any position'. Pappus observes that this last is the most fundamental principle of the theory of the centre of gravity, the elementary propositions of which are proved in Archimedes' *On Equilibriums* and in Heron's *Mechanics*. Pappus again says that it was proved in Archimedes' book *On balances* or *levers* (περὶ ζυγῶν), and in Philon's and Heron's *Mechanics*, that 'greater circles overpower lesser circles when they revolve about the same centre'. It seems possible that there was originally a larger work *On Equilibriums* of which the extant two Books were only a part; and περὶ ζυγῶν and κεντροβαρικά may be only alternative titles.

Theon of Alexandria quotes a proposition from a work of Archimedes entitled *Catoptrica* (theory of mirrors) to the effect that things thrown into water look larger and still larger the farther they sink (cf. p. 268 above). A scholiast to the Pseudo-Euclid's *Catoptrica* quotes a proof, which he attributes to Archimedes, of the equality of the angles of incidence and reflection in a mirror.

The Arabians attributed yet other works to Archimedes, (1) *On the Circle*, (2) *On the Heptagon in a Circle*, (3) *On*

Circles touching one another, (4) *On Parallel Lines*, (5) *On Triangles*, (6) *On Properties of Right-angled Triangles*, (7) a Book of *Data*, (8) *De clepsydris*. We shall return later to the question of the heptagon in a circle.

Since Heron, Pappus, and Theon of Alexandria all cite works of Archimedes which no longer survive, those works must have been in existence at Alexandria as late as the third and fourth centuries A.D. Attention came, however, to be concentrated on two treatises only, *On the Sphere and Cylinder* and *Measurement of a Circle*, on which Eutocius (fl. about A.D. 520) wrote commentaries afterwards revised by Isidorus of Miletus. Eutocius commented on the *Plane Equilibriums* also, but does not seem even to have been acquainted with the *Quadrature of the Parabola* and the book *On Spirals*.

In the ninth century Leon, who restored the University of Constantinople, collected all the works he could find there, and had the archetype MS. written from which all the others derive, except the Constantinople MS. discovered by Heiberg in 1906 and containing, among other things, the *Method*. The archetype MS. (A) was presented to the Pope by Charles of Anjou in 1266 and was in the Papal Library for some time after 1269, but passed into private hands at some date after 1368 and finally (between 1544 and 1564) disappeared, leaving no trace. Fortunately it was in 1269 translated into Latin by William of Moerbeke, whose translation, preserved at Rome, was so literal that, as a source, it is almost as good as a Greek manuscript. William of Moerbeke added a translation of the two Books *On Floating Bodies* from another manuscript containing works not included in A. The *editio princeps* of the Greek text was published at Basel in 1544 by Thomas Gechauff Venatorius. A Latin translation of five of the treatises by

Commandinus appeared at Venice in 1558. Torelli's edition in Greek and Latin (Oxford 1792), like Commandinus' translation, followed the *editio princeps* in the main, but Torelli also collated E, the earliest of the extant manuscripts, and Abram Robertson, who brought out the book after Torelli's death, five more. Heiberg's definitive Greek text (with Latin translation) of all the works of Archimedes with Eutocius' commentaries, &c., includes (in the second edition, 1910–15) the *Method*, the fragment of the *Stomachion*, and so much of the Greek text of the treatise *On Floating Bodies* as could be restored from the newly discovered Constantinople MS.

Since Archimedes explains, in the preface to the *Method*, that it was by a certain mechanical method that he discovered the solution of the most important problems in quadrature and cubature, and the *Method* itself is entitled *On Mechanical Theorems, Method* (addressed) *to Eratosthenes*, it will be appropriate to begin with this work in describing the various treatises. Premising certain mechanical propositions and a Lemma ($=$ *On Conoids and Spheroids*, Prop. 1), Archimedes finds, in the *Method*, (1) the area of any segment of a section of a right-angled cone (a parabola); (2) the ratio ($1\frac{1}{2}$) of the volume of the right cylinder circumscribing a sphere or a spheroid of revolution, and having the same length of axis, to the volume of the sphere or spheroid itself; (3) the volume of any segment cut off, by a plane at right angles to the axis, from any right-angled conoid (paraboloid of revolution), sphere, spheroid, and obtuse-angled conoid (hyperboloid of revolution) in terms of the volume of the cone having the same base as the segment and equal height; (4) the centre of gravity of a segment of a paraboloid of revolution, a sphere, and a spheroid respectively. These results are obtained in Propositions 1–11. Propositions 12–15 and Proposition 16 give the cubature of two special figures.

(*a*) Suppose a prism with a square base to have a cylinder inscribed in it, the bases of the cylinder being circles inscribed in the square bases of the prism, and suppose a plane drawn through one side of one base of the prism and that diameter of the circle inscribed in the opposite base which is parallel to the said side. This plane cuts off a solid bounded by two planes and by part of the curved surface of the cylinder (a solid shaped like a part of a *hoof* cut off by a plane); it is proved that the volume of this figure is one-sixth of that of the prism.

(*b*) Suppose a cylinder inscribed in a cube, the circular bases of the cylinder being circles inscribed in two opposite faces of the cube, and suppose another cylinder similarly inscribed with reference to another pair of opposite faces. The two cylinders enclose a certain solid which is actually made up of eight 'hoofs' like that of Props. 12–15. The volume of the solid is proved to be two-thirds of that of the cube.

In accordance with his dictum that the mechanical method only serves to indicate results, which have then to be proved rigorously by the orthodox geometrical method of exhaustion, Archimedes gave the rigorous proof of the last two results in Propositions 15, 16; part of the text is missing, but restoration of the argument is not difficult.

The method of the treatise can be best understood by taking a particular case. It must be sufficient here to describe the easiest. We take first the problem of finding the area of a *parabolic segment*.

Let ABC be the given segment, BD its diameter, CF the tangent at C. Let P be any point on the arc of the segment, and let AKF, $OPNM$ be drawn parallel to BD. Join CB, and produce it to meet MO in N and FA in K, and let KH on CK produced be made equal to KC.

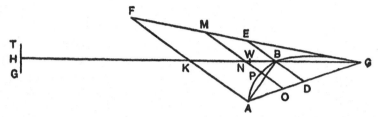

Now, by a proposition 'proved in a lemma' (cf. *Quadrature of the Parabola*, Prop. 5),

$$MO : OP = CA : AO$$
$$= CK : KN$$
$$= HK : KN.$$

Also, by the property of the parabola, $EB = BD$, so that $MN = NO$, and $FK = KA$.

It follows that, if HC be regarded as the bar of a balance, a line TG equal to PO and placed with its middle point at H will balance, about K, the straight line MO placed where it is, i.e. with its middle point at N.

Similarly with all lines parallel to BD, such as MO, PO, in the triangle CFA and in the segment CBA respectively.

And there are the same number of the lines in each; therefore the whole segment of the parabola acting at H balances the triangle CFA placed where it is.

But the centre of gravity of the triangle CFA is at the point W on the straight line CK such that $CW = 2WK$ [and the whole triangle may be taken as acting as one mass at W].

Therefore (segmt. ABC) : ($\triangle CFA$) $= WK : KH$
$$= 1 : 3,$$

that is, $\quad$ (segmt. ABC) $= \frac{1}{3} \triangle CFA$
$$= \frac{4}{3} \triangle ABC.$$

It will be observed that Archimedes takes the segment and the triangle to be *made up* of the infinity of parallel straight lines that may be drawn in the figures respectively. The weights are, of course, taken to be proportional to the lengths of the lines. Archimedes also, without mentioning *moments*, in effect assumes that the sum of the moments of each particle of a figure, acting where it is, is equal to the moment of the whole figure applied as one mass at its centre of gravity.

When the volumes of solids (spheres, spheroids, and paraboloids) are being dealt with, the comparison is with cones and cylinders, and the elements are parallel *plane sections* of each solid instead of the parallel *lines* drawn in the plane figures compared; the weights are taken to be proportional to the areas of the plane sections.

We may illustrate by the case of any right segment of a *spheroid of revolution*.

The ellipse with axes AA', BB' is a section made by the plane of the paper of a spheroid with AA' as axis of revolution. It is required to find the volume of any right segment ADC of the spheroid in terms of the right cone with the same base and height.

DC is the diameter, and G the centre, of the circular base of the segment. Join AB, AB', and produce them to meet the tangent at A' to the ellipse in K, K', and DC produced in E, F.

Conceive a cylinder described with axis AA' and base the circle on KK' as diameter, and imagine cones described with AG as axis and bases the circles on EF, DC as diameters.

Let N be any point on AG, and let $MOPQNQ'P'O'M'$ be drawn through N parallel to BB' or DC, as shown in the figure. Produce $A'A$ to H so that $HA = AA'$.

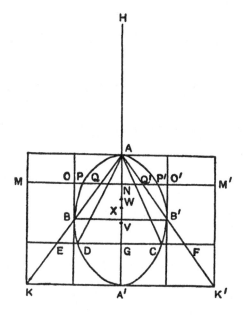

Now
$$HA : AN = A'A : AN$$
$$= KA : AQ$$
$$= MN : NQ$$
$$= MN^2 : MN \cdot NQ. \quad \ldots \quad (1)$$

Archimedes next proves that $MN \cdot NQ = NP^2 + NQ^2$, thus.

By the property of the ellipse,
$$AN \cdot A'N : NP^2 = (\tfrac{1}{2}AA')^2 : (\tfrac{1}{2}BB')^2$$
$$= AN^2 : NQ^2;$$

therefore $NQ^2 : NP^2 = AN^2 : AN \cdot A'N$
$= NQ^2 : NQ \cdot QM,$
whence $NP^2 = NQ \cdot QM.$

Add NQ^2 on both sides, and we have
$$NP^2 + NQ^2 = MN \cdot NQ.$$

If follows, by (1) above, that
$$HA : AN = MN^2 : (NP^2 + NQ^2). \quad \ldots \quad (2)$$

And MN^2, NP^2, NQ^2 are proportional to the areas of the sections perpendicular to AA' of the cylinder with axis AG, the spheroid ABB' and the cone AEF respectively.

Therefore, if we regard HAA' as a lever, the element of the cylinder, placed where it is, balances, about A, the sum of the corresponding elements of the spheroid and the cone AEF, both placed at H.

The same applies to *all* the elements so taken, and it follows that the whole cylinder with axis AG, where it is, balances the segment of the spheroid and the cone AEF taken together when both are placed with their centres of gravity at H.

Regarding the cylinder as acting as one mass at its centre of gravity W, midway between A and G, and taking moments (as we should say), we thus find the sum of the volumes of the segment of the spheroid and the cone AEF, and thence (since the volume of the cone is known) the volume of the segment of the spheroid.

The volume of the segment being thus known, Archimedes is in a position to find its centre of gravity by the same procedure slightly modified (Prop. 10). The variation is caused by the fact that in this case we must have the segment acting *where it is*, not at H.

Archimedes proves that, with the above figure,
$$HA : AN = (NP^2 + NQ^2) : NQ^2.$$
For, as above,
$$MN \cdot NQ = NP^2 + NQ^2,$$
whence $MN^2 : (NP^2 + NQ^2) = (NP^2 + NQ^2) : NQ^2;$
therefore, by (2) above,
$$HA : AN = (NP^2 + NQ^2) : NQ^2.$$
It results, by a like argument, that the segment and the cone AEF taken together where they are balance, about A, the cone AEF acting as one mass at H. Taking moments as before, we can (since we know the position of the centre of gravity of the cone AEF as well as its volume) find the centre of gravity of the segment of the spheroid.

Incidentally, in Proposition 13, Archimedes finds the centre of gravity of the half of a right cylinder cut off by a plane through its axis, or, in other words, the centre of gravity of a semicircle.

We pass on to speak of the other treatises in the order in which they appear in the editions.

ON THE SPHERE AND CYLINDER. I, II

In his preface to this work Archimedes himself states shortly the main results obtained, explaining that they are new and are being published for the first time. They are: (1) that the surface of a sphere is four times that of a great circle in it (i.e. $= 4\pi r^2$); (2) that the surface of a segment of a sphere is equal to a circle with radius equal to the straight line drawn from the vertex of the segment to a point on the circumference of the base; (3) that, if about a sphere there be circumscribed a cylinder with height equal to the diameter of the sphere, then (*a*) the volume, (*b*) the surface (including the bases), of the cylinder are one and a half times the volume and surface respectively of the sphere. Although the proposition about the surface of a sphere precedes the proposition about its volume, we now know from the *Method* that the latter was discovered first, and the fact stated in the former was originally inferred from the latter.

'From the theorem', says Archimedes, 'that a sphere is four times as great as the cone with a great circle of the sphere as base and with height equal to the radius of the sphere I conceived the notion that the surface of any sphere is four times as great as a great circle in it; for, judging from the fact that any circle is equal to a triangle with base equal to the circumference and height equal to the radius of the circle, I apprehended that, in like manner, any sphere is equal to a cone with base equal to the surface of the sphere and height equal to the radius'.

Book I begins with definitions, and these are followed by five Assumptions, two of which are well known, namely, *Of all lines which have the same extremities, the straight line is the least* and (the 'Axiom of Archimedes') *Of unequal magnitudes, the greater exceeds the less by such a magnitude as, when added to itself* [continually], *can be made to exceed any assigned magnitude of the same kind.* Though known by the name of Archimedes, the latter assumption is practically equivalent to Eucl. V, Def. 4, and is closely connected with the theorem of Eucl. X. 1 (cf. pp. 193-4 above).

In applying the method of exhaustion Euclid uses inscribed figures only, approaching only from *below* the area or volume to be measured. Archimedes, on the other hand, uses circumscribed figures as well, approaching the area or volume to be measured from above as well as from below and, as it were, *compressing* the inscribed and circumscribed figures into coalescence with the figure to be measured. For the solution in the present treatise of his

ON THE SPHERE AND CYLINDER. I

main problems he requires to know the surface of a right cone and a cylinder. For this purpose he inscribes in the bases, and circumscribes to them, regular polygons with a large number of sides, thereby obtaining pyramids (or prisms) with many faces inscribed in and circumscribed to the cone (or cylinder). He thus finds the surfaces of the inscribed and circumscribed pyramids (or prisms), and then proves that the surfaces of the circumscribed figures (excluding the bases) are greater, and the surfaces of the inscribed figures (excluding the bases) less, than the surfaces (excluding the bases) of the cone and cylinder respectively (Props. 7–12). In Propositions 13, 14 follow the proofs, by exhaustion, of the theorems (1) that the curved surface of the cylinder is equal to a circle the radius of which is a mean proportional between the 'side' (i.e. a generator) of the cylinder and the diameter of the base; (2) that the curved surface of the cone is equal to a circle the radius of which is a mean proportional between the 'side' (i.e. generator) and the radius of the base. That is, if s be a generator and r the radius of the base, the curved surface of the cone is (as we should say) $\pi r s$ and that of the cylinder $2\pi r s$.

We will illustrate Archimedes' working of the method of exhaustion by the case of the cone.

Let A be the base of the cone, C a straight line equal to its radius, D a straight line equal to a generator of the cone, E a mean proportional to C, D, and B a circle with radius equal to E.

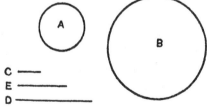

If S is the curved surface of the cone, we have to prove that $S = B$. For, if not,

I. Suppose $B < S$.

Circumscribe a regular polygon about B and inscribe a similar polygon in it such that the former may have to the latter a ratio less than $S : B$ (Prop. 5). Describe about A a similar polygon and set up from it a pyramid circumscribing the cone.

Then (polygon about A) : (polygon about B)
$= C^2 : E^2$
$= C : D$
$=$ (polygon about A) : (surface of pyramid excluding base).

Therefore (surface of pyramid) = (polygon about B).

But (polygon about B) : (polygon in B) $< S : B$;
therefore (surface of pyramid) : (polygon in B) $< S : B$.

This is impossible, since (surface of pyramid) $> S$, while (polygon in B) $< B$.

Therefore B is not less than S.

II. Suppose $B > S$.

Circumscribe and inscribe to B similar regular polygons such that the former has to the latter a ratio less than $B : S$. Inscribe in A a similar regular polygon and set up on it a pyramid inscribed in the cone.

Then (polygon in A) : (polygon in B) $= C^2 : E^2$
$= C : D$
$>$ (polygon in A) : surface of pyramid.

[The latter inference is clear because the ratio $C : D$ is greater than the ratio of the perpendiculars from the centre of A and from the vertex of the pyramid respectively on any side of the polygon in A; in other words, if $\beta < \alpha < \frac{1}{2}\pi$, $\sin \alpha > \sin \beta$.]

Therefore (surface of pyramid) $>$ (polygon in B).

ON THE SPHERE AND CYLINDER. I

But (polygon about B) : (polygon in B) $< B : S$;
therefore, *a fortiori*,

(polygon about B) : (surface of pyramid) $< B : S$.

This is impossible, since (polygon about B) $> B$, while surface of pyramid $< S$.

Therefore B is not greater than S.

Since then B is neither greater nor less than S, B must be equal to S.

Before passing to the propositions about the surface and volume of a sphere or a segment thereof, Archimedes interpolates some propositions about what he calls a 'solid rhombus', a figure made up of two right cones, equal or unequal, with bases coincident but with their vertices in opposite directions. The use of this figure will shortly appear.

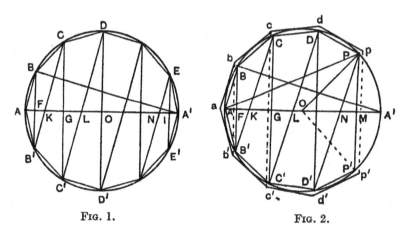

Fig. 1.   Fig. 2.

To find the surface of a sphere or a segment thereof, Archimedes takes a great circle of the sphere and first inscribes, and then circumscribes, a regular polygon of an even number of sides bisected by the diameter of the circle or segment (in the case of the segment APP' the polygon has PP' for one side and only the rest of the sides are equal to one another).

Then, by making the polygons revolve about the axis AA', he obtains surfaces, one greater, and the other less, than that of the sphere or segment, and by measuring these surfaces respectively he approximates to the surface of the sphere or segment itself.

Construct the inscribed polygons as shown in the figures. Joining BB', CC' ... CB', DC' ... we see that BB', CC' ... are all parallel, and so are AB, CB', DC' ... Join $A'B$.

Then, by similar triangles,
$$A'B : BA = BF : FA$$
$$= B'F : FK$$
$$= CG : GK$$
$$= C'G : GL$$
$$\cdots\cdots$$
$$= E'I : IA' \text{ in Fig. 1,}$$
$$\text{or } = PM : MN \text{ in Fig. 2.}$$

Adding antecedents and consequents, we have
$$A'B : BA = (BB' + CC' + \ldots + EE') : AA' \text{ (Fig. 1)}, \quad (1)$$
$$\text{or } = (BB' + CC' + \ldots + \tfrac{1}{2}PP') : AM \text{ (Fig. 2)}. \quad (2)$$

When we make the polygon revolve about AA', the inscribed figure so obtained has a surface made up of the surfaces of cones and frusta of cones; and Archimedes has proved (Prop. 14) that the surface of the cone ABB' is what we should write as $\pi . AB . BF$, and (Prop. 16) that the surface of the frustum BC' is $\pi . BC(BF + CG)$.

Therefore, since $AB = BC = \ldots$,
the surface of the inscribed solid of revolution is
$$\pi . AB \{ \tfrac{1}{2} BB' + \tfrac{1}{2}(BB' + CC') + \ldots \}$$
$$= \pi . AB(BB' + CC' + \ldots + EE'), \text{ (Fig. 1)}$$
$$\text{or } \pi . AB(BB' + CC' + \ldots + \tfrac{1}{2}PP'). \text{ (Fig. 2)}$$

Hence, from above, the surface of the inscribed solid is $\pi \cdot A'B \cdot AA'$ in Fig. 1 and $\pi \cdot A'B \cdot AM$ in Fig. 2.

Thus the surfaces of the inscribed figures are less than $\pi \cdot AA'^2$ or $\pi \cdot A'A \cdot AM$, i.e. $\pi \cdot AP^2$, in the two cases respectively, but approximate thereto as the number of the sides in the polygons is increased.

Similar propositions prove that, by means of like solids circumscribed to the sphere or segment, we approximate from *above* to the same values.

Then, by the method of exhaustion, it is proved that the surfaces of the sphere and segment are neither greater nor less than, and must therefore be *equal to*, $\pi \cdot AA'^2$ and $\pi \cdot AP^2$ respectively.

We can express the above results in trigonometrical form. If $4n$ is the number of the sides of the polygon inscribed in the great circle of the sphere, and $2n$ the number of the equal sides of the polygon inscribed in the section of the segment, and if α be the angle AOP, then the proportions in (1) and (2) above are equivalent to

(1) $\quad \sin\dfrac{\pi}{2n} + \sin\dfrac{2\pi}{2n} + \ldots + \sin(2n-1)\dfrac{\pi}{2n} = \cot\dfrac{\pi}{4n}$,

and (2) $\quad 2\left\{\sin\dfrac{\alpha}{n} + \sin\dfrac{2\alpha}{n} + \ldots + \sin(n-1)\dfrac{\alpha}{n}\right\} + \sin\alpha$

$$= (1 - \cos\alpha)\cot\dfrac{\alpha}{2n}.$$

Thus in effect Archimedes gives the summation of the series

$$\sin\theta + \sin 2\theta + \ldots + \sin(n-1)\theta$$

for any case where $n\theta$ is not greater than π.

The surfaces of the solids of revolution as obtained by Archimedes are the equivalent of the products of the first of the above trigonometrical series by $\pi \cdot AB \cdot AA'$ or (if r be the radius of the sphere) by $4\pi r^2 \cdot \sin\dfrac{\pi}{4n}$, and of the second by $\pi \cdot AB \cdot \tfrac{1}{2}AA'$ or $\pi r^2 \cdot 2\sin\dfrac{\alpha}{2n}$; the surfaces are therefore $4\pi r^2 \cdot \cos\dfrac{\pi}{4n}$ and $\pi r^2 \cdot 2\cos\dfrac{\alpha}{2n}(1-\cos\alpha)$ respectively. Archimedes' results for the surfaces of the sphere and segment respectively, namely, $4\pi r^2$ and $2\pi r^2(1-\cos\alpha)$, are in effect the limiting values of these expressions when n is indefinitely increased.

Again, the effect of multiplying the above expressions by $4\pi r^2 \cdot \sin\dfrac{\pi}{4n}$ and $\pi r^2 \cdot 2\sin\dfrac{\alpha}{2n}$ respectively and increasing n indefinitely is precisely what we should represent by
$$4\pi r^2 \cdot \tfrac{1}{2}\int_0^\pi \sin\theta\, d\theta, \text{ or } 4\pi r^2,$$
and
$$\pi r^2 \cdot \int_0^\alpha 2\sin\theta\, d\theta, \text{ or } 2\pi r^2(1-\cos\alpha),$$
so that Archimedes' procedure is in essence equivalent to an *integration*.

In order to find the volumes of a sphere and a segment thereof, Archimedes makes the circumscribed and inscribed polygons revolve about the axis as before, finds the volumes of the circumscribed and inscribed solids of revolution respectively, and then, by increasing the number of sides in the polygons respectively, compresses the solids into coalescence with one another and with the included sphere or segment, finally confirming his results by exhaustion as usual. The contents of the inscribed and circumscribed figures are obtained by adding the content of 'solid rhombi' and differences between such. Thus, in the case of the whole sphere, Archimedes takes for convenience an inscribed regular polygon of $4n$ sides as shown in Fig. 1. If BO be joined, the revolution of the triangle ABO about AA' produces a solid rhombus the content of which is known. If CO be joined and CB be produced to meet $A'A$ produced in T, the revolution of the triangle BOC about AA' produces an extinguisher-shaped figure the content of which is the difference between the volumes of the solid rhombi $TCOT$ and $TBOT$. And so on.

Archimedes proves that the volumes of the elements of the solid are equal respectively to a series of cones of height h, where h is the length of the perpendicular from O on any one of the sides $AB, BC \ldots$. The cone which is equal to the solid rhombus $ABOB'$ has for base a circle equal to the surface of the cone ABB'. The base of the cone which is equal to the first 'extinguisher' is equal in area to the surface of the frustum described by the revolution of BC about AO; and so on. Hence the volume of the whole inscribed solid of revolution is equal to that of a cone with height h and base equal to the whole surface of the said solid. The volume of the sphere is, of course, greater than that of the inscribed solid. Similarly, Archimedes proves that it is less than that of a similar figure circumscribed to it, which is again equal to that of the cone which has for base a circle equal to the surface of the circumscribed figure, and for height, not h, but the radius of the great circle. By increasing indefinitely the number of sides in the polygons, and so compressing the outer and inner solids into coincidence with the sphere or segment, Archimedes proves, by exhaustion, that the volume of the sphere is equal to that of a cone having for height the radius of the sphere and for base a circle equal in area to the surface of the sphere and is, therefore, $\frac{1}{3}r \cdot 4\pi r^2$ or $\frac{4}{3}\pi r^3$. It follows that the volume of the cylinder circumscribing the sphere is $\frac{3}{2}$ of that of the sphere.

In the case of the segment of the sphere, the sum of the elements makes up, not the segment, but the *sector* included in it or including it according as the segment is greater or less than a hemisphere. The *sector* is thus equal to the cone with height r and base $\pi \cdot AP^2$; and the volume of the *segment* is the sum or difference of the volumes of this cone and the cone OPP'.

In Proposition 2 of Book II Archimedes reduces the

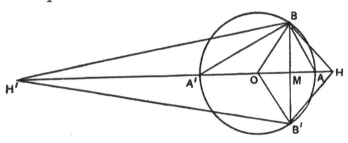

sum or difference of the two cones which has been proved equal to the segment APP' to one cone with base the same as that of the segment. We have simply, in the case of the segment $A'BB'$ greater than a hemisphere, to add to the axis OM of the cone OBB' a length OH' along OA' such that

$$H'O : OA = A'B^2 : BM^2$$
$$= AA' : AM.$$

(Since $H'O \cdot BM^2 = OA \cdot A'B^2$, the effect of this is to transform the cone with height OA and base $\pi \cdot A'B^2$ into an equivalent cone with height $H'O$ and base $\pi \cdot BM^2$, which base is the same as that of the cone OBB'.)

The height of the required single cone with $\pi \cdot BM^2$ for base which is equal to the sum of the two cones is proved to be $H'O + OM = H'M$, where

$$H'M : A'M = (OA + AM) : AM.$$

Similarly, the height of the cone with base $\pi \cdot BM^2$ which is equal to the segment ABB' less than a hemisphere is proved to be HM, where
$$HM : AM = (OA' + A'M) : A'M.$$

Upon these results follows the most important proposition (4) of Book II, namely, the problem *To divide the sphere by a plane into segments the volumes of which are in a given ratio*.

Let AM, $A'M$, the heights of the required segments, be denoted by h, h', and let m/n be the ratio of their volumes. The nature of the problem is seen in this way.

From the above proportions, if r be the radius of the sphere, we have
$$HM : h = (r + h') : h',$$
$$H'M : h' = (r + h) : h,$$
whence $\quad \dfrac{h(r+h')}{h'} = \dfrac{m}{n} \cdot \dfrac{h'(r+h)}{h}.$

Eliminate h' by means of the relation $h + h' = 2r$.

Thus $\quad nh^2(3r - h) = m(2r - h)^2(r + h)$
$$= m(4r^2 - 4rh + h^2)(r+h)$$
$$= m(h^3 - 3h^2r + 4r^3);$$
therefore $\quad h^3(m+n) - 3h^2r(m+n) + 4mr^3 = 0,$

or $\quad h^3 - 3h^2r + \dfrac{4m}{m+n}r^3 = 0.$

Archimedes in effect reduces his problem to the solution of this cubic equation. It may be written in the form
$$(2r-h)^2(r+h) = 4r^3 \dfrac{n}{m+n},$$
and Archimedes treats it as a particular case of a more general equation which he expresses in the form
$$(r+h) : b = c^2 : (2r-h)^2,$$

where b is a given length and c^2 a given area. The solution is not given in the text of Archimedes as it stands, but Eutocius found it in an Archimedean fragment which he reproduces, along with other solutions of our problem by Dionysodorus and Diocles.

If we put x for $2r-h$ and a for $3r$, our equation reduces to

$$x^2(a-x) = bc^2.$$

In the Archimedean fragment this equation is solved by means of the intersections of a parabola and a rectangular hyperbola, the equations of which may be written thus,
$$x^2 = \frac{c^2}{a}y, \qquad (a-x)y = ab.$$

The fragment includes the διορισμός, or the determination of the conditions of possibility of a real solution. This takes the form of investigating the maximum possible value of $x^2(a-x)$, which is found to be that corresponding to the value $x = \frac{2}{3}a$. This is established by showing that, if $bc^2 = \frac{4}{27}a^3$, the curves touch at the point for which $x = \frac{2}{3}a$. If $bc^2 < \frac{4}{27}a^3$, there are two solutions.

Book II contains the following further problems: To find a sphere equal to a given cone or cylinder (Prop. 1); to cut a sphere by a plane into segments having their *surfaces* in a given ratio (Prop. 3); given two segments of spheres, to find a third segment similar to one and having its surface equal to that of the other (Prop. 6); the same problem with volume substituted for surface (Prop. 5); from a given sphere to cut off a segment having a given ratio to the cone with the same base and equal height (Prop. 7). The problems of Props. 1 and 5 are reduced to that of finding two mean proportionals. Two interesting theorems complete the Book. (1) If a sphere be cut by a plane into segments with surfaces S, S' and volumes V, V'

MEASUREMENT OF A CIRCLE 305

respectively, S, V being the surface and volume of the greater segment, then

$$V : V' < S^2 : S'^2 \text{ but } > S^{\frac{3}{2}} : S'^{\frac{3}{2}} \text{ (Prop. 8).}$$

(2) Of all segments of spheres which have equal surfaces, the hemisphere is the greatest in volume (Prop. 9).

The *Measurement of a Circle* contains only three propositions and is not in its original form; like the treatise *On the Sphere and Cylinder*, it has lost all trace of Archimedes' Doric dialect, and Proposition 2 is at all events not in its proper place. Perhaps we have only a fragment of a longer treatise. The important propositions are 1 and 3; the first proves that the area of a circle is equal to that of a right-angled triangle having for perpendicular the radius of the circle, and for base its circumference; the third proves that the ratio of the circumference of any circle to its diameter (i.e. what we call π) is $< 3\frac{1}{7}$ but $> 3\frac{10}{71}$. Proposition 1 is proved by inscribing and circumscribing polygons, beginning from a square and continually doubling the number of sides, and then applying the method of exhaustion. Proposition 3 is proved by means of a regular polygon of 96 sides inscribed in the circle and a similar polygon circumscribed to it. Geometry does its part first, and the rest is sheer calculation.

Archimedes takes the circumscribed figure first. Let CA be the tangent at A to a circular arc with centre O.

Make the angle AOC equal to one-third of a right angle. Let OD bisect the angle AOC, OE the angle AOD, OF the angle AOE, and OG the angle AOF.

Produce GA to H so that $GA = AH$.

Thus the angle GOH is equal to the angle AOF and is therefore $\frac{1}{24}$th of a right angle, so that GH is the side of a circumscribed regular polygon with 96 sides. Archimedes has to find a value for GH (or GA) in terms of the radius AO of the circle.

We know the values of CO, CA, and we have to find successively values for DA, EA, FA, GA. Archimedes uses the well-known theorem that, since OD bisects the angle COA,
$$CO : OA = CD : DA.$$

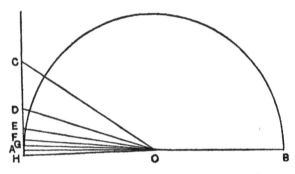

In order to apply this theorem continually, we must also know the values of DO, EO, FO, GO, which are obtained from $DO^2 = DA^2 + AO^2$, &c.

First, to find DA, we have
$$OA : AD = CO : CD$$
$$= (CO + OA) : CA.$$
Now, since $\angle COA = 30°$, $OC : CA = 2 : 1$, and
$$OA : AC = \sqrt{3} : 1.$$

Archimedes has now to use an approximate value for $\sqrt{3}$, and he assumes as known, without proof, that $\sqrt{3} : 1 > 265 : 153$.

Thus $\qquad OA : AC > 265 : 153.$
But $\qquad OC : CA = 306 : 153;$
therefore $\quad OA : AD = (CO + OA) : CA$
$$> 571 : 153.$$

MEASUREMENT OF A CIRCLE

whence $\quad OD^2 : AD^2 > (571^2 + 153^2) : 153^2$
$$> 349450 : 23409,$$
and (says Archimedes)
$$OD : AD > 591\tfrac{1}{8} : 153.$$

Next $\quad OA : AE = (DO + OA) : DA$
$$> (591\tfrac{1}{8} + 571) : 153, \text{ or } 1162\tfrac{1}{8} : 153,$$
and $\quad OE^2 : AE^2 > \{(1162\tfrac{1}{8})^2 + 153^2\} : 153^2,$
whence Archimedes derives
$$OE : AE > 1172\tfrac{1}{8} : 153.$$

In like manner
$$OA : AF > (1172\tfrac{1}{8} + 1162\tfrac{1}{8}) : 153, \text{ or } 2334\tfrac{1}{4} : 153,$$
and $\quad\quad\quad OF : AF > 2339\tfrac{1}{4} : 153;$

Lastly,
$$OA : AG > (2339\tfrac{1}{4} + 2334\tfrac{1}{4}) : 153, \text{ or } 4673\tfrac{1}{2} : 153.$$

The perimeter of the polygon $= 96 \cdot GH = 192 \cdot GA$, and we have to compare this with the diameter $2OA$.

We have
(perimeter of polygon) : (diameter) $< 96 \times 153 : 4673\tfrac{1}{2}$.

Now $\quad \dfrac{96 \times 153}{4673\tfrac{1}{2}} = 3\dfrac{667\tfrac{1}{2}}{4673\tfrac{1}{2}}$, and $\dfrac{667\tfrac{1}{2}}{4672\tfrac{1}{2}} = \tfrac{1}{7}$.

Therefore, *a fortiori*, $\pi < 3\tfrac{1}{7}$.

To obtain the inscribed polygon, Archimedes takes a semicircle ABC and makes the angle BAC equal to one-third of a right angle. He then bisects the angle BAC by AD, the angle BAD by AE, the angle BAE by AF, and the angle BAF by AG.

The chord GB is then the side of an inscribed regular polygon of 96 sides.

Archimedes has now successively to calculate CB, DB,

EB, FB, GB, and incidentally CA, DA, EA, FA, GA, in terms of the diameter of the circle.

We have $\qquad AC:CB = \sqrt{3}:1$.

Archimedes assumes, again without proof, that
$$\sqrt{3}:1 < 1351:780,$$
and proceeds thus.

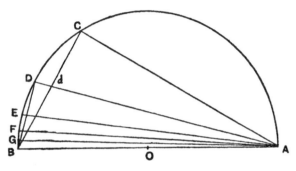

$$AC:CB < 1351:780,$$
while $\qquad AB:CB = 2:1 = 1560:780$.

Let AD, CB meet in d. Then, since AD bisects the angle CAB, the triangles ADB, ACd, BDd are similar.

Therefore $\quad AD:DB = AC:Cd$,
$\qquad\qquad\qquad = AB:Bd$, since AD bisects $\angle BAC$,
$\qquad\qquad\qquad = (AB+AC):CB$.

It follows, from above, that
$$AD:DB < 2911:780,$$
and $\quad AB^2:BD^2 < \{(2911)^2+(780)^2\}:(780)^2$
$\qquad\qquad\qquad < 9082321:608400$.

Therefore (says Archimedes)
$$AB:BD < 3013\tfrac{3}{4}:780.$$

Next he proves, in like manner, that

$$AE:EB < 5924\tfrac{3}{4} : 780$$

$$< 1823 : 240, \text{ in lower terms,}$$

whence $\quad AB:BE < 1838\tfrac{9}{11} : 240.$

Again $\quad AF:FB < 3661\tfrac{9}{11} : 240$

$$< 1007 : 66, \text{ in lower terms,}$$

and we derive $\quad AB:BF < 1009\tfrac{1}{6} : 66.$

Finally, $\quad AG:GB < 2016\tfrac{1}{6} : 66,$

whence $\quad AB:BG < 2017\tfrac{1}{4} : 66.$

Since the polygon has 96 sides,

(perimeter of polygon) : (diameter) $> 96 \times 66 : 2017\tfrac{1}{4}.$

That is, says Archimedes, the ratio in question (and, *a fortiori*, π) $> 3\tfrac{10}{71}$. (Archimedes may have obtained the latter figure by developing the ratio $96 \times 66 : 2017\tfrac{1}{4}$ as a continued fraction, since $3\tfrac{10}{71} = 3 + \dfrac{1}{7+} \dfrac{1}{10}.$)

The above working shows that Archimedes was able to reckon with large numbers, and to approximate to the square roots of such numbers when they are not exact squares.

It remains to consider his assumption that

$$\frac{265}{153} < \sqrt{3} < \frac{1351}{780}.$$

Many have been the conjectures as to how Archimedes arrived at these particular approximations. Some have suggested methods which are the equivalent of the use of continued fractions more or less disguised. But the simplest supposition is that of Hunrath and Hultsch that the formula used was

$$a \pm \frac{b}{2a} > \sqrt{(a^2 \pm b)} > a \pm \frac{b}{2a \pm 1},$$

where a is the nearest square number above or below $\sqrt{(a^2 \pm b)}$. The first part of this formula was used by Heron, as we know from the *Metrica*; the second part was used by the Arabian algebraist al-Karkhī (eleventh century), who drew from Greek sources.

Applying the above formula, we find, in the first place,

$$2 - \tfrac{1}{4} > \sqrt{3} > 2 - \tfrac{1}{3},$$

or $\qquad \tfrac{7}{4} > \sqrt{3} > \tfrac{5}{3}.$

Next, clearing of fractions, and considering 5 as an approximation to $\sqrt{(3 \cdot 3^2)}$ or $\sqrt{27}$, we have

$$5 + \tfrac{2}{10} > \sqrt{27} > 5 + \tfrac{2}{11},$$

whence we obtain $\qquad \tfrac{26}{15} > \sqrt{3} > \tfrac{19}{11}.$

Again clearing of fractions, and treating 26 as an approximation to $\sqrt{(3 \cdot 15^2)}$ or $\sqrt{675}$, we have

$$26 - \tfrac{1}{52} > 15\sqrt{3} > 26 - \tfrac{1}{51},$$

which reduces to $\qquad \dfrac{1351}{780} > \sqrt{3} > \dfrac{265}{153}.$

We learn from Heron's *Metrica* that Archimedes himself improved on the above approximations to π (cf. p. 146 above).

ON CONOIDS AND SPHEROIDS

The treatise has a preface addressed to Dositheus stating the problems dealt with, and begins as usual with definitions of the solids in question and various parts of

ON CONOIDS AND SPHEROIDS 311

them. The solids are (1) the *right-angled conoid* (paraboloid of revolution), (2) the *obtuse-angled conoid* (hyperboloid of revolution), and (3) the *spheroids*, namely (a) the *oblong*, or *prolate*, spheroid described by the revolution of an ellipse about its *major* axis, (b) the *flat* or *oblate*, generated by revolution about the *minor* axis.

The first propositions (a Lemma and Props. 1-18) are preliminary. They include theorems connected with the summation of series required for the purpose of finding the sum of all the small elements making up the volumes of the figures measured.

It is proved that

$$2(a+2a+3a+ \ldots +na) > n.na$$
$$> 2\{a+2a+3a+ \ldots +(n-1)a\}$$

(this is clear from our formula $S_n = \tfrac{1}{2}n(n+1)a$); also that

$$(n+1)(na)^2+a(a+2a+3a+ \ldots +na)$$
$$= 3\{a^2+(2a)^2+ \ldots +(na)^2\},$$

whence it follows that

$$3\{a^2+(2a)^2+(3a)^2+ \ldots +(na)^2\} > n(na)^2$$
$$> 3\{a^2+(2a)^2+ \ldots +(\overline{n-1}\ a)^2\};$$

and, lastly, Proposition 2 gives limits for the sum of n terms of the series

$$ax+x^2,\ a.2x+(2x)^2,\ \ldots$$

in the form of inequalities of ratios, thus:

$$n\{a.nx+(nx)^2\} : \Sigma_1^{n-1}\{a.rx+(rx)^2\}$$
$$> (a+nx) : (\tfrac{1}{2}a+\tfrac{1}{3}nx)$$
$$> n\{a.nx+(nx)^2\} : \Sigma_1^n\{a.rx+(rx)^2\}.$$

Proposition 3 proves that, if QQ' be a chord of a parabola bisected at V by the diameter PV, and if the length of PV be constant, the areas of the triangle PQQ' and of the segment PQQ' are also constant, whatever be the direction of QQ'. Propositions 4–6 find the area of an ellipse in comparison with that of the auxiliary circle. Propositions 7, 8 show how, given elliptic sections of cones and cylinders (generally oblique), to find the circular sections. Propositions 11–18 give simple properties of the conoids and spheroids which are easily deduced from the properties of the respective conics.

With Proposition 19 begins the real business of the treatise, namely the investigation of the volumes of segments (right or oblique) of each conoid and spheroid. The nature of the method followed will be understood from the three diagrams opposite and the explanations appended.

Segments of each of the three solids (the paraboloid and hyperboloid of revolution and the spheroid) are cut by a plane (represented by the plane of the paper) through the axis of the segment and at right angles to the plane section which is the base of the segment, and which is a circle or an ellipse according as the said base is or is not at right angles to the axis of the segment; BB', the section by the plane of the paper of the base of the segment, is the diameter of the circle or an axis of the ellipse as the case may be.

The axis AD of the segment is divided into a large number of equal parts, and through the points of division planes are drawn parallel to the base of the segment, making sections similar to the base in which PP', QQ' ... are axes. By drawing element-cylinders (or 'frusta of cylinders', as Archimedes calls them when oblique) through the various sections, with generators parallel to the axis of the segment, we obtain circumscribed and inscribed solids

ON CONOIDS AND SPHEROIDS 313

consisting of the sums of two series of short cylinders, as shown in the figures. If we begin from A, the first inscribed

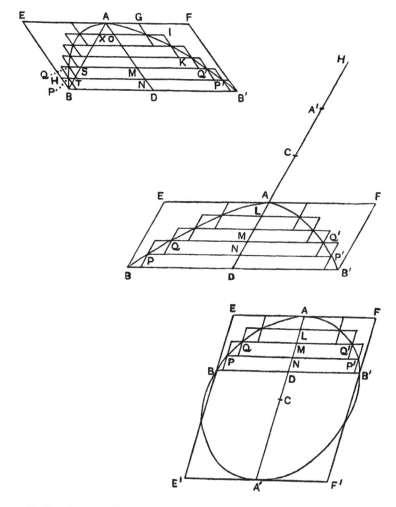

cylinder is equal to the first circumscribed cylinder, the second to the second, and so on, but there is one more cylinder in the circumscribed figure. The difference between the circumscribed and inscribed figures is therefore equal to the last element-cylinder of which BB' is the base and ND the axis. As the last element-cylinder can be made as small as we please by increasing the number of equal parts into which AD is divided, we have the necessary condition for applying the method of exhaustion.

In order to find the volumes of the inscribed and circumscribed solids, Archimedes considers each element-cylinder in relation to the corresponding element-cylinder in the cylinder EB' described with BB' as base and AD as axis.

Suppose that $AD = h$, and that AD is divided into n equal parts of length k. Let M, any one of the points of division, be taken, and suppose that $AM = rk$. Consider the element-cylinder on QQ' as base in relation to the corresponding element of the cylinder EB' or, in other words, to the equal element-cylinder HB'. The elements are in the ratio of QM^2 to BD^2.

In the case of the paraboloid,
$$QM^2 : BD^2 = AM : AD$$
$$= rk : nk.$$

In the case of the hyperboloid,
$$QM^2 : BD^2 = AM \cdot A'M : AD \cdot A'D$$
$$= rk(a+rk) : nk(a+nk),$$
where a denotes the transverse axis AA'.

That is, $QM^2 : BD^2 = \{a \cdot rk + (rk)^2\} : \{a \cdot nk + (nk)^2\}$.

To obtain the volume of the inscribed figure, we have to find the sum of terms in which r is given all values from 1 to $(n-1)$, while in the case of the circumscribed figure r has to be given all values from 1 to n.

For the paraboloid Archimedes obtains the equivalent of the proportion

(inscribed figure) : (cylinder EB')
$$= \{1+2+3+ \ldots +(n-1)\}k : n^2k,$$
and (circumscribed figure) : (cylinder EB')
$$= (1+2+3+ \ldots +n)k : n^2k.$$

But, by the Lemma,
$$k+2k+ \ldots +(n-1)k < \tfrac{1}{2}n^2k < (k+2k+ \ldots +nk);$$
therefore

(cylinder EB') : (inscribed figure) > 2
$\qquad >$ (cylinder EB') : (circumscribed figure).

This indicates the desired result, which is then confirmed by the method of exhaustion, namely that

$\qquad$ cylinder $EB' = 2$(segment of paraboloid),

or, if V be the volume of the 'segment of a cone' with vertex A and base the same as that of the segment,

$\qquad$ volume of segment $= \tfrac{3}{2}V.$

It will be seen that Archimedes' procedure proves in effect that, if k be indefinitely diminished and n indefinitely increased, while nk remains equal to h, then

$\qquad$ limit of $k\{k+2k+3k+ \ldots +(n-1)k\} = \tfrac{1}{2}h^2,$

which is equivalent in our notation to
$$\int_0^h x \, dx = \tfrac{1}{2}h^2.$$

In the case of the *hyperboloid*, Archimedes proves in like manner the equivalent of the proportion

(inscribed figure) : (cylinder EB')
$$= \Sigma_1^{n-1}\{a \cdot rk+(rk)^2\} : n\{a \cdot nk+(nk)^2\}$$

and (circumscribed figure) : (cylinder EB')
$$= \Sigma_1^n\{a \cdot rk+(rk)^2\} : n\{a \cdot nk+(nk)^2\}.$$
But he has proved (Prop. 2) that
$$n\{a \cdot nk+(nk)^2\} : \Sigma_1^{n-1}\{a \cdot rk+(rk)^2\}$$
$$> (a+nk) : (\tfrac{1}{2}a+\tfrac{1}{3}nk)$$
$$> n\{a \cdot nk+(nk)^2\} : \Sigma_1^n\{a \cdot rk+(rk)^2\}.$$

From these relations it is inferred, and then proved by the method of exhaustion, that

(volume of segment) : (volume of cylinder BE')
$$= (\tfrac{1}{2}a+\tfrac{1}{3}nk) : (a+nk),$$
whence (volume of segment) : (volume of cone ABB')
$$= (\tfrac{3}{2}a+nk) : (a+nk)$$
$$= (AD+3CA) : (AD+2CA).$$

The result obtained is equivalent to a proof of the fact that, if
$$S_n = a(k+2k+3k+ \ldots +nk)+\{k^2+(2k)^2+ \ldots +(nk)^2\},$$
and if we suppose k to be indefinitely diminished and n indefinitely increased, while nk remains always equal to h, then

limit of $n(ah+h^2)/S_n = (a+h)/(\tfrac{1}{2}a+\tfrac{1}{3}h)$,

or $\qquad$ limit of $\dfrac{h}{n} S_n = h^2(\tfrac{1}{2}a+\tfrac{1}{3}h)$,

In other words, the limit of kS_n or
$$ak(k+2k+ \ldots +nk)+k\{k^2+(2k)^2+ \ldots (nk)^2\}$$
is $h^2(\tfrac{1}{2}a+\tfrac{1}{3}h)$.

This is equivalent in our notation to the integration
$$\int_0^h (ax+x^2)\,dx = h^2(\tfrac{1}{2}a+\tfrac{1}{3}h).$$

The case of the *spheroid* is made a little more complicated by the fact that (in order probably to avoid the multiplication of auxiliary propositions) Archimedes makes the *same* integration serve in this case also.

ON SPIRALS

The treatise *On Spirals* has a preface addressed to Dositheus beginning with an allusion to the death of Conon and summarizing the main results, first of the works *On the Sphere and Cylinder* and *On Conoids and Spheroids*, and then of the treatise itself. The spiral is thus defined:

If a straight line one extremity of which remains fixed be made to revolve at a uniform rate in a plane until it returns to the position from which it started, and if, at the same time as the straight line is revolving, a point move at a uniform rate along the straight line starting from the fixed extremity, the point will describe a spiral in the plane.

As usual, we have a series of propositions preliminary to the main subject, first two propositions about uniform motion, then (5–9) geometrical propositions the object of which is to show the possibility of certain constructions. Proposition 10 repeats the Lemma to Proposition 2 of *On Conoids and Spheroids* involving the summation of the series $1^2 + 2^2 + \ldots + n^2$, while Proposition 11 proves that

$$(n-1)(na)^2 : \{a^2 + (2a)^2 + \ldots + (\overline{n-1}a)^2\}$$
$$> (na)^2 : \{na \cdot a + \tfrac{1}{3}(na-a)^2\}$$
$$> (n-1)(na)^2 : \{(2a)^2 + (3a)^2 + \ldots + (na)^2\}.$$

This proposition is true of the series

$$a^2, (a+b)^2, (a+2b)^2 \ldots (a+\overline{n-1}b)^2$$

and is assumed in Propositions 25, 26 for that series: thus

$(n-1)\{a+(n-1)b\}^2 : \{a^2+(a+b)^2+(a+2b)^2+ \ldots$
$\qquad\qquad\qquad\qquad\qquad\qquad +(a+\overline{n-2}b)^2\}$
$> (a+\overline{n-1}b)^2 : \{(a+\overline{n-1}b)a+\tfrac{1}{3}(\overline{n-1}b)^2\}$
$> (n-1)\{a+(n-1)b\}^2 : \{(a+b)^2+(a+2b)^2+ \ldots$
$\qquad\qquad\qquad\qquad\qquad\qquad +(a+\overline{n-1}b)^2\}.$

The definitions are now introduced, namely, those of the *spiral*, the *origin*, and the *initial line* respectively. The annexed figure represents the spiral between O, the origin, and the point A where, at the end of the first complete turn, it meets the initial line OA. Let OP, OQ, two radii vectores, meet the circle described with O as centre and OA as radius in P', Q'.

Propositions 12, 14, 15 give the fundamental property of the spiral. For the points P, Q on it we have

$$OP:OQ = (\text{arc } AKP'):(\text{arc } AKQ').$$

This is, of course, equivalent to the equation in polar co-ordinates, $\rho = a\theta$, where θ is the circular measure of the angle through which the revolving line has turned from its original position.

As Archimedes does not speak of any angles greater than π or 2π, he has to distinguish between the first turn of the spiral, the second turn, &c. He calls OA, the radius vector at the end of the first revolution, the 'first distance'. The 'second distance' is the equal length added to the radius vector in the second revolution, and so on. The 'first area' is the area bounded by the portion of the spiral described in the first revolution and the 'first distance', the 'second area' the area added in the second revolution and bounded by the 'second distance', and so on. The 'first circle' is that described with OA, the 'first distance', as radius, the 'second circle' that described with O as centre and radius equal to the sum of the 'first' and 'second' distances ($=2OA$), and so on.

We can express the relation of points on *any* turn by means of the 'first circle' only. If P, Q are points on the nth turn, and c is the circumference of the 'first circle',

$$OP:OQ = \{(n-1)c + \text{arc } AKP'\} : \{(n-1)c + \text{arc } AKQ')\}.$$

Tangents to the spiral come in Propositions 13-20, the last three proving the property of the tangent. This is given in the form of theorems stating for different cases the length of the *subtangent* at any point P, that is to say, the distance between the origin and the point in which the tangent is intersected by the line drawn from O at right angles to the radius vector OP. Archimedes deals separately with the tangents at the end of the first turn, the second turn, &c., and then with the tangent at any point on the first and any subsequent turn.

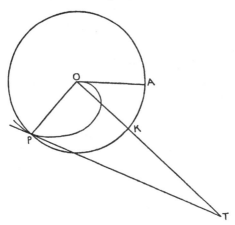

If P be a point on the first turn, O the origin, OA the initial line, and AKP the circle described with O as centre and OP as radius, and if OT drawn at right angles to OP meets the tangent at P in T, then

(subtangent OT) = (arc AKP).

If P is a point on any subsequent turn, say the nth, and AKP is the circle drawn with O as centre and OP as radius, and cutting the initial line at A, and if c be the circumference of this circle, then

(subtangent OT) = $(n-1)c +$ (arc AKP).

The rest of the book (Props. 21–8) is devoted to finding the areas of portions of the spiral and its several turns cut off by the initial line, or between two radii vectores. We will indicate the procedure in the general case which, as usual, amounts to a genuine integration.

Take $OB(=b)$ and $OC(=c)$, any two radii vectores. With centre O and radius OC describe a circle. Divide the angle BOC into any number of equal parts by radii OP, $OQ \ldots OY$, OZ meeting the spiral in P, $Q \ldots Y$, Z. OB, OP, $OQ \ldots OZ$, OC are then in arithmetical progression. Let the common difference be h, and suppose the number of radii to be n, so that the number of equal angles (and sectors of the circle) is $n-1$. Through B, P, $Q \ldots Y$, Z draw arcs of circles with O as centre as shown in the figure.

We thus have figures (1) inscribed, (2) circumscribed, to the figure bounded by OB, OC and the spiral, each consisting of the sum of a number of element-sectors of circles. By increasing n indefinitely we can compress the circumscribed and inscribed figures into coincidence with the area bounded by the spiral.

Each element-sector of the inscribed and circumscribed figures is compared with the corresponding sector of the large circle. The areas of the sectors are proportional to the squares on their radii.

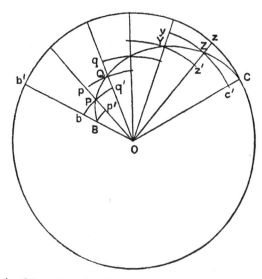

Thus Archimedes obtains

(whole sector $Ob'C$) : (circumscribed figure)
$$= (n-1)OC^2 : (OP^2 + OQ^2 + \ldots + OC^2),$$
and (whole sector $Ob'C$) : (inscribed figure)
$$= (n-1)OC^2 : (OB^2 + OP^2 + OQ^2 + \ldots + OZ^2).$$

Now $OB = b, OP = b+h, OQ = b+2h\ldots, OZ = b+(n-2)h$, $OC = b+(n-1)h$, and, by applying the result of Propositions 25, 26 as generalized, we see that the ratio of the sector $Ob'C$ to the circumscribed figure is less, and that of the sector to the inscribed figure greater, than the ratio of $OC^2 : \{OC \cdot O\widetilde{B} + \tfrac{1}{3}(OC-OB)^2\}$.

Hence, by the method of exhaustion,

(sector $Ob'C$) : (area of spiral OBC)
$$= OC^2 : \{OC \cdot OB + \tfrac{1}{3}(OC-OB)^2\}.$$

This is the equivalent of saying that, when h is diminished, and n increased, indefinitely but so that $(n-1)h$ remains equal to $c-b$,

$$\text{limit of } h\{b^2+(b+h)^2+(b+2h)^2+\ldots+(b+\overline{n-2}h)^2\}$$
$$= (c-b)\{cb+\tfrac{1}{3}(c-b)^2\}$$
$$= \tfrac{1}{3}(c^3-b^3),$$

that is, with our notation,

$$\int_b^c x^2\,dx = \tfrac{1}{3}(c^3-b^3).$$

As particular cases Archimedes finds:

(1) the area included between the first complete turn and the initial line, i.e. between the radii vectores 0 and $2\pi a$; the said area is to the circle with radius $2\pi a$ as $\tfrac{1}{3}(2\pi a)^2$ is to $(2\pi a)^2$, i.e. it is equal to one-third of the said circle (Prop. 24);

(2) the area bounded by the *second* turn and the initial line; this is to the circle with radius $4\pi a$ as $\{r_2 r_1+\tfrac{1}{3}(r_2-r_1)^2\}$ to $r_2{}^2$, where $r_2=4\pi a$, $r_1=2\pi a$, i.e. as $(8+\tfrac{4}{3})(\pi a)^2$ is to $16(\pi a)^2$, or as 7 to 12.

If R_1 is the area of the first turn bounded by the initial line, R_2 the area of the ring added by the second complete turn, and so on, then (Prop. 27)

$$R_3 = 2R_2,\ R_4 = 3R_2,\ R_5 = 4R_2\ldots,\ R_n = (n-1)R_2,$$

while $R_2 = 6R_1$.

Lastly (Prop. 28), if OBC be a spiral and circles be described with O as centre and OB, OC as radii, the radii OB, OC meeting the circles with OC, OB as radii respectively in b, c, and if E, F be the areas of the parts into which the arc BC of the spiral divides the area $bBcC$, then

$$E:F = \{OB+\tfrac{2}{3}(OC-OB)\} : \{OB+\tfrac{1}{3}(OC-OB)\}.$$

In the work on *Plane Equilibriums* Archimedes establishes the fundamental theorems of mechanics on the basis

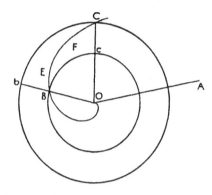

of a few simple postulates suggested, no doubt, by common experience. His postulates are these. Equal weights at equal distances balance; equal weights at unequal distances do not balance but incline towards the weight which is at the greater distance. If, when weights at certain distances balance, something be added to one of the weights, equilibrium will not be maintained, but there will be inclination on the side of the weight to which the addition was made; similarly, if anything be taken away from one of the weights, there will be inclination on the side of that weight from which nothing was taken. When equal and similar plane figures coincide if placed on one another, their centres of gravity similarly coincide; and in figures which are unequal but similar the centres of gravity will be 'similarly situated'.

Simple propositions (1–5) proved by *reductio ad absurdum* lead up to the fundamental theorem, proved first for commensurable and then by *reductio ad absurdum* for incommensurable magnitudes that *Two magnitudes, whether commensurable or incommensurable, balance at distances reciprocally proportional to the magnitudes* (6, 7). The main problems of Book I follow, namely, the finding of the centres of gravity of a parallelogram (9, 10), a triangle (13, 14), and a parallel-trapezium (15). The correctness of the determination in the first two cases is confirmed by *reductio ad absurdum*; and alternative proofs of both cases are added, depending on the postulate that in similar figures the centres of gravity are 'similarly situated' (10, 11, 12, and 13 (2)). In the case of the trapezium the result is expressed thus: If AD, BC are the parallel sides (AD being the smaller), and EF is the straight line joining their middle points, the centre of gravity is at a point G on EF such that

$$GE : GF = (2BC + AD) : (2AD + BC).$$

Book II is entirely devoted to finding the centres of gravity (1) of any parabolic segment, (2) of a portion of it cut off by a chord parallel to the base. Archimedes uses for the purpose of the former problem a series of triangles inscribed to the segment 'in the well-known manner' (γνωρίμως, as he calls it). BB' being the chord which is the base of the segment, we inscribe the triangle having the same base and height; the common vertex A is the point of contact of the tangent parallel to BB'. The diameter AO bisects BB' at O. If triangles be similarly inscribed in the segments AQB, $AQ'B'$, the diameters through the vertices Q, Q' bisect the chords AB, AB' and, if produced to meet BB', will, with the diameter AO, divide it into four equal parts. If, again, triangles be inscribed 'in the well-known manner' in the segments cut off by AQ, QB, AQ', $Q'B'$, the diameters through the vertices R, P, P', R' will, with the diameters through Q, Q', A, divide BB' into eight equal parts, and so on; and Archimedes easily proves that, if PP', QQ', RR' meet the diameter AO in L, M, N, the parts AL, LM, MN, NO are in the ratio of the odd numbers 1, 3, 5, 7. If the number of sides in the polygon

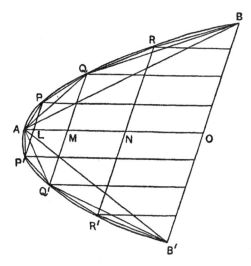

so inscribed in the segment be increased to any extent, the segments of AO are always in the ratio 1 : 3 : 5 : 7, &c. The centre of gravity of the inscribed figure necessarily lies on OA (Prop. 2), and if there be two parabolic segments, and two figures inscribed 'in the well-known manner' and with the same number of sides, the centres of gravity divide the diameters in the same ratio, since the ratio depends on the same ratio of odd numbers 1 : 3 : 5 : 7, &c.

The centre of gravity of the segment is nearer to A than the centre of gravity of the inscribed figure is: but the interval between the two can be made as small as we please by increasing sufficiently the number of the sides of the inscribed figure (Props. 5, 6). When we inscribe in any segment the triangle with the same base and height, the triangle is greater than half the segment; hence each time that we double the number of sides (other than the base) in the inscribed figure, we take away more than half of the area of the segments remaining over. Corresponding segments on opposite sides of the diameter AO, e.g. AQB, $AQ'B'$ have their diameters equal and are therefore equal in area (Prop. 5). Lastly (Prop. 7), if there be two parabolic segments, their centres of gravity divide their diameters in the same ratio (this is enunciated of similar segments only, but is true of any two segments and is so assumed in Proposition 8, where the centre of gravity of any segment is found).

Let H, H' be the centres of gravity of the segments AQB, $AQ'B'$, and G that of the whole segment ABB'. Therefore
$$QH : QD = Q'H' : Q'D' = AG : AO = m : 1, \text{ say.}$$

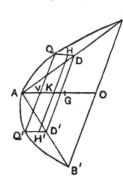

Let HH' (which is parallel to QQ' and DD') meet the diameter in K, so that K is the centre of gravity of the two small segments taken together.

Since the segment
$$ABB' = \tfrac{4}{3}(\triangle ABB'),$$
the sum of the two segments
$$AQB,\ AQ'B' = \tfrac{1}{3}(\triangle ABB').$$
Now $AO = 4AV$,
and $QD = \tfrac{1}{2}AO - AV = AV$.

Therefore $VK = QH = m \cdot QD = m \cdot AV$.

Equate the moments about A of (1) the segment ABB', and (2) the triangle ABB' and the two segments AQB, $AQ'B'$ taken together.

Giving each of the areas its value in terms of $\triangle ABB'$, we can divide out by $\triangle ABB'$, and, if we regard AV as of unit length, we have
$$\tfrac{4}{3} \cdot 4m = \tfrac{1}{3}(1+m) + \tfrac{2}{3} \cdot 4,$$

whence $15m = 9$, and $m = \frac{3}{5}$.

That is, $AG = \frac{3}{5}AO$, or $AG:GO = 3:2$.

Proposition 10 finds the centre of gravity of the portion of a parabola between two parallel chords. This area being the difference between two segments, the problem is easy to work out with the help of our algebraical notation. Archimedes can only use geometry, and it is difficult enough to express the result in geometrical language but more difficult still to prove it. Archimedes uses for this purpose a remarkable Lemma (Prop. 9). If a, b, c, d, x, y be straight lines satisfying the conditions

$$\frac{a}{b} = \frac{b}{c} = \frac{c}{d} (a > b > c > d),$$

$$\frac{d}{a-d} = \frac{x}{\frac{3}{5}(a-c)},$$

and
$$\frac{2a+4b+6c+3d}{5a+10b+10c+5d} = \frac{y}{a-c},$$

then
$$x+y = \frac{2}{5}a.$$

The proof amounts to eliminating, by geometry, three quantities b, c, d from the above four equations.

The *Sand-reckoner* (*Psammites* or *Arenarius*) is a *tour de force* of another kind. In it Archimedes undertakes to prove that, notwithstanding the inadequacy of the ordinary Greek numerical notation, it is possible to express in language enormous numbers, as large as we are likely to have any use for, as, for instance, a number as large as the number of grains of sand which the universe, on any reasonable view of the size to be attributed to it, could hold. The system for expressing such numbers had already been explained in a work now lost, Ἀρχαί, *Principles*, addressed to Zeuxippus, and is again set out in this treatise (for an account of it see pp. 19–20 above). In order to effect his object, Archimedes has to lay down what he thinks a reasonable maximum limit to the size of the universe, and this leads him to cite certain results obtained by his predecessors. It is to this fortunate circumstance that we owe our knowledge that Aristarchus of Samos, in a book of *Hypotheses*, put forward a heliocentric explanation of the universe. Archimedes quotes earlier estimates of sizes and distances; Eudoxus, he says, made the diameter of the sun nine times that of the moon, his own father Phidias twelve times, and Aristarchus between eighteen and twenty times. The perimeter of the earth had also to be assumed; some, he says, had tried to prove that it was about 300,000 stades. In order to be on the safe side, he will assume it to be ten times as much (3,000,000 stades), and the ratio of the sun's diameter to that of the moon to be 30:1, but not more. The earth's diameter he assumes to be greater than that of the moon, but less than that of the sun. Aristarchus, he says, had discovered the angular diameter of the sun to be 1/720th part of a great circle (i.e. half a degree); he himself, by means of a rough instrument consisting of a small cylinder or disk movable along a straight stick and remaining always at right angles to it, estimated the said angular diameter at the moment when the sun has just risen to be between $\frac{1}{164}$th and $\frac{1}{200}$th of a right angle. Aristarchus had assumed that the size of the circle in which the earth revolves about the sun is negligible in comparison with the distance of the fixed stars. This will not do for Archimedes' argument; so he suggests that a reasonable view would be to take the sphere of the fixed stars to bear the same ratio to the sphere with radius equal to the distance of the sun from

THE SAND-RECKONER

the earth as that sphere bears to the earth itself. But since, on this view, the size of the earth is not negligible in comparison with the sun's circle, he has to allow for parallax and to find limits to the angle subtended by the sun *at the centre of the earth* instead of at the point on its surface from which the observer sees the sun when it has just risen. This he does by clever geometry very much after the manner of Aristarchus' work *On the sizes and distances of the Sun and Moon*, and proves that the diameter of the sun is greater than the side of a regular polygon of 812 sides, and *a fortiori* greater than the side of a regular polygon of 1,000 sides, inscribed in the circle in which the sun revolves about the centre of the earth.

Now the diameter of the sun's orbit, being less than one-third of the perimeter of the chiliagon inscribed in it, is *a fortiori* less than one-third of 1,000 times the sun's diameter and therefore again *a fortiori* less than one-third of 30,000 times the earth's diameter, that is to say, less than 10,000,000,000 stades. But, on our assumptions, the diameter of the universe is in the same ratio to the diameter of the sun's orbit as the diameter of the sun's orbit is to the diameter of the earth; therefore the content of the universe, or the sphere of the fixed stars, is less than $10,000^3$ times the sphere in which the sun's orbit is a great circle.

Archimedes has now to work up to this from the size of a grain of sand. He assumes that a poppy seed is not greater in content than 10,000 grains of sand, and that the diameter of a poppy seed is not less than 1/40th of a fingerbreadth. Therefore a sphere of diameter one fingerbreadth is not greater than 64,000 poppy seeds and contains not more than 640,000,000 grains of sand ('6 units of the *second order plus* 40,000,000 of the *first order*') and *a fortiori* not more than 1,000,000,000 ('10 units of *second order* of numbers'). Gradually increasing the diameter of the sphere by multiplying it each time by 100 (making the sphere 1,000,000 times larger in content each time), and substituting for 10,000 finger-breadths one stade ($<$ 10,000 finger-breadths), Archimedes finds the number of grains of sand in a sphere of diameter 10,000,000,000 stades to be less than '1,000 units of the *seventh order* of numbers' (or 10^{51}), and the number in a sphere $10,000^3$ times this size to be less than '10,000,000 units of the *eighth order* of numbers (or 10^{63})'.

The *Quadrature of a Parabola* has a preface addressed to Dositheus after the death of Conon, in which Archimedes claims originality for his quadrature of any segment of a parabola, observing that he first discovered it by means of mechanics and then confirmed it by geometry, using the lemma already mentioned (the 'Axiom of Archimedes'). The treatise contains two solutions, the first of which corresponds to this description. The mechanical solution does not differ essentially from that given in the *Method* (see pp. 289–90 above), but to a Greek it would seem more orthodox because Archimedes does not here, as in the *Method*, take an infinity of parallel *lines* as making up the segment, but inscribes and circumscribes *trapezia* proportional to the straight lines, and then, by increasing indefinitely the number of trapezia and diminishing indefinitely their breadth, he compresses the inscribed and circumscribed figures formed by them into coincidence with the segment and with each other. The procedure is, as in other cases, equivalent to integration, and the area of the segment is found in a form corresponding to

$$\frac{1}{A^2}\int_0^A X^2\, dX = \tfrac{1}{3}A,$$

where A is the area of a triangle in the figure which is equal to four times the triangle with the same base and height as those of the segment.

The second method is purely geometrical and simply 'exhausts' the area of the segment by inscribing successive sets of triangles 'in the well-known manner' (cf. pp. 324-5 above). For this purpose it is necessary to find, in terms of the area of the inscribed triangle with the same base and height as the segment has, the area added to the inscribed figure each time that the sides of it (other than the base) are doubled in number.

Let QPq be the triangle inscribed 'in the well-known manner' in the segment QPq, PV the diameter bisecting Qq. If QV, Vq be bisected at M, m, and MR, mr be drawn parallel to VP to meet the curve in R, r, the latter points are the vertices of the next triangles inscribed in the same manner, namely PRQ, Prq, for RY, ry are diameters bisecting QP, Pq respectively.

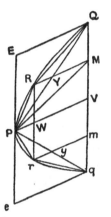

Now $QV^2 = 4RW^2$, so that $PV = 4PW$, and $RM = 3PW$.

But $YM = \frac{1}{2}PV = 2PW$, so that
$$YM = 2RY.$$

Therefore $\triangle PRQ = \frac{1}{2}\triangle PQM = \frac{1}{4}\triangle PQV$.

Similarly $\triangle Prq = \frac{1}{4}\triangle PVq$; whence
$$\triangle PRQ + \triangle Prq = \frac{1}{4}(\triangle PQq). \quad \text{(Prop. 21.)}$$

In like manner it can be proved that the next addition of the same kind to the inscribed figure adds
$$\tfrac{1}{4}(\triangle PRQ + \triangle Prq)$$
to its area, and so on.

Therefore the area of the inscribed figure is
$$\{1+\tfrac{1}{4}+(\tfrac{1}{4})^2+ \ldots \} \cdot \triangle PQq. \quad \text{(Prop. 22.)}$$
Further, each addition to the inscribed figure is greater than half of the segments of the parabola left over immediately before such addition. For, if we draw the tangent at P and complete the parallelogram $QEeq$, the triangle QPq is half of this parallelogram and therefore greater than half the segment, and so on (Prop. 20).

Archimedes now sums any number of terms of the above series in this form. Given any number of areas A, B, C, $D \ldots Z$ of which A is the greatest and each is four times the next in order, then (Prop. 23)
$$A+B+C+ \ldots +Z+\tfrac{1}{3}Z = \tfrac{4}{3}A.$$
The algebraical equivalent is
$$1+\tfrac{1}{4}+(\tfrac{1}{4})^2+ \ldots +(\tfrac{1}{4})^{n-1} = \tfrac{4}{3}-\tfrac{1}{3}(\tfrac{1}{4})^{n-1}$$
$$= \frac{1-(\tfrac{1}{4})^n}{1-\tfrac{1}{4}}.$$

Archimedes, instead of taking the *limit* when n is indefinitely increased, proves by *reductio ad absurdum* that the area of the segment can neither be greater nor less than $\tfrac{4}{3}(\triangle PQq)$.

In *On Floating Bodies*, I, II, Archimedes lays down the foundations of hydrostatics, premising two Postulates only. The first of these is that 'a fluid at rest is of such a nature that, of the parts of it which lie evenly and are continuous, that which is pressed the less is driven along by that which is pressed the more; and each of its parts is pressed by the fluid which is perpendicularly above it except when the fluid is shut up in anything and pressed by something else'; the second, placed after Proposition 7, is that, 'of bodies which are borne upwards in a fluid, each is borne upwards along the perpendicular drawn through its centre of gravity'.

Proposition 2 boldly states and proves that '*the surface of any fluid at rest is a sphere the centre of which is the same as that of the earth*', and in the whole of Book I the diagrams show the surface of the fluid as spherical. It is next proved that a solid which, size for size, is of equal weight with a fluid will, if let down into it, sink till it is just covered but not lower (3); a solid lighter than the fluid will, if let down into it, be only partially immersed, and in fact just so far that the weight of the solid is equal to the weight of the fluid displaced (4, 5), and, if it is forcibly immersed, it will be driven upwards by a force equal to the difference between its weight and that of the fluid displaced (6). The important proposition (7) follows that a solid heavier than a fluid will, if placed in it, sink to the bottom of the fluid, and the solid will, when weighed in the fluid, be lighter than its true weight by the weight of the fluid displaced.

The last-cited proposition gives a method of solving the Problem of the Crown, the solution of which (according to the story) occurred to Archimedes in the bath so that in his excitement he ran home naked shouting 'Εὕρηκα, εὕρηκα.' The crown, made for Hieron, was believed to contain some silver as well as gold, and the problem was to determine the actual proportions of each. If W be the weight of the crown, w_1 and w_2 the weights of gold and silver contained in it, $W = w_1 + w_2$. The method is to take the crown itself, an equal weight of pure gold, and an equal weight of pure silver, and to weigh them separately in the fluid, noting the loss of weight in each case, which of course represents the weight of the fluid displaced by the object weighed.

If the weight W of gold loses F_1 in weight, a weight w_1 of gold would lose $\frac{w_1}{W} \cdot F_1$.

If the weight W of silver loses F_2 in weight, a weight w_2 of silver would lose $\frac{w_2}{W} \cdot F_2$.

If the crown itself loses F in weight, F is equal to the loss in weight by w_1 of gold and w_2 of silver, that is,

$$\frac{w_1}{W} F_1 + \frac{w_2}{W} F_2 = F,$$

whence $\qquad w_1 F_1 + w_2 F_2 = (w_1 + w_2) F,$

and $\qquad \dfrac{w_1}{w_2} = \dfrac{F_2 - F}{F - F_1}.$

According to Vitruvius, however, Archimedes used a slightly different method and, on the whole, having regard to the story of the bath, this seems more likely to have been the way in which he discovered the solution. The method is simpler in that we only need to know the volumes of the fluid displaced by the crown, an equal weight of gold and an equal weight of silver respectively. If, as before, W is the weight of the crown, and w_1, w_2 the weights of gold and silver contained in it respectively, suppose that a weight W of gold displaces a volume V_1 of the fluid; then w_1 of gold will displace a volume $\frac{w_1}{W} \cdot V_1$ of the fluid. Similarly, if a weight W of silver displaces a volume V_2 of the fluid, a weight w_2 of silver will displace a volume $\frac{w_2}{W} \cdot V_2$ of the fluid. If, then, the crown itself displaces a volume V of the fluid,

$$\frac{w_1}{W} \cdot V_1 + \frac{w_2}{W} \cdot V_2 = V,$$

whence $\quad w_1 V_1 + w_2 V_2 = V(w_1 + w_2),$

and $\quad \dfrac{w_1}{w_2} = \dfrac{V_2 - V}{V - V_1}.$

The last Propositions (8, 9) of Book I deal with the case of any segment of a sphere of substance lighter than a fluid and immersed in it in such a way that either (1) the curved surface is downwards and the base entirely outside the fluid, or (2) the curved surface is upwards and the base is entirely submerged. Archimedes proves that, in either case, 'if the figure be forced into such a position that the base of the segment touches the surface of the fluid (at one point), the figure will not remain inclined but will return to the upright position'; in other words, the upright position is one of stable equilibrium.

Book II is a *tour de force*, and must be read in full to be appreciated. It contains a full investigation of the conditions of rest and stability of a right segment of a paraboloid of revolution floating in a fluid (either way up, but so placed that the base is either entirely above or entirely below the surface of the fluid) for different values of what we call the specific gravity of the solid in relation to the fluid, and for different ratios between the length of the axis of the segment and the principal parameter of the generating parabola. I will quote only one of the results obtained, by way of example. Suppose that h is the length of the axis, or the height, of the segment, p the principal parameter of the generating parabola, and $s\,(<1)$ the ratio of the specific gravity of the solid to that of the fluid, then (Props. 8, 9), if $h > \tfrac{3}{4}p$ but $h/\tfrac{1}{2}p < 15/4$, and if s is less than $(h-\tfrac{3}{4}p)^2/h^2$ in the case where the curved surface is downwards, or greater than $\{h^2-(h-\tfrac{3}{4}p)^2\}/h^2$ in the case where the curved surface is upwards, the position of stability is one in which the axis of the segment is not vertical but is inclined to the surface of the fluid at a certain angle, the construction for which, as given by Archimedes, is equivalent to the solution of the following equation in θ,

$$\tfrac{1}{4}p \cot^2\theta = \tfrac{2}{3}(h-k) - \tfrac{1}{2}p,$$

where k is the axis of the segment of the paraboloid cut off by the surface of the fluid. This example will serve to show the advanced nature of the whole investigation.

The *Cattle-Problem* attributed to Archimedes is a difficult problem in indeterminate analysis. There are eight unknown quantities (the respective numbers of bulls and cows of four different colours) to be determined from seven simple equations coupled with two other conditions. If W, w be the number of white bulls and cows respectively and $(X, x), (Y, y), (Z, z)$ represent the numbers of bulls and cows of the other three colours, we are given the seven equations:

(I) $\quad W = (\tfrac{1}{2}+\tfrac{1}{3})X + Y,$
$\quad\quad\quad X = (\tfrac{1}{4}+\tfrac{1}{5})Z + Y,$
$\quad\quad\quad Z = (\tfrac{1}{6}+\tfrac{1}{7})W + Y,$
(II) $\quad w = (\tfrac{1}{3}+\tfrac{1}{4})(X+x),$
$\quad\quad\quad x = (\tfrac{1}{4}+\tfrac{1}{5})(Z+z),$
$\quad\quad\quad z = (\tfrac{1}{5}+\tfrac{1}{6})(Y+y),$
$\quad\quad\quad y = (\tfrac{1}{6}+\tfrac{1}{7})(W+w),$

with the further conditions that

$\quad\quad W + X = $ a square number,

and $\quad\quad Y + Z = $ a triangular number,

though an ambiguity in the text makes it just possible that $W + X$ need only be the product of two whole numbers,

CATTLE-PROBLEM. SEMI-REGULAR SOLIDS 337

instead of a square. If $W+X$ must be a square, this, with the rest of the conditions involves (as was shown by Amthor in the *Zeitschrift für Math. und Physik*, 1880) the solution of the 'Pellian' equation

$$t^2 - 4729494 u^2 = 1,$$

which leads to prodigious figures; one of the eight unknown quantities would alone have over 206,500 digits!

Pappus, in the part of his *Collection* devoted to the five regular solids, tells us that Archimedes discovered thirteen other polyhedra which are *semi-regular*, being contained by equilateral and equiangular, but not all similar, polygons. If, for short, we denote by $m_\alpha, n_\beta \ldots$ a polyhedron contained by m regular polygons of α sides, n regular polygons of β sides, &c., the thirteen Archimedean polyhedra, which we will call $P_1, P_2 \ldots P_{13}$, are as follows:

Figure with 8 faces: $P_1 \equiv (4_3, 4_6)$.
Figures with 14 faces: $P_2 \equiv (8_3, 6_4), P_3 \equiv (6_4, 8_6),$
$P_4 \equiv (8_3, 6_8).$
Figures with 26 faces: $P_5 \equiv (8_3, 18_4), P_6 \equiv (12_4, 8_6, 6_8).$
Figures with 32 faces: $P_7 \equiv (20_3, 12_5), P_8 \equiv (12_5, 20_6),$
$P_9 \equiv (20_3, 12_{10}).$
Figure with 38 faces: $P_{10} \equiv (32_3, 6_4).$
Figures with 62 faces: $P_{11} \equiv (20_3, 30_4, 12_5),$
$P_{12} \equiv (30_4, 20_6, 12_{10}).$
Figure with 92 faces: $P_{13} \equiv (80_3, 12_5).$

Kepler in his *Harmonice Mundi* showed how these figures can be obtained respectively. Most of them can be produced by truncating, symmetrically and uniformly all round, (1) the corners, (2) both the corners and the edges of the regular solids. P_2, P_3, P_4 (two of which are said to have been known to Plato) have been described above (pp. 176–7). Four of the semi-regular solids can be obtained from the cube, namely, P_2 and P_4 by the first kind of truncation, P_5 and P_6 by the second. Similarly, by cutting off corners only, we can obtain P_1 from the tetrahedron, P_2 and P_3 from the octahedron, P_7 and P_8 from the icosahedron, and P_7 and P_9 from the dodecahedron, while by suitably cutting off both corners and edges from the icosahedron we can obtain P_{11} and P_{12} respectively. P_{10} and P_{13} do not yield to this method; but they are reproduced in the annexed figures taken from Kepler. P_{10} is the *snub cube*, in which each solid angle is formed by the

 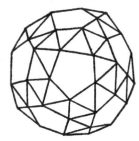

angles of four equilateral triangles and one square; P_{13} is the *snub dodecahedron*, each solid angle of which is formed by the angles of four equilateral triangles and one regular pentagon.

The *Liber Assumptorum*, which has come to us through the Arabic, contains some things which are so elegant that they may well go back to Archimedes, and notably those relating to two figures which had special names. The first of these is the ἄρβηλος, 'shoemaker's knife', a figure formed by three semicircles drawn with collinear and coterminous diameters as shown on p. 339. If PN is the straight line perpendicular to AB which touches both the smaller semicircles, it is proved that the area of the ἄρβηλος included between the three semicircles is equal to the circle on PN as diameter (Prop. 4). If we draw (1) a circle in the space

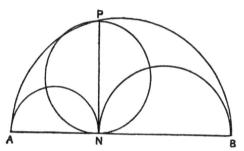

between the arcs AN, AP and the straight line PN and touching all three, (2) a circle on the other side of PN and similarly touching PN and the arcs PB, NB, the two circles are equal (Prop. 5). If one circle be drawn in the ἄρβηλος touching all three semicircles, and if the ratio of AN to NB be given, we can find the relation between the diameter of the circle so inscribed and AB; in particular, if $AN = \frac{3}{2}BN$, the diameter of the circle is $\frac{6}{19} AB$ (Prop. 6).

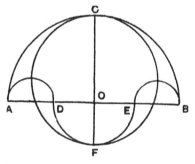

The second figure, called the *Salinon* (perhaps 'salt-cellar'), consists of four semicircles with their diameters in one straight line and drawn as in the annexed diagram, where $AD = EB$; and it is proved that, if the perpendicular to AB drawn through O, the centre of the semicircles ACB, DFE, meets those semicircles in C, F, the area included between the semicircles is equal to the circle on CF as diameter.

Proposition 8 has already been mentioned as being of interest in connexion with the problem of trisecting any angle (pp. 151-2).

The Arabian author of a book on the finding of chords in a circle, Abū'l Raihān al-Bīrūnī, mentions Archimedes as the author of solutions of two problems hitherto generally attributed to Heron, (1) the finding of the length of the perpendiculars of a triangle when the sides are given, (2) the finding of the area of a triangle in terms of the sides, namely as $\sqrt{\{s(s-a)(s-b)(s-c)\}}$, where s is half the sum of the sides. These things may have appeared in a work such as that 'On Triangles' attributed to Archimedes by the Arabs (cf. pp. 285-6 above).

The Arabian tradition that Archimedes also wrote a book *On the Heptagon in a Circle* may be said to have acquired more substance since a theorem on the subject handed down by Thābit b. Qurra has become known. This highly interesting theorem is to the following effect:

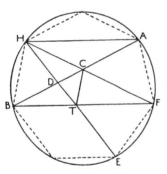

Given a straight line AB, we mark upon it two points C, D such that $\quad AD.CD = DB^2$
and $\quad CB.DB = AC^2$;

the segments AC, DB will both be greater than CD, so that we can construct on CD a triangle CHD such that $CH = AC$, $DH = DB$. About the triangle AHB circumscribe a circle, and produce HC, HD to meet this circle again in F, E. Let BF, HE intersect in T. Then C, T, B, H will be concyclic, the arcs BH, AF, FE will be equal, while the two remaining arcs AH, BE will be double of each of the latter arcs, so that the circle is divided into seven equal parts.

ON THE REGULAR HEPTAGON IN A CIRCLE

We can prove the result in this way.

Since $AD \cdot CD = DB^2$, $AD:DB = DB:CD$, or (since, by construction, $DB = DH$) $AD:DH = HD:DC$, and the triangles ADH, HDC are similar, so that $\angle CHD = \angle DAH = \angle AHC$ (since $CA = CH$). Hence the subtended arcs EF, BH, AF are all equal.

Again, since $CB \cdot DB = AC^2$, $CB:AC = AC:DB$, or (since $CA = CH$, and $DB = DH$) $BC:CH = CH:HD$.

But, since $\angle BCH = 2 \angle CAH$, and $\angle BTH = 2 \angle BFH$, it follows that $\angle BCH = \angle BTH$, and B, T, C, H are concyclic. And, since $BD = DH$, it follows that $CD = DT$, so that $HT = BC$.

Therefore, from above, $TH:HC = CH:HD$, and the triangles THC, CHD are similar.

Therefore $\angle HCD = \angle CTH = \angle HBA$ (in the same segment).

And $\angle HCD = 2 \angle BAH$, so that $\angle HBA = 2 \angle BAH$, and the arc HA is double of the arc BH. Also, since $DH = DB$, $\angle BHE = \angle HBA$, so that the arc BE is equal to the arc HA. If, then, we mark the middle points of the arcs HA, BE, we have a regular heptagon inscribed in the circle. Q.E.D.

When the author speaks of marking points C, D on AB such that $AD \cdot CD = DB^2$ and $CB \cdot DB = AC^2$, he does not, of course, imply that this is a *plane* problem.

Archimedes, in fact, according to our authority, reduced this auxiliary problem to a kind of νεῦσις solved by means of a ruler, without troubling to show how it might alternatively be solved by means of conics or otherwise.

The νεῦσις, which Archimedes apparently gave without any hint of how he arrived at it, is a remarkable example of geometrical ingenuity.

Let $ABDC$ be a square, with one side BA produced to H. 'We draw the diagonal BC' (said Archimedes);

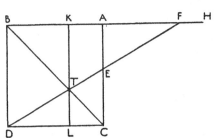

'then, bringing the end of the ruler to the point D, we direct the other end to a point F on AH so chosen that (if DF meets the diagonal BC in T) the triangle FAE shall be equal to the triangle CTD.

Through T draw KTL parallel to AC (meeting BA in K). Then I say that
$$AB \cdot KB = AF^2,$$
and $$FK \cdot AK = KB^2.$$'

Since $CD \cdot LT = AF \cdot AE$, and $LT = LC = AK$, it follows that $AF > AK$.

Again, since $CD \cdot LT = AF \cdot AE$, while $AB = CD$,
$$AB : AF = AE : LT.$$
But, since the triangles FAE, DLT are similar,
$$AE : LT = AF : LD = AF : BK;$$
therefore $$AB : AF = AF : BK,$$
or $$AB \cdot KB = AF^2. \quad \ldots \ldots \quad (1)$$

Again, since the triangles DLT, FKT are similar,
$$TL : KT = DL : KF;$$
i.e. $$AK : KB = KB : KF,$$
or $$AK \cdot FK = KB^2, \quad \ldots \ldots \quad (2)$$
while $$KB > AK. \quad\quad Q.E.D.$$

Thus BF is a line divided at K and A in the way in which, in the first figure, BA has to be divided at D, C.

ERATOSTHENES of Cyrene, who was very little younger than Archimedes, was a man of great distinction in all branches of knowledge, if not quite of the first rank in any; thus he was called *Beta* as well as πένταθλος (an all-round athlete). It was to Eratosthenes that Archimedes addressed the *Method*. When about forty years of age, Eratosthenes was invited by Ptolemy Euergetes to be tutor to his son (Philopator), and he became librarian at Alexandria; it was in recognition of his obligation to Ptolemy that he erected a column and inscribed upon it the epigram already mentioned relating to the solutions of the problem of the two mean proportionals including his own (cf. p. 154 above).

Eratosthenes wrote a book entitled Πλατωνικός, which evidently dealt with the fundamental ideas of mathematics in connexion with Plato's philosophy. It seems to have begun with the story of the problem of Delos, and to have treated of definitions in geometry and arithmetic, of proportions, and also of the principles of music.

Eratosthenes contributed to arithmetic by his device known as the *sieve* (κόσκινον) for finding prime numbers (cf. p. 63 above), and to geometry by his solution of the problem of finding any number of mean proportionals between two given straight lines and by an independent work *On Means*. The latter work was important enough to be included by Pappus in the so-called *Treasury of Analysis*; Pappus also alludes to certain loci which Eratosthenes called 'loci with reference to means', but the nature of these can only be conjectured.

Eratosthenes' most famous achievement, however, was his measurement of the earth. Archimedes tells us that 'some' had tried to prove that the circumference of the earth is about 300,000 stades. The origin of this was an observation that Lysimachia (on the Hellespont) and Syene were on the same meridian, and that two stars which were in the zenith at the two places respectively were at an angular distance of 1/15th of a complete circle; the distance between the two places being estimated at 20,000 stades, it was inferred that the whole circumference of the earth was 300,000 stades. Eratosthenes improved upon this estimate. Observing that at Syene, at the summer solstice, the sun cast no shadow from an upright gnomon, while at Alexandria (taken to be on the same meridian with Syene) the inclination of the sun's rays to the vertical was 1/50th of a complete circle, or of four right angles, he calculated the circumference of the earth from the known distance (5,000 stades) between Alexandria and Syene, making it accordingly 50 times 5,000, or 250,000, stades. Cleomedes gives this as the figure arrived at, while Theon of Smyrna and Strabo quote it as 252,000. If the latter figure is correct, Eratosthenes must have altered 250,000 into 252,000 for some reason, perhaps to make it divisible by 60. Pliny says that Eratosthenes made 40 stades equal to the Egyptian σχοῖνος, and, taking this at 12,000 Royal cubits of 0·525 metres, we obtain for the circumference of the earth 24,662 miles and for the diameter 7,850 miles, which is only 50 miles shorter than the true polar diameter.

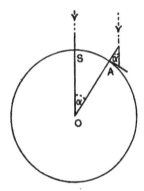

The above calculation was given in a separate work, *On the Measurement of the Earth*, which is said to have dealt with mathematical geography generally, including questions such as the size of the equator, the distance of the tropic and polar circles, the extent of the polar zone, as well as the sizes and distances of the sun and moon, eclipses, and changes in the length of the day according to the latitude and season. Eratosthenes is said to have estimated the distance of the moon from the earth at 780,000 stades, and that of the sun at 804,000,000 stades, and to have obtained these results by means of lunar eclipses.

Eratosthenes is supposed to have estimated the distance between the tropic circles, or twice the obliquity of the ecliptic, at 11/83rds of a great circle, i.e. 47° 42′ 39″; but the passage of Ptolemy mentioning this estimate in connexion with Eratosthenes is not quite conclusive as to whether Eratosthenes actually used this figure or was content with the value of 24° for the obliquity of the ecliptic discovered before Euclid's time.

An observation by Eratosthenes is recorded which was the basis of an estimate of the size of the sun made by Posidonius. Eratosthenes observed that at Syene (on the summer tropic), and throughout a circle round it with a radius of 300 stades, the sun throws no shadow at noon. Using this datum, and assuming that the circle of the sun about the earth is 10,000 times the size of a circular section of the earth, Posidonius arrived at 3,000,000 stades as the sun's diameter.

Eratosthenes wrote works, mythological and descriptive, on the constellations. One was a poem of which only fragments remain. Another apparently bore the title *Catasterismi*, 'placings of stars', or alternatively Ἀστρονομία or Ἀστροθεσίαι; the extant *Catasterismi* can, however, hardly be genuine in its present form.

Eratosthenes is also famous as the first to attempt a scientific chronology starting with the siege of Troy. Clement of Alexandria gives a short résumé of the main results of his Χρονογραφίαι, with which must be connected the separate 'Ολυμπιονῖκαι in several Books; both works were largely used by Apollodorus. The *Geographica* of Eratosthenes in three Books began with a history of geography down to his own time, and then treated of mathematical geography, the spherical form of the earth, the neglibility for this purpose of the unevennesses caused by mountains and valleys, changes of features due to floods, earthquakes, and the like. We gather from Theon of Smyrna that Eratosthenes' estimate of the height of the highest mountain was 10 stades, or about 1/8000th part of the diameter of the earth.

XII

CONIC SECTIONS: APOLLONIUS OF PERGA

THE discovery of the conic sections is attributed to Menaechmus, who is said to have used two of them, the parabola and the rectangular hyperbola, for solving the problem of the two mean proportionals (cf. pp. 158-61 above).

How Menaechmus came to think of producing curves by cutting a cone is uncertain. We learn on the authority of Geminus that the ancients knew of no other cones but right cones, but that they distinguished three kinds, namely those in which the vertical angle is acute, right, and obtuse, and produced the three conics as sections of the three kinds of cone respectively by planes which were in all cases at right angles to a generator. Hence the ellipse was called a 'section of an acute-angled cone', the parabola a 'section of a right-angled cone', and the hyperbola a 'section of an obtuse-angled cone', which names were used by Euclid and Archimedes, and persisted until Apollonius gave the curves the names by which we know them.

But the Greeks may easily have had their attention attracted, in the first instance, to the shape of the curve produced by cutting a right cylinder by a plane obliquely inclined to its axis, when this occurred in real life; they would then naturally try to investigate, and find geometrical expression for, the elongation of the curve so produced (an ellipse) in comparison with the form of the circle. The next step might be that they found that a curve of the same nature could be produced by cutting right through a right circular *cone*. Euclid actually says in his *Phaenomena* that, 'if a *cone* or a cylinder be cut by a plane not parallel to the base, the resulting section is a section of an acute-angled cone, which is similar to a θυρεός (shield)'. Then, when a right cone had been cut by a plane cutting all the generators, and the resulting curve had been investigated, it would be natural to inquire what sort of curves are produced if a right cone is cut by a plane which does not cut through it completely and is either parallel or not parallel to a generator.

If we actually take a right-angled cone and cut it by a plane at right angles to a generator, it is quite easy to show that the curve so produced has the property of the parabola used by Menaechmus, namely, that which we express by $PN^2 = p \cdot AN$, where p is the principal parameter.

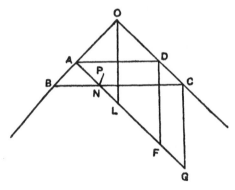

Let the section perpendicular to the generator OB cut the axial triangle BOC in AFG. If P be any point on the section, draw PN (the ordinate) perpendicular to AG. Draw BC through N at right angles to the axis OL of the cone, CG parallel to OL meeting AG in G, and AD parallel to BC.

Then P is on the circular section of the cone which has BC as a diameter.

Since OL bisects the right angle AOD, it bisects AD, BC.
Now $PN^2 = BN.NC$, by the property of the circle.
But, since B, A, C, G are concyclic,
$$BN.NC = AN.NG.$$
And $NG = AF = 2AL$;
therefore $PN^2 = 2AL.AN.$

The point A can be chosen so that $2AL$ is of any given length; thus a parabola can be drawn with principal parameter equal to any given straight line.

The particular *hyperbola* used by Menaechmus is a rectangular hyperbola, and it is remarkable that the property used is the fundamental *asymptote*-property. This suggests a deeper study of the curve than would be necessary in the case of the parabola, and it is less easy to suggest what Menaechmus' procedure would be in this case. The easiest way of obtaining a *rectangular* hyperbola would be to use the same right-angled cone, and to cut it by a plane parallel to the axis. The central property could be easily obtained in this way, and the asymptote-property could be deduced from it.

If the conics were really discovered by Menaechmus, whose date we must place about 360 or 350 B.C., the theory must have developed rapidly, since by the end of the century or thereabouts there were in existence two works on the subject considerable enough to be included by Pappus in his list of treatises belonging to the *Treasury of Analysis*. Euclid wrote a systematic treatise on *Conics* in four Books, the scope of which was much the same as that of Apollonius' first four Books, though the treatment was less general; and Euclid's work was preceded by the 'five Books of *Solid Loci* connected with the conics' written by Aristaeus. If we may trust the text of Pappus,

Euclid based himself, in places, on the work of Aristaeus; in particular, in dealing with what Apollonius calls the 'locus with respect to three or four lines', Euclid (so we are told) wrote so much about the locus as was possible by means of the conics of Aristaeus, without claiming completeness for his demonstrations. The title of Aristaeus' work, *Solid Loci*, requires a word of explanation. The Greeks (cf. p. 139 above) distinguished three classes of loci corresponding to the three kinds of problems which they called 'plane', 'solid', and 'linear' respectively. (1) 'Plane loci' are not merely loci in a plane; they are actually only straight lines and circles; (2) 'Solid loci' are conics pure and simple, the idea being that they arise from plane sections of solid figures, though not solid figures in general but only cones and cylinders; finally (3) 'Linear loci' are curves of degree higher than conics, e.g. the *quadratrix* or the conchoid. We may take it, therefore, that Aristaeus' work was on loci which proved, on investigation, to be conics, and included some discussion of the 'three- and four-line locus'.

Euclid's *Conics* and Aristaeus' *Solid Loci* were no doubt lost because they were immediately superseded by Apollonius' *Conics*, just as Euclid's *Elements* at once superseded all earlier Elements.

It is curious that the focus-directrix property of the conics does not appear in Apollonius' *Conics*. Apollonius obtains the foci of central conics, and gives focal properties of the ellipse and hyperbola, but makes no mention of the directrix, or of the focus of a *parabola*. Euclid's *Conics* would, therefore, probably not contain the focus-directrix property. As we have seen, however, Pappus gives a complete statement and proof of it for all three conics in a lemma to another work of Euclid's, the *Surface-Loci*. It is a fair inference that in that treatise Euclid had assumed the property as well known. Perhaps it had a place in Aristaeus' *Solid Loci*, to which it would be appropriate.

We may assume Euclid's *Conics* to have contained most, if not all, of the propositions in conics assumed by Archimedes as known or 'proved' (as he says) 'in the Elements of Conics'. It is evident that Euclid's work contained the ordinary chord and tangent properties of all three curves. The fundamental property was the ordinate-property referred to the principal axes. In the case of the hyperbola only a single branch was considered. The hyperbola, therefore, was not regarded as having a centre; what we call the centre was the point of intersection of the asymptotes, which Archimedes called 'the nearest lines to the section of an obtuse-angled cone', and the half of the transverse axis (our CA) was the 'line adjacent to the axis'. The constant ratio $PN^2 : AN . A'N$ was, in the case of the hyperbola, not equated to $CB^2 : CA^2$, as it was in the case of the ellipse, and a similar remark applies to the constant ratio $QV^2 : PV . P'V$. The main asymptote-properties were included, namely, (1) that the portion intercepted between the asymptotes of any tangent to a hyperbola is bisected at the point of contact, and (2) that, if x, y are straight lines drawn from any point on a hyperbola in fixed directions to meet the asymptotes, the rectangle xy is constant.

The chord and tangent properties included the property of the rectangles contained by the segments of two intersecting chords, namely, that they are in the ratio of TP^2 to TP'^2, where TP, TP' are tangents parallel to the chords respectively and intersecting in T.

Archimedes' investigations compelled him to specialize in properties of the parabola. He has, therefore, himself to prove certain propositions which would not come into an ordinary treatise on conics. But he assumes as known: (1) the property of the subnormal; (2) the proposition that, if the diameter PV bisects a chord QQ' in V and the tangent at Q meets the diameter in T, $PV = PT$; (3) a theorem by no means easy to prove, namely that, if the diameter PV bisects the chord QQ' in V, and QD is drawn perpendicular to PV, then

$$QV^2 : QD^2 = p : p_a,$$

where p is the parameter of the ordinates to PV, and p_a the parameter of the principal ordinates.

Although Euclid and Archimedes no doubt still regarded the conics as 'sections of a right-angled, acute-angled, and obtuse-angled cone' respectively, they knew that at all events the ellipse could be produced as a section of other than acute-angled cones or of cylinders, right or oblique. But Apollonius was the first to produce all three conics, including the double-branch hyperbola, with complete generality from any oblique circular cone, and to give them the new names *parabola, ellipse, hyperbola*, adapted from the form of the fundamental property of each as obtained by so generating the curves respectively.

Apollonius of Perga, the 'great geometer', as he was called, studied with the successors of Euclid at Alexandria, and flourished in the reigns of Ptolemy Euergetes (247–221 B.C.) and Ptolemy Philopator (221–203 B.C.). He dedicated the fourth and following Books of his *Conics* to King Attalus I of Pergamum (241–197 B.C.). He would, therefore, presumably be some twenty-five years younger than Archimedes. The *Conics* was at once recognized as the authoritative treatise on the subject, and superseded all others. Serenus and (so we are told by Suidas) Hypatia wrote commentaries on it; Eutocius (fl. about A.D. 520) prepared an edition of the first four Books, and added a commentary which survives and is included in Heiberg's Greek text of the four Books. Only those Books survive in Greek; but of the original eight Books seven exist in Arabic, the eighth being lost. The first four were translated from Eutocius' edition by Hilāl b. Abī Hilāl al-Ḥimṣī (died 883/4), and Books V–VII by Thābit b. Qurra (826–901) from some other MS. The first important edition of Books I–IV is the Latin translation by Commandinus (1566), which included the lemmas of Pappus and the commentary of Eutocius. Books V–VII were first published (1661) in a Latin translation by Abraham Echellensis and Giovanni Alfonso Borelli from a reproduction of the Books written in 983 by Abū'l Fatḥ al-Iṣfahānī.

The *editio princeps* of the Greek text is the monumental edition by Halley (Oxford, 1710). Books I–IV were to have been edited in Greek and Latin, with Eutocius' commentary, by Gregory, but Gregory died while the work was proceeding, and Halley took over the responsibility for the whole undertaking, adding a translation of Books V–VII from the Arabic and an attempted restoration of Book VIII. Heiberg's definitive text of Books I–IV (Teubner, 1891–3) contains, besides Eutocius' commentary, the fragments of Apollonius, &c.

According to the prefaces of Apollonius himself, the first four Books form an elementary introduction, the main new features being that the treatment was more general than in earlier works, and that the third Book contained a number of new propositions 'useful for the synthesis of solid loci and for *diorismi* (determinations of limits of possibility)', which also enabled the theory of the conic as the 'locus with respect to three or four lines' to be worked out more fully than it was in Euclid.

Books V–VII contain more specialized, and (so far as Book V is concerned) much more difficult, investigations. Book VI deals mainly with equal and similar conics and segments of such, and Book VII mainly with what we may describe as the values of certain linear and quadratic functions of conjugate diameters, including their variations and maximum and minimum values, I mean such functions as $PP' \pm DD'$, $PP'^2 \pm DD'^2$, and $PP' \pm p$, $PP'^2 \pm p^2$, where p is the parameter to the ordinates of PP'. VII. 31 in particular gives the important theorem that, if PP', DD' be conjugate diameters in an ellipse or in conjugate hyperbolas, and if the tangents at their extremities form the parallelogram $LL'MM'$, then

parallelogram $LL'MM'$ = rectangle $AA' \cdot BB'$.

Book V is a wonderful *tour de force*. The subject is in general what we call *normals* and their construction, but normals regarded, not as perpendiculars to tangents, but as maximum and minimum straight lines drawn to the curve from points inside or outside the conic. Apollonius of course gives the property of the subnormals in each of the curves; he discusses the number of normals that can be drawn to the conic from any internal or external point; and, most remarkable of all, he gives two difficult propositions (V. 51, 52) which lead directly to the determination of the *evolute* of a conic. The latter propositions naturally contain nothing but pure geometry; nevertheless, from the results obtained it is easy to deduce the equations of the evolute of a parabola and a central conic respectively, namely $\qquad 27ay^2 = 4(x-2a)^3,$

and $\qquad (ax)^{\frac{2}{3}} \mp (by)^{\frac{2}{3}} = (a^2 \pm b^2)^{\frac{2}{3}}.$

By way of detail as regards Books I–IV forming the text-book portion, as it were, of the treatise, it must suffice here to show the way in which Apollonius produces the three conics and obtains their fundamental property in each case. This fundamental property is, as we shall see, equivalent to the equation of the conic in Cartesian coordinates referred to axes which are, in general, not the principal axes, but any diameter and the tangent to the conic at its extremity. First of all, Apollonius generates a (double) cone, in general oblique or (as Apollonius calls it) 'scalene', by taking a circle and any point outside its plane and then making an endless straight line move in such a way that, while always passing through the point, it passes successively through all the points of the circle.

Apollonius defines the *axis* of the cone as the straight line drawn from the vertex of the cone through the centre of the circular base, and he first proves that all sections of the cone parallel to that base are also circles, and that there is another set of circular sections parallel to one another and 'sub-contrary' to the first set. Next, says Apollonius, let ABC be any triangle through the axis, and let any section whatever cut the base in a straight line DE at right angles to BC. If, then, PM be the intersection of the axial triangle and the cutting plane, and if QQ' be any chord in the section parallel to DE and cutting PM in V, Apollonius proves that QQ' is bisected at V.

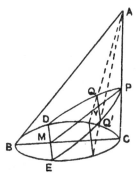

Therefore PM is a *diameter* of the section in accordance with his general definition of a diameter as a straight line drawn from the curve which bisects every one of a series of chords parallel to a given straight line, i.e. parallel to one another. Apollonius then explains that in general the chords QQ', though parallel to DE, which is at right angles to BC, will not be at right angles to PM, and will only be at right angles to PM if either (1) the cone is a right cone, or (2) the plane of the axial triangle is at right angles to the base of the cone. Thus Apollonius works out the properties of the sections with reference to a diameter the ordinates of which are in general obliquely inclined to it.

We will now take the three cases: (1) where the diameter PM is parallel to AC; (2) where PM meets CA *produced* in P', so that the cutting plane makes sections in both halves of the double cone; (3) where PM cuts AC in P'; and we have to prove that the three curves so produced have the properties of the parabola, hyperbola, and ellipse respectively. In cases (2) and (3) AF is drawn parallel to PM to meet BC in F.

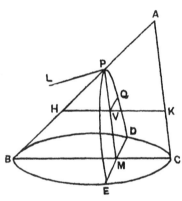

The property of the curve is in all three cases expressed by means of a certain straight line PL drawn in the plane of the section at right angles to PM.

In case (1) Apollonius takes PL perpendicular to PM and of length such that

$$PL : PA = BC^2 : BA \cdot AC,$$

and in cases (2) and (3) such that

$$PL : PP' = BF \cdot FC : AF^2.$$

In the latter two cases he joins $P'L$ and draws VR parallel to PL to meet $P'L$ or $P'L$ produced in R.

Draw HK through V parallel to BC. HK is then the diameter of a circular section parallel to the base.

Therefore $QV^2 = HV \cdot VK$.

Then (1) for the *parabola* we have, by similar triangles,
$$HV : PV = BC : CA,$$
$$VK : PA = BC : BA.$$

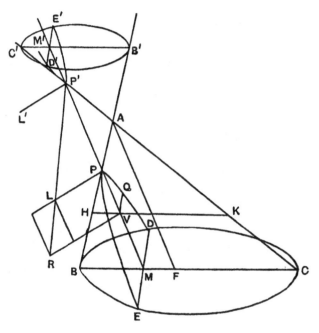

Therefore
$$QV^2 : PV \cdot PA = HV \cdot VK : PV \cdot PA$$
$$= BC^2 : BA \cdot AC$$
$$= PL : PA, \text{ by hypothesis,}$$
$$= PL \cdot PV : PV \cdot PA,$$
whence $QV^2 = PL \cdot PV$.

In the case (2) of the *hyperbola* and (3) of the *ellipse*
$$HV : PV = BF : FA,$$
$$VK : P'V = CF : FA.$$

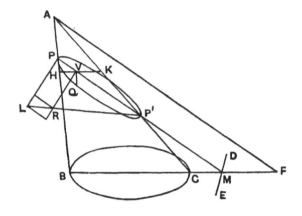

Therefore
$$QV^2 : PV . P'V = HV . VK : PV . P'V$$
$$= BF . FC : FA^2$$
$$= PL : PP', \text{ by hypothesis,}$$
$$= VR : P'V$$
$$= PV . VR : PV . P'V,$$
whence $\quad QV^2 = PV . VR.$

Accordingly (1), in the case of the *parabola*, the square on the ordinate (QV^2) is equal to the rectangle *applied* to PL with width equal to the abscissa PV.

(2) In the case of the *hyperbola* QV^2 is equal to the rectangle applied to PL with width equal to the abscissa but *exceeding* by the small rectangle LR, which is similar and similarly situated to the rectangle contained by PL, PP'.

(3) In the case of the ellipse the corresponding rectangle which is equal to QV^2 and is applied to PL *falls short* by the rectangle LR, which is similar and similarly situated to the rectangle contained by PL, PP'.

The properties are thus expressed in the precise language of the Pythagorean 'application of areas', and Apollonius named the three conics accordingly: *parabola* (παραβολή, *application*), *hyperbola* (ὑπερβολή, application with *excess*), *ellipse* (ἔλλειψις, application with *falling-short*).

Apollonius calls PL the *latus rectum* (ὀρθία, erect) or the *parameter of the ordinates* (ἡ παρ' ἥν δύνανται αἱ καταγόμεναι τεταγμένως, the line to which are applied the rectangles equal to the squares of the straight lines drawn ordinatewise); the *diameter* PP' of the central conic he calls ἡ πλαγία (*transverse*), and he often speaks of the diameter and parameter together as the *transverse side* (πλαγία πλευρά) and the *erect side* (ὀρθία πλευρά) of the *figure* (εἶδος).

If we call the parameter p, and the corresponding diameter d (in the case of the central conics), the properties of the curves, as stated by Apollonius, are the equivalent of the Cartesian equations referred to the diameter and the tangent at its extremity as axes (generally oblique), namely

$y^2 = px$ (the parabola)

$y^2 = px \pm \dfrac{p}{d}x^2$ (the hyperbola and ellipse respectively).

Apollonius was the first to treat fully the double-branch hyperbola as one curve like the ellipse; he calls it *the opposites*, αἱ ἀντικείμεναι (τομαί). A necessary preliminary is the proposition I. 14, where he proves that the opposite branches of a hyperbola have the same diameter and equal *latera recta* corresponding thereto. The properties of the single-branch hyperbola are commonly included with those of the ellipse and circle in one enunciation, which begins with the words 'If in a hyperbola, an ellipse or the circumference of a circle' . . .; and sometimes the 'opposites' are included with an ellipse: thus, 'If in an ellipse or the opposites a straight line be drawn through the centre meeting the section on both sides of the centre, it will be bisected at the centre' (I. 30).

Conjugate diameters in an ellipse are dealt with in I. 15 in relation to the diameter of reference PP' and its conjugate DD', and it is proved that the ellipse has the same property in relation to DD' and its parameter that it has in relation to PP' and its parameter PL, the parameter to DD' being DL' where $DL'.DD' = PP'^2$ (just as $PL.PP' = DD'^2$, since the constant ratio to which $QV^2 : PV.P'V$ is equal, namely $PL : PP'$, is equal to $DD'^2 : PP'^2$). Similarly I. 16 introduces the *secondary* diameter of the double-branch hyperbola, which is defined as that straight line drawn through the centre parallel to the ordinates of the transverse diameter which is bisected at the centre and is equal in length to the mean proportional between PP' and PL, i.e. $DD'^2 = PP'.PL$.

Next we have a series of propositions leading up to and containing the tangent properties. If the tangent at any point Q meets the diameter through P in T, and if QV is the ordinate to the diameter, then (1) for the parabola $PV = PT$ (I. 33, 35), and (2) for the central conics $TP : TP' = PV : VP'$ (I. 34, 36). The latter property is then used (I. 37, 39) to prove that $CV.CT = CP^2$ and $QV^2 : CV.VT = p : PP'$ [i.e. $CD^2 : CP^2$].

The section of Book I from Proposition 41 to Proposition 50 gives a series of theorems leading up to what amounts to a transformation of co-ordinates from the original diameter and the tangent at its extremity as axes to *any* diameter and the tangent at its extremity; Apollonius shows that, if *any* other diameter than the original one be taken, the property of the curve with reference to the new diameter and the corresponding parameter is of the same form as it was with reference to the original diameter and parameter. The principal *axes* only appear as particular cases; they are introduced for the first time in Propositions 52–8, where Apollonius shows how to construct each of the conics, having given (1) a diameter, (2) the length of the corresponding parameter, and (3) the inclination of the ordinates to the diameter. The construction in each case takes the form of finding the cone of which the required conic is a section. The angle between the diameter and the ordinates to it is first assumed to be a right angle; and then the case where the angle is oblique is reduced to that in which it is right. The problem of finding the axis of a parabola and the centre and axes of a central conic when the conic (and not merely the elements as here) is given comes later (II. 47).

The first section of Book II contains the asymptote-properties. The asymptotes are first constructed in this way (II. 1). Given any diameter of reference PP' with the corresponding parameter PL, and the angle of inclination between the diameter and the ordinates to it, Apollonius draws the tangent at P (which is of course parallel to the ordinates to PP'), and measures lengths PE, PE' along it in both directions such that

$$PE^2 = PE'^2 = \tfrac{1}{4}p \cdot PP' \; [= CD^2],$$

where p is the parameter; he then proves that, if, C being the centre, CE, CE' be joined and produced indefinitely they will not meet the curve in any finite point and are therefore *asymptotes*. II. 2 proves further that no straight line through C drawn within the angle between the asymptotes can be an asymptote. The asymptote-properties follow (II. 8–14). Opposite branches of a hyperbola and the 'conjugate opposites' have the same asymptotes (II. 15, 17). Propositions about conjugate hyperbolas are contained in II. 18–23; e.g. if a chord Qq in one branch of a hyperbola meet the asymptotes in R, r and the conjugate hyperbola in Q', q', then $Q'Q \cdot Qq' = 2CD^2$ (II. 23). The Book concludes with problems of drawing tangents to conics in various ways (II. 49–53).

Book III begins with a series of propositions leading up to the well-known theorems about the rectangles contained by the segments of intersecting chords and the harmonic properties of the pole and polar (III. 16–23 and 37–40, with 30–6 as special cases). Next we have propositions about intercepts made by two tangents on a third, e.g. (III. 41), if the tangents at three points of a parabola form a triangle, all three tangents will be cut by the points of contact into parts in the same ratio, that is to say, if the tangents at points P, Q, R form the triangle pqr,

$$Pr : rq = rQ : Qp = qp : pR,$$

a property from which it is not difficult to deduce the Cartesian equation of a parabola referred to two fixed tangents, namely

$$\left(\frac{x}{h}\right)^{\frac{1}{2}} + \left(\frac{y}{k}\right)^{\frac{1}{2}} = 1.$$

The focal properties of central conics (other than those which bring in the directrix) are proved in III. 45–52. The foci are called 'the points arising out of the application' (τὰ ἐκ τῆς παραβολῆς γινόμενα σημεῖα), the meaning of which is that S, S' are taken on the axis AA' such that $AS.SA' = AS'.S'A' = \frac{1}{4}p.AA'$ [$=CB^2$], i.e. in the phraseology of application of areas, rectangles are applied to AA' and $A'A$ equal to CB^2 and *exceeding* (in the case of the hyperbola) or *falling short* (in the case of the ellipse) by a square figure. There is no mention of the directrix of any conic or of the focus of a parabola.

Book III ends with propositions which are of use with reference to conics as 'loci with respect to three or four lines'. In particular, Propositions 54, 56 lead immediately to the 'three-line locus', from which it is easy to obtain the Cartesian equation of a conic with reference to two fixed tangents where the lengths of the tangents are h, k, namely

$$\frac{x}{h} + \frac{y}{k} - 1 = 2\lambda \left(\frac{xy}{hk}\right)^{\frac{1}{2}}.$$

Book IV is on the whole dull. Propositions 1–23 prove, for a large number of cases, the converse of propositions in Book III about the harmonic properties of pole and polar; the rest of the Book deals with conics which intersect or touch one another or both.

Of other works of Apollonius Pappus mentions six as having formed part of the *Treasury of Analysis*. Only one has survived, and that only in the Arabic; Halley published a Latin translation of it in 1706. The Greek title was λόγου ἀποτομή, '*Cutting-off of a ratio*', *Sectio Rationis*, and it dealt, in two Books, with the general problem, 'Given two straight lines, parallel or intersecting, and a fixed point on each line, to draw through a given point a straight line which shall cut off segments from each line (measured from the fixed points on them respectively) bearing a given ratio to one another.' The two Books discussed the various possible cases arising according to the relative positions of the given straight lines and points, with the conditions of possibility in certain cases. Book I discusses some particular cases, (1) in which the lines are parallel, (2) in which, the lines intersecting, one of the fixed points from which the segments are measured is at the intersection of the two lines. Book II deals with the general case and first shows how it can be reduced to the second of the foregoing cases, thus.

Let A, B be fixed points on the straight lines AC, BK respectively, and O the given point. It is required to draw through O a straight line, as ONM, cutting the straight lines AC, BK in points M, N respectively such that AM is to BN in a given ratio.

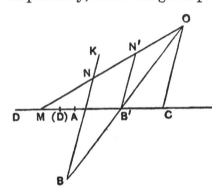

To reduce the problem, join OB meeting AC in B', and draw $B'N'$ parallel to BN meeting OM in N'.

Now the ratio $B'N' : BN$ is equal to the ratio $OB' : OB$, and is therefore constant. Therefore, since the ratio $BN : AM$ is given, the ratio $B'N' : AM$ is also given ($= \lambda$, say). The problem is, therefore, reduced to the case in which the two given straight lines are AC and $B'N'$, and the fixed point (B') on $B'N'$ is the point in which $B'N'$, AC intersect.

Apollonius proceeds by analysis. Suppose the problem solved, and $ON'M$ drawn through O so that $B'N' : AM = \lambda$.

Draw OC parallel to BN or $B'N'$. Take D on AM such that $OC : AD = \lambda = B'N' : AM$.

Then
$$AM : AD = B'N' : OC$$
$$= B'M : CM;$$

therefore $MD:AD = B'C:CM$,
or $CM.MD = AD.B'C$.

But AD, $B'C$ are given, by construction; therefore the rectangle $AD.B'C$ is given. Hence the problem is that of finding M such that $CM.MD$ is equal to a given rectangle, or, in other words, of *applying to CD a rectangle equal to a given rectangle and falling short by a square figure*.

Another treatise entitled *Cutting-off of an area* (χωρίου ἀποτομή), in two Books, dealt with the like problem in which the intercepted segments are required, instead of being in a given ratio, to contain a given rectangle. With the same figure as before, we have to draw ONM through O in such a way that the rectangle $AM.BN$ is equal to a given area. We reduce the problem, as before, to the case where one of the fixed points is B', the point of intersection of $B'N'$ and AC; and we have to draw $ON'M$ meeting $B'N'$ and AC in points N', M such that the rectangle $B'N'.AM$ is equal to a given area.

Take a point D on AM such that

$OC.AD =$ the given area

(D must in this case lie between A and M).

Suppose the problem solved; then,
since $B'N'.AM = OC.AD$,
$B'N':OC = AD:AM$.
But, by parallels, $B'N':OC = B'M:MC$;
therefore $AM:CM = AD:B'M$
$= MD:B'C$,
so that $B'M.MD = AD.B'C$.

But the rectangle $AD.B'C$ is given (by the construction); therefore the problem is here one of applying a rectangle equal to a given rectangle but *exceeding* by a square figure. The determination of the conditions of possibility in different cases (διορισμοί) would present no difficulty to Apollonius.

A third treatise was *On determinate section* (διωρισμένη τομή) in two Books. The general problem is, 'Given four points A, B, C, D on a given straight line, to determine another point (P) on it such that $AP.CP$ is to $BP.DP$ in a given ratio.' From Pappus' account of the contents of this work we may infer that the subject was very exhaustively discussed. To determine P by means of the equation

$$AP.CP = \lambda.BP.DP$$

is not in itself a difficult matter, because the problem can at once be put into the form of a quadratic equation, which the Greeks would have no difficulty in reducing to the usual *application of areas*. But from Pappus' statements about the separate cases dealt with we may conclude that the discussion included the determination of limits of possibility, the number of solutions, &c., and that the treatise amounted to a sort of Theory of Involution.

The treatise on *Contacts* or *Tangencies* (ἐπαφαί) in two Books is also described by Pappus. The following enunciation shows the scope of the work. *Given three things, each of which may be a point, a straight line, or a circle, to draw a circle which shall pass through the point or points (if such are given) and touch the straight lines or circles, as the case may be.* The three things given may be (1) three points, (2) three straight lines, (3) two points and a straight line, (4) two straight lines and a point, (5) two points and a circle, (6) two circles and a point, (7) two straight lines and a circle, (8) two circles and a straight line, (9) a point, a straight line, and a circle, (10) three circles. Two of these cases are treated in Eucl. IV; (3), (4), (5), (6), (8), (9) came in Book I of Apollonius' work, while (7), the case of two straight lines and a circle, and (10) that of three circles, occupied the whole of Book II.

The case of the three circles has exercised the ingenuity of many distinguished geometers, including Vieta and Newton. Vieta (1540-1603) set the problem to Adrianus Romanus (van Roomen, 1561-1615), who solved it by means of a hyperbola. Vieta rejoined with *Apollonius Gallus* (1600), in which he solved it by 'plane' methods, as did Newton afterwards in his *Arithmetica universalis* (Prob. xlvii). There is no doubt that Apollonius solved it by 'plane' methods, i.e. by means of straight lines and circles, and his solution can be restored with the aid of Pappus' lemmas. It is a highly characteristic example of Greek geometry and deserves to be reproduced here. Three preliminary propositions are required. (1) If two circles touch internally or externally, any straight line through the point of contact divides the circles into segments respectively similar. (2) Given three circles, their six centres of similitude (external and internal) lie three by three on four straight lines. Pappus uses the first of these propositions, and it is easily proved. The second must have been known to the Greeks. (3) is a problem solved by Pappus: Given a circle *ABC*, and given three points *D*, *E*, *F* in a straight line, to 'inflect (the broken line) *DAE* (to the circle) in such a way that *BC* may be in a straight line with *CF*', in other words, to inscribe in the circle a triangle such that the sides may pass respectively through three points lying in a straight line. The solution is a typical example of procedure by analysis and synthesis.

Suppose the problem solved, i.e. suppose DA, EA drawn to a point A on the circle and cutting it at two points B, C such that BC produced passes through F.

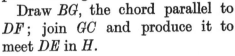

Draw BG, the chord parallel to DF; join GC and produce it to meet DE in H.

Then $\angle BAC = \angle BGC$
$= \angle CHF$, by parallels,
$=$ supplement of $\angle DHC$.

Therefore A, D, H, C are concyclic, and

rect. $DE \cdot EH =$ rect. $AE \cdot EC$.

And the rectangle $AE \cdot EC$ is given, being equal to the square on the tangent from E to the circle; therefore $DE \cdot EH$ is given, and, since DE is given, HE is given, and therefore H.

But F is also given; and the problem is therefore reduced to that of drawing HC, FC through a point C on the circle cutting the circle again at points B, G such that BG is parallel to HF.

Suppose *this* done, and draw BK, the tangent to the circle at B, meeting HF in K.

Then $\angle KBC = \angle BGC$, in the alternate segment,
$= \angle CHF$, by parallels.

But the triangles FBK, FHC also have the angle CFH common; therefore they are similar, and

$$CF : FH = KF : FB,$$
or $$HF \cdot FK = BF \cdot FC.$$

Now $BF \cdot FC$ is given, being equal to the square on the tangent from F to the circle. And HF is given; therefore FK is given, so that K is given.

The synthesis is as follows. We have first to find H and K. Take H on DE such that $DE.EH$ is equal to the square on the tangent from E to the circle; and take K on HF such that $HF.FK$ is equal to the square on the tangent from F to the circle.

Next draw from K a tangent to the circle, and let B be the point of contact. Join BF meeting the circle in C. Join HC, and produce it to meet the circle again in G.

We have now to prove, first, that BG is parallel to HF.

Since
$$HF.FK = BF.FC,$$
$$HF:FC = BF:FK;$$
therefore the triangles FHC, FBK are similar, and
$\angle CHF = \angle FBK = \angle BGC$, in the alternate segment; therefore BG, HF are parallel.

Now join EC and produce it to meet the circle at A. Join AB, BD; and it only remains to prove that AB, BD are in a straight line. This is easy, for $DE.EH = AE.EC$, so that A, D, H, C are concyclic and, if AD be joined, the angle DAC is equal to the angle CHF, while the angle BAC, being equal to the angle BGC, is also equal to the angle CHF; therefore the angles BAC, DAE are equal, and ABD is a straight line.

Take now the case of Apollonius' problem in which the required circle is to touch the three given circles externally, as shown in the figure overleaf. Let the radii of the circles be a, b, c and their centres A, B, C respectively. Let D, E, F be the three external centres of similitude, so that $BD:DC = b:c$, &c., and D, E, F are in a straight line.

Suppose the problem solved and the circle PQR drawn touching the given circles in P, Q, R. Join PQ and produce it both ways to meet again in K, L the circles with centres A and B respectively. Then, by the first of the proposi-

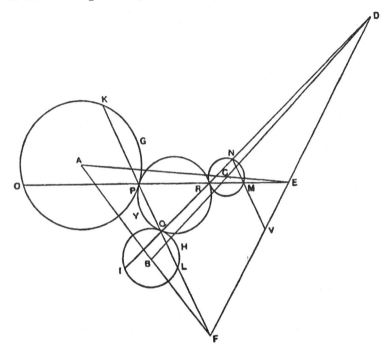

tions cited above, the segments of circles KGP, QHL are similar to the segment PYQ and therefore to one another. Therefore PQ produced beyond L passes through F. Similarly QR, PR produced pass through D, E respectively.

Let PE, QD meet the circle with centre C again in M, N. Then, the segments PQR, RNM being similar, the angles PQR, RNM are equal, so that NM is parallel to PF. Produce NM to meet FD in V.

Then $EV : EF = EM : EP = EC : EA = c : a$;

therefore EV, and therefore V, is given.

PLANE LOCI

The problem thus reduces itself to this: Given three points V, E, D in a straight line, and the circle with centre C, to draw DR, ER to a point R on the circle meeting the circle in two other points M, N such that NM may pass through V. This is Pappus' problem just solved.

Thus R is found, and ER, DR produced pass through P, Q, the other required points of contact, respectively.

Another treatise by Apollonius was on *Plane Loci* in two Books. Pappus described its contents in detail sufficient to enable three distinguished geometers to attempt restorations, namely Fermat, van Schooten, and Simson. 'Plane loci' are, as we have seen, straight lines and circles only; hence, as we should expect, the object of the treatise is to prove that various loci which we can denote in co-ordinate geometry either by simple equations or by certain quadratic equations are straight lines or circles respectively. Everything is, of course, expressed in geometrical language; hence the enunciations are often unwieldy. I will quote only one example:

If from any point straight lines be drawn to meet at given angles two straight lines either parallel or intersecting, and if the straight lines so drawn have a given ratio to one another, or if the sum of one of them and a line to which the other has a given ratio be given (in length), the point will lie on a straight line given in position.

This includes the equivalent of saying that, if x, y be the co-ordinates of the point, each of the equations $x = my$, and $x + my = c$ represents a straight line. Another involved enunciation is the equivalent of the statement that any linear equation between trilinear or multilinear co-ordinates represents a locus which is a straight line. In Book II there were propositions equivalent to saying that the equations $x^2 + y^2 = ax$ and $x^2 + y^2 = a(x-b)$ in Cartesian co-ordinates represent circles. Another is tantamount to saying that, if $A, B, C \ldots$ be any number of fixed points and $\alpha, \beta, \gamma \ldots$ any constants, the locus of a point P such that $\alpha \cdot AP^2 + \beta \cdot BP^2 + \gamma \cdot CP^2 + \ldots = $ a given area is a circle. The Book of course contained the theorem, quoted by Eutocius but already known to and quoted by Aristotle, that, if A, B be fixed points and AP and BP are in a given ratio, the locus of P is a straight line or a circle according as the given ratio is or is not one of equality.

We have had occasion to notice a type of problem called by the special name νεῦσις (*verging* or *inclinatio*). The object of such a problem is to place between two straight lines, or a straight line and a curve, or two curves, a straight line of given length in such a way that it *verges* towards a given point, i.e. will, if produced, pass through the point. Apollonius wrote a treatise in two Books on the classes of νεύσεις which can be solved by 'plane' methods. According to Pappus, the following were the cases dealt with.

I. Given (*a*) a semicircle and a straight line at right angles to the diameter, or (*b*) two semicircles with their bases in one straight line, to insert between them a straight line of given length verging to an angle of the semicircle or one of the semicircles respectively.

II. Given a rhombus with one side produced, to insert a straight line of given length in the exterior angle so that it verges to the opposite angle.

III. In a given circle to insert a chord of given length verging to a given point.

Book I of Apollonius' treatise contained four cases of I (*a*), two cases of (III), and two of (II); Book II contained ten cases of I (*b*).

ΝΕΥΣΙΣ IN RHOMBUS

Restorations were attempted by Marino Ghetaldi (*Apollonius Redivivus*, 1607, and *Apollonius Redivivus . . . Liber secundus*, 1613), Alexander Anderson (in *Supplementum Apollonii Redivivi*, 1612), and Samuel Horsley (Oxford, 1770), the last being the most complete.

The case of the rhombus is interesting and characteristic. A lemma by Pappus affords the key to Apollonius' solution.

He would naturally begin with analysis. Suppose the side AC of the rhombus $ACDB$ produced beyond C, and suppose KH, a straight line of given length ($=k$, say), inserted between CD and AC produced in such a way that BKH is a straight line.

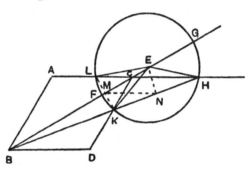

Bisect KH in N, and draw NE at right angles to KH meeting BC produced in E. Draw KM perpendicular to BC and produce it to meet AC in L, so that, by the property of the rhombus, $KM = ML$.

Then, since $KM = ML$, and $KN = NH$, MN is parallel to LH.

And, since the angles at M, N are right, M, K, N, E lie on a circle.

Therefore $\angle CEK = \angle MNK = \angle CHK$, so that C, K, H, E are also concyclic.

Therefore $\angle EHK =$ supplement of $\angle ECK = \angle DCB$, so that the isosceles triangles EKH, DCB are similar.

Lastly $\angle EBK = \angle EKH - \angle CEK = \angle EHK - \angle CHK$
$= \angle EHC = \angle EKC.$

The triangles EBK, EKC also have the angle BEK common; they are therefore similar, and
$$BE : EK = EK : EC,$$
or $$BE . EC = EK^2.$$

But, by the similar triangles EKH, DCB,
$$EK : KH = DC : CB,$$
and, the ratio $DC : CB$ being given, while KH is given in length $(= k)$, EK is given in length $(= p$, say).

The problem is therefore reduced to that of finding E on BC produced such that $BE . EC$ is equal to p^2.

The construction is therefore as follows:

Find a length p such that $p : k = DC : CB$.

Then apply to BC a rectangle $BE . EC$ equal to p^2 and exceeding by a square figure.

Lastly, with E as centre and radius equal to p, describe a circle meeting DC in K and AC produced in H.

Then HK is the straight line required; and it only remains to verify the fact, i.e. to prove that $HK = k$, and that HK, when produced, passes through B.

How to prove this is shown by Pappus in his lemma, which proves more generally that, if E be *any* point on BC produced, and F a point taken on BC such that $EF^2 = BE . EC$, and if a circle be described with E as centre and EF as radius cutting CD in K and AC produced in H, then B, K, H are in one straight line.

We use the same figure as above.

Since BCE bisects the angle ACD, and KL is perpendicular to BC, $CL = CK$, and $EL = EK$.

Therefore the triangles ECK, ECL are equal in all respects.

Therefore $\angle CKE = \angle CLE = \angle CHE$.

It follows that C, K, H, E are concyclic, so that $\angle KCE$ is supplementary to $\angle EHK$ and therefore to $\angle EKH$.

Now, by hypothesis, $BE \cdot EC = EF^2$,

whence $BE : EF = EF : EC$,

or $BE : EK = EK : EC$.

And the triangles EBK, EKC have the angle BEK common; therefore the triangles are similar and

$$\angle BKE = \angle KCE.$$

But $\angle KCE$ was proved supplementary to $\angle EKH$; therefore $\angle$s BKE, EKH are supplementary, and accordingly BKH is a straight line.

Lastly, since $\angle EKH = \angle ECH$, in the same segment,

$= \angle DBC$, by parallels,

the triangles DBC, EKH are similar,

whence $EK : KH = DC : CB$.

But, by Apollonius' construction, $EK = p$, where $p : k = DC : CB$.

Therefore $KH = k$.

We hear of yet other works by Apollonius which are lost.

1. In a *Comparison of the dodecahedron with the icosahedron* (inscribed in the same sphere) he proved that the perpendiculars from the centre of the circumscribing sphere to a pentagonal face of the dodecahedron and to a triangular face of the icosahedron are equal, so that the volumes of the solids are to one another as their surfaces.

2. A *General Treatise* alluded to by Marinus (on Euclid's *Data*) must have dealt in great measure with the funda-

mental principles of geometry, definitions, axioms, &c. To this work must be referred certain things quoted by Proclus, the elucidation of the notion of a line, a definition of an angle (plane and solid), and attempts to prove the axioms, as well as alternative (but scarcely better) constructions for the problems of Eucl. I. 10, 11, 23.

3. *On the Cochlias* or cylindrical helix.

4. *On unordered irrationals*, a work referred to in Pappus' commentary on Euclid's Book X as containing some extensions or generalizations of the theory of irrationals as expounded in that Book.

5. *On the burning-mirror.* In this work Apollonius is said to have discussed the spherical form of mirror among others. Anthemius of Tralles, the architect (died about A.D. 534), in a work on burning-mirrors of which we possess a fragment, deals with elliptic and parabolic mirrors as reflecting certain pencils of rays to a focus, and says that the ancients knew of the properties of mirrors in the shape of the conic sections. It is quite possible that Apollonius was one of 'the ancients' referred to, and that he knew of the focal properties of the parabola, though his *Conics* is silent on the subject.

6. The Ὠκυτόκιον ('quick-delivery') seems to have been a handbook teaching quick methods of calculation. In it Apollonius is said to have calculated a nearer, though less handy, approximation to the value of π than that made by Archimedes.

Lastly, Apollonius was a distinguished astronomer. We know from Ptolemy that he was a master of the theories of epicycles and eccentric circles as means of accounting for the movements of the planets.

XIII

THE SUCCESSORS OF THE GREAT GEOMETERS

WITH Apollonius the golden age of Greek geometry comes to an end. There remained details to be filled in, and no doubt geometers of the requisite calibre might have found in such a work, for instance, as Apollonius' *Conics* propositions containing the germs of theories which were capable of independent development. But further progress was practically barred by the restrictions of method and form which were inseparable from the classical Greek geometry. Want of notation such as our algebraical symbols provide was the greatest obstacle. As we have seen, the Greeks had only a *geometrical* algebra to work with, and this meant that they could not go beyond what could be represented in terms of geometry, that is, by lines, areas, or solid figures, supplemented by the unlimited use of proportions. Heron, in his proof that the area of a triangle is what we represent by $\sqrt{\{s(s-a)(s-b)(s-c)\}}$, felt it necessary to apologize for the use of a product of *four* linear factors, seeing that only three can be represented in a diagram. An irrational quantity could only be represented by a straight line, which again meant that a product of two irrationals must be represented by a *rectangle*, and so on.

Theoretical geometry being thus practically at the end of its resources, mathematicians could only occupy themselves (1) in multiplying solutions of problems which could be solved by means of conics, problems such as those which we can reduce to cubic equations; (2) in trying to discover new curves with a like object; (3) in pursuing the *applications* of geometry. (1) and (2) offered little scope, having regard to the restrictions of method. Under (3) comes *mensuration* in general, including the calculation of areas and volumes of figures of different shapes, and arithmetical approximations to the true values in cases where incommensurables such as surds and the ratio (π) of the circumference of a circle to its diameter are involved; but one special branch of mensuration, that of triangles, plane and spherical, in other words, trigonometry, was immediately required for astronomy. Mensuration in general is represented by Heron of Alexandria, trigonometry by Hipparchus, Menelaus, and Ptolemy.

We have, however, in this chapter to deal with the geometers of note who continued the true geometrical tradition.

The first is NICOMEDES, whose place is here, although he seems to have been intermediate in date between Eratosthenes and Apollonius. He is famous for his discovery of the conchoid already described (pp. 150–1), which gave a means of solving any νεῦσις where one of the lines between which an intercept of given length on a line 'verging' to a given point has to be placed is a straight line. The conchoid was used both for the finding of two mean proportionals and for trisecting any angle. As we have seen, its equation referred to polar co-ordinates is $\rho = a + b\sec\theta$. Nicomedes proved that the edge of the so-called 'ruler' in the instrument for constructing the curve is an asymptote, and that any straight line drawn in the space between the asymptote and the conchoid must, if produced, cut the curve.

Nicomedes is associated with DINOSTRATUS, the brother of Menaechmus, as having applied to the squaring of the circle the *quadratrix* of Hippias.

DIOCLES is known as the discoverer of the *cissoid* ('ivy-leaf-shaped'), which he used to solve the problem of the two mean proportionals (see pp. 168–9 above). He also solved the problem of Archimedes *On the Sphere and Cylinder*, II. 4, by means of the intersection of an ellipse and a rectangular hyperbola. Eutocius, who gives these solutions, tells us that they were contained in a work by Diocles *On burning-mirrors*. Its contents are otherwise unknown, but an Arabian writer attributes to Diocles the discovery of the property of a mirror having the form of a parabola. Ibn al-Haitham, on the other hand, who wrote a work on burning-mirrors which is still extant, and whose proofs are after the Greek model, does not mention Diocles but only Archimedes, Apollonius, and Anthemius; and he attributes to Apollonius the proof of the main property of the parabola required, namely, that the tangent at any point on the curve makes equal angles with the focal distance of the point and the diameter passing through the point. Very interesting in this connexion is the *Fragmentum mathematicum Bobiense*, which evidently came from some treatise on the parabolic form of mirror, and which contains a proof amounting to the following.

APR is a parabola with axis AN, P a point on the curve, PN the ordinate, PT the tangent at P meeting the axis in T, S a point on AN such that $AS = \tfrac{1}{4}AL$, where AL is the parameter.

Draw AY at right angles to AN meeting PT in Y, and join SY.

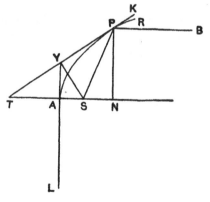

Now $PN^2 = AL \cdot AN$
$\qquad = 4AS \cdot AN$
$\qquad = 4AS \cdot AT.$

But (since $AN = AT$) $PN = 2AY$;
therefore $AY^2 = AS.AT$,
so that the angle TYS is a right angle.

And, since $TA = AN$, $TY = YP$;
therefore the triangles SYT, SYP are equal in all respects, and $SP = ST$, so that the angles STY, SPY are equal.

If then PB be the diameter through P,
$$\angle KPB = \angle STP = \angle SPT.$$

(The author is not content with this, but argues that the 'mixed' angles RPB, SPA are equal, since the rectilineal angles SPT, KPB are equal, and the parts of them, the mixed angles YPA, KPR made by the tangent with the curve on either side of P are equal. He also speaks of the curve as a 'section of a right-angled cone'.)

It was at one time conjectured that we might have here a fragment of Diocles' work, but it is hardly likely that Diocles, living a century after Apollonius, would have used the 'mixed' angles or spoken of a 'section of a right-angled cone' instead of a 'parabola'. The fragment would seem rather to belong to a date not later than Apollonius or even Archimedes.

PERSEUS is only known as the discoverer and investigator of the so-called *spiric sections*. His date is uncertain, but he was probably older than Apollonius. Our informant about him is Proclus, who distinguishes three sections of the *spire* (σπεῖρα). The *spire* in one of its forms was what we call a *tore* or anchor-ring. Imagine a circle and a straight line outside it but in its plane. Then, if the circle revolves bodily about the straight line, it describes a solid ring or *tore*. Archytas had already used such a solid in his solution of the problem of the two mean proportionals, namely, the solid produced by the revolution of a semicircle about the tangent at one extremity of its diameter, that is, the half of a tore with inner diameter *nil*. This latter kind of tore was called *continuous* (συνεχής); the kind first mentioned (a real anchor-ring) was called *open* (διεχής). Yet a third kind of *spire* was distinguished, namely, that in which the axis of revolution cuts the circle, so that the figure produced crosses itself, as it were, and is *interlaced* (ἐμπεπλεγμένη).

Suppose that we have an open tore, and let c be the distance of the centre of the generating circle from the axis of revolution, and a the radius of the circle, so that $c > a$.

The sections described by Proclus are made by planes parallel to the axis of revolution. Suppose the tore cut by such a plane at a distance d from the axis of revolution. Then there are the following cases:

(1) $c+a > d > c$. Here the curve of section is an oval.

(2) $d = c$: the limiting case of (1).

(3) $c > d > c-a$. The curve of section is now a closed curve narrowest in the middle.

(4) $d = c-a$. In this case the section touches the ring on the inner side, and the section is the *hippopede* (horse-fetter) in the shape of a figure of eight. The lemniscate of Bernoulli is a particular case of this curve, that namely in which $c = 2a$.

(5) $c-a > d > 0$. In this case the section consists of two ovals symmetrical with one another.

The three curves described by Proclus correspond to (1), (3), and (4).

In an epigram by which Perseus celebrated his discovery he spoke of 'three lines (i.e. curves) upon five sections' (ἐπὶ πέντε τομαῖς). Paul Tannery took the 'three lines' to be *additional* to the 'five sections' and to refer to sections of the third variety of tore, the interlaced or overlapping; this conjecture is, however, unconfirmed by any positive evidence.

According to Proclus, Perseus worked out the property of his curves just as Nicomedes did for the conchoid, Hippias for the quadratrix, and Apollonius for the three conics.

ZENODORUS wrote, at some date between (say) 200 B.C. and A.D. 90, a treatise Περὶ ἰσομέτρων σχημάτων, *On isometric figures*, by which expression he meant figures with equal perimeter or contour, but of different shapes. The comparison of the content of such figures would interest Greek mathematicians if only on account of popular misconceptions on the subject, examples of which are not infrequent in classical literature. Thus Thucydides estimates the size of Sicily by the time required for circumnavigating it. About 130 B.C. Polybius observed that there were people who could not understand that camps of equal periphery might be of different area. Pliny compared the size of different parts of the earth by adding their lengths to their breadths. On the theorems of Eucl. I. 36, 37, proving that triangles or parallelograms on equal bases and between the same parallels are equal, Proclus observes that to the ordinary person it seemed paradoxical that the area should be the same however long the sides other than the base might be. He mentions also that certain writers estimated the size of cities by their circumferences; and again that certain members of communistic societies in his own time cheated their fellow-members by giving them land of greater perimeter but less area than the plots they took themselves, and thereby got a reputation for greater honesty, while in fact they took more than their share of the produce.

A number of Zenodorus' propositions are preserved in the commentary of Theon of Alexandria on Book I of Ptolemy's *Syntaxis*. Hultsch reproduced them in Latin in the third volume of his edition of Pappus, for the purpose of comparison with Pappus' own version of the same propositions in his fifth Book, which seems to have followed Zenodorus pretty closely, though with some changes in detail.

The important propositions proved by Zenodorus and Pappus include the following:

(1) *Of all regular polygons of equal perimeter that is the greatest in area which has the most angles.*

(2) *A circle is greater than any regular polygon of equal perimeter.*

(3) *Of all polygons with the same number of sides and equal perimeter the equilateral and equiangular polygon is the greatest in area.*

Pappus added the further proposition that:

Of all segments of a circle having equal circumferences the semicircle is the greatest in area.

Nor were Zenodorus' theorems confined to plane figures; he gave also the theorem that

Of all solid figures the surfaces of which are equal, the sphere is the greatest in solid content.

HYPSICLES (second half of second century B.C.) has already been mentioned (p. 254) as the author of the continuation of Euclid's *Elements* known as 'Book XIV'. He appears to have written on Polygonal Numbers, though no such work by him is extant; for Diophantus cites him as having defined a polygonal number in this way:

Given as many numbers as we please beginning from 1 and increasing by the same common difference, then, if the common difference is 1, the sum of all the numbers is a triangular number; if 2, a square; if 3, a pentagonal number [and so on]. The number of angles is called after the number which exceeds the common difference by 2, and the side after the number of terms including 1.

This is tantamount to saying that the nth a-gonal number (1 counting as the first) is $\frac{1}{2}n\{2+(n-1)(a-2)\}$.

A tract of astronomical content entitled Ἀναφορικός (= *Ascensiones*) purporting to be by Hypsicles survives in Greek as well as in Arabic. The problem dealt with is that of finding, on certain assumptions, the time that a given sign of the zodiac will take to rise. The treatment, however, is crude, and there is nothing of real interest in the tract except that it seems to be the first Greek work in which we find the division of the ecliptic or zodiac circle into 360 parts or *degrees*; it distinguishes a degree *in space* (μοῖρα τοπική) and a degree *in time* (μοῖρα χρονική). It is certain that the division of the ecliptic into 360 degrees and that of the νυχθήμερον into 360 time-degrees were adopted by the Greeks from Babylon. While the Chaldaeans thus divided the ecliptic into 360 degrees, it does not appear that they so divided the equator or any other circle. They measured angles in general by *ells*, an ell representing 2°, so that the complete circle contained 180 such parts. The explanation of this may be that they divided the diameter of a circle into 60 parts in accordance with their usual sexagesimal division, and then took the circumference to have 180 parts on the ground that the circumference is roughly three times the diameter. The measurement of angles by *ells* and *dactyli* (24 to the ell) survives in Hipparchus' work *On the Phaenomena of Eudoxus and Aratus*. It was Hipparchus who first divided the circle in general into 360 parts or degrees, the occasion being the invention of trigonometry by him.

DIONYSODORUS. POSIDONIUS

DIONYSODORUS is known as the author of a solution, by means of a parabola and a rectangular hyperbola, of the cubic equation to which Archimedes reduces the problem of *On the Sphere and Cylinder*, II. 4, 'To cut a sphere by a plane in such a way that the volumes of the segments are to one another in a given ratio'; the solution is given by Eutocius. This Dionysodorus may have been Dionysodorus of Amisene in Pontus, whom Suidas describes as 'a mathematician worthy of mention in the domain of education', but is more likely to have been Dionysodorus of Caunus, a contemporary of Apollonius or very little later. Heron also mentions a Dionysodorus as the author of a tract *On the Spire* (or tore), in which he proved that, if d be the diameter of the circle which by its revolution generates the tore, and c the distance of its centre from the axis of revolution, the (volume of the) tore is to the cylinder with height d and radius c as the generating circle is to half the parallelogram cd, that is to say, that

$$(\text{volume of tore}) : \pi c^2 . d = \tfrac{1}{4}\pi d^2 : \tfrac{1}{2} cd,$$

or $\qquad (\text{volume of tore}) = \tfrac{1}{2}\pi^2 . cd^2,$

which is, of course, equal to the product of the area of the generating circle and the length of the path of its centre of gravity.

POSIDONIUS, a Stoic, the teacher of Cicero, is known as Posidonius of Apamea (where he was born) or of Rhodes (where he taught); he lived from about 135 to 51 B.C. In pure mathematics he is mainly quoted as the author of certain definitions, or for views on technical terms, such as 'theorem' and 'problem', and things connected with elementary geometry. He defined a 'figure' as 'confining limit' (πέρας συγκλεῖον), and 'parallels' as 'those lines which, being in one plane, neither converge nor diverge, but have all the perpendiculars equal which are drawn from points of the one line to the other'. He wrote a separate work in refutation of the Epicurean Zeno, who had questioned the very beginnings of the *Elements* on the ground that they contained unproved assumptions; Zeno argued, for instance, that even Eucl. I. 1 requires the preliminary assumption that 'two straight lines cannot have a common segment', and that the usual 'proof' of this fact by means of a circle bisected by its diameter again has to assume that two arcs of circles cannot have a common part.

More important were Posidonius' contributions to mathematical geography and astronomy. He gave his great work on geography the title *On the Ocean*, using the word which had always had such a fascination for the Greeks; the contents of the book are known to us through the copious quotations from it in Strabo; it dealt with physical as well as mathematical geography, the zones, the tides, and their connexion with the moon, ethnography, and all sorts of observations made during extensive travels. His astronomical work was entitled Περὶ μετεώρων or *Meteorologica*; Geminus wrote a commentary on it, and various views quoted from Posidonius by Cleomedes (*De motu circulari corporum caelestium*) may safely be referred to it. Posidonius also wrote a separate tract on the size of the earth.

His estimate of the circumference of the earth was less accurate than that of Eratosthenes. According to Cleomedes, he based himself on observations of the star Canopus showing that, while it just grazed the horizon at Rhodes, its meridian altitude at Alexandria was 'a fourth part of a sign' = $7\frac{1}{2}$ degrees; this gave 240,000 stades or 180,000 stades for the circumference of the earth according as the distance from Rhodes to Alexandria (supposed to be on the same meridian) was taken to be 5,000 or 3,750 stades; but the $7\frac{1}{2}°$ was an inaccurate estimate of the difference of latitude, which is in fact only $5\frac{1}{4}°$.

We have already mentioned Posidonius' estimate of the size of the sun and the basis of his calculation (p. 345 above). The figure of 3,000,000 stades for the diameter of the sun thus arrived at was much nearer the truth than the estimates given by Aristarchus, Hipparchus, and Ptolemy. The latter estimates make the diameter of the sun about $6\frac{3}{4}$, $12\frac{1}{3}$, and $5\frac{1}{2}$ times the diameter of the earth respectively; Posidonius' estimate made it about $39\frac{1}{4}$ times the diameter of the earth (the true figure being 108·9 times).

GEMINUS may properly be dealt with in this chapter because of his relation to Posidonius. He wrote a commentary on or exposition of Posidonius' work Περὶ μετεώρων, and many views stated in extracts made by Proclus and others from his works on subjects connected with elementary geometry suggest the influence of Posidonius. It is a fair inference that he was a Stoic and a pupil of Posidonius. His birthplace is uncertain, but was probably the island of Rhodes; and it is assumed that he wrote about 73–67 B.C. It is not even yet quite clear whether his name was Gemīnus or the Latin Gemĭnus. Manitius, his latest editor, satisfied himself that the name was Γεμῖνος and suggested that it was derived from the root γεμ, as Ἐργῖνος from ἐργ and Ἀλεξῖνος from ἀλεξ; he compared also the undisputed Greek names Ἰκτῖνος, Κρατῖνος. Certain it is that it appears consistently with the properispomenon accent in Greek (Γεμῖνος), while it is found in inscriptions with the spelling Γεμεῖνος. Tittel, however (in Pauly-Wissowa's *Real-Encyclopädie*), maintains that the name is after all Gemĭnus.

Geminus is an important person in the history of mathematics because he wrote a comprehensive work on the subject which is extensively cited by Proclus and others. The title of the work was probably *On the Doctrine*, or *Theory, of Mathematics*. Eutocius quotes from the sixth Book of this treatise some remarks about the way in which the first writers on conic sections produced the three curves (cf. p. 347 above). Part of the work (perhaps the first Book) dealt with the classification of the mathematical sciences, the division of mathematics into arithmetic, geometry, mechanics, astronomy, optics, geodesy, canonic (musical harmony), and logistic. Since Book VI seems to have treated of conics, it is probable that quotations made by Proclus of passages about higher curves came from later Books; the work must therefore have been of considerable size; indeed it would seem to have been an attempt to give a complete view of the whole science of mathematics, or a kind of encyclopedia of the subject. Other collections of extracts besides those given by Proclus are found (1) in the scholia to Book I of Euclid's *Elements*, (2) in the Arabic commentary on the *Elements* by an-Nairīzī (about A.D. 900), where long extracts from Geminus under the name of 'Aganis' are given on the authority of Greek commentators on Euclid and especially Simplicius, (3) to some extent in the *Anonymi variae collectiones* appended to the edition of Heron of Alexandria by Hultsch, and (4) in Eutocius, who gives one extract in his commentary on Archimedes' *Plane Equilibriums*, and another in his commentary on Apollonius' *Conics*.

The purpose of Geminus' work was evidently the examination of the first principles, the logical building-up of mathematics on the base of those admitted principles, and the defence of the whole structure against the attacks of the declared enemies of the science, the Epicureans and

Sceptics. In geometry he begins by distinguishing between the principles or hypotheses which must be taken for granted, though not admitting of proof, and the things which must not be assumed but are matter for demonstration.

Geminus treated the definitions historically, giving the various alternative definitions which had been suggested for each fundamental concept such as 'line', 'surface', 'figure', 'body', 'angle', and adding instructive classifications of the different species of 'lines' (which of course include curves both in a plane and 'on surfaces'), and of 'surfaces'. He observed that only three 'lines' and two surfaces are *homoeomeric* or uniform, i.e. such that any one part will fit on any other part, the lines being the straight line, the circle, and the cylindrical helix, and the surfaces being the plane and the sphere. Posidonius' definitions of 'figure' and 'parallels' and his division of quadrilaterals into seven kinds, which are cited by Proclus, were evidently taken from Geminus' work.

He devoted special attention to the distinction between postulates and axioms, giving the views of earlier philosophers and mathematicians (Aristotle, Euclid, Archimedes, Apollonius, the Stoics) as well as his own. This was with reference to the special stumbling-block, the fifth Postulate of Euclid, and in a lesser degree to the fourth (that all right angles are equal). Geminus observed that the *converse* of the fifth Postulate is proved by Euclid in a theorem (I. 17); hence the fifth Postulate itself is a theorem which ought equally to be proved; and it is as incorrect to assume what really requires proof as it is to attempt to prove the indemonstrable (as Apollonius did when he tried to prove the Axioms).

'So in this case', says Geminus, 'the fact that, when the right angles are lessened, the straight lines converge is true and necessary; but the statement that, since they converge more and more as they are produced, they will sometime meet is plausible but not necessary, in the absence of some argument showing that this is true in the case of straight lines. For the fact that some lines exist which approach one another indefinitely but yet remain non-secant (ἀσύμπτωτοι), although it seems improbable and paradoxical, is nevertheless true and fully ascertained with reference to other species of lines [the hyperbola and its asymptotes and the conchoid and its asymptote]. May not then the same thing be possible in the case of straight lines which happens in the case of the other lines referred to? ... It is clear from this that we must seek a proof of the present theorem, and that it is alien to the special character of Postulates.'

A modern geometer could hardly have put the dilemma more forcibly. Geminus himself attempted a remedy by first substituting a definition of parallels based, like that of Posidonius, on *equidistance*, and then adding the statement that the said 'distance' between the lines is the shortest straight line that can be drawn between them. Premising these things, he tried to prove the Postulate by a series of propositions which an-Nairīzī reproduces from Simplicius. The 'proof' naturally fails; close examination of it shows that in one place Geminus makes a tacit assumption which amounts to 'Playfair's Axiom' to the effect that through a given point only one straight line can be drawn parallel to a given straight line.

In astronomy, as we have seen (p. 387), Geminus wrote an exposition or elucidation (ἐξήγησις) of Posidonius' *Meteorologica*. An important extract from this work is given by Simplicius, after Alexander Aphrodisiensis, in his Commentary on Aristotle's *Physics*. The extract is on the distinction between physical and astronomical inquiry. Astronomy, he says, deals, not with causes, but with facts; hence it often proceeds by hypotheses, stating various expedients by which the phenomena may be saved. For example, why do the sun, moon, and planets appear to move irregularly? To explain the observed facts, we may assume, for instance, that the orbits are eccentric circles or that the stars describe epicycles on a carrying circle; and then we have to go further and examine other ways in which it is possible for the phenomena to be brought about. '*Hence we actually find a certain person coming forward and saying that, even on the assumption that the earth moves in a certain way, while the sun is in a certain way at rest, the apparent irregularity with reference to the sun may be saved.*' The text has after 'a certain person' the words 'Heraclides Ponticus', but it is practically certain that they were interpolated under a misapprehension, and that Geminus wrote τις, 'a certain person', simply, meaning thereby Aristarchus of Samos.

A useful astronomical handbook which has come down to us in Greek is entitled Γεμίνου εἰσαγωγὴ εἰς τὰ Φαινόμενα, Geminus' *Introduction to the Phaenomena*. This is clearly not the original Commentary on Posidonius' *Meteorologica*, for it does not contain the famous extract just quoted. Nor can it be by Geminus in its present form, since it contains views which we cannot attribute to him. But it may be either a genuine treatise by Geminus disfigured by later interpolations and mistakes of copyists, or a compilation from an original *Isagoge* with foreign elements added by inferior hands. In any case it is a tolerable elementary treatise, suitable for teaching purposes and containing the most important doctrines of Greek astronomy from the standpoint of Hipparchus. It treats of the zodiac, the solar year, the irregularity of the sun's motion, and the order and periods of revolution of the planets and the moon (c. 1); the twelve signs of the zodiac, the constellations, the axis of the universe and the poles, the circles on the heavenly sphere (cc. 2–5); day and night, risings and settings (cc. 6, 7, 13); the moon's phases, eclipses of the sun and moon, and the problem of accounting for the motions of the sun, moon, and planets (cc. 9–12); physical geography, zones, &c. (cc. 15, 16); weather indications (c. 17). Chapter 8 is a clear, interesting, and valuable disquisition on the calendar, the lengths of months and years, and the various cycles, the octaëteris, the 16-years and 160-years cycles, the 19-years cycle of Euctemon and Meton, and the cycle of Callippus (76 years). Chapter 18 is on the ἐξελιγμός, the shortest period which contains an integral number of synodic months, of days, and of anomalistic revolutions of the moon; this period is three times the Chaldaean period of 223 lunations used for predicting eclipses.

XIV

TRIGONOMETRY: HIPPARCHUS, MENELAUS, AND PTOLEMY

SPHAERIC, or the geometry of the sphere, with special reference to astronomy, was one of the subjects of the Pythagorean *quadrivium*. The subject was so far advanced before Euclid's time that there already existed a text-book or text-books containing the principal propositions about great and small circles on the sphere, from which both Autolycus and Euclid quoted the propositions as generally known. These propositions, with others of purely astronomical interest, were afterwards collected in a work entitled *Sphaerica*, in three Books, by Theodosius.

Theodosius was at one time commonly described as Theodosius of Tripolis, through confusion with a poet of that name who came from Tripolis. He was, however, probably the Theodosius of Bithynia mentioned by Strabo in a reference to certain Bithynians distinguished in their particular sciences, 'Hipparchus, Theodosius, and his sons, mathematicians'; and, as Vitruvius (say 20 B.C.) mentions a Theodosius who invented a sun-dial for any climate, it is not unlikely that he is the same person, and lived about the time of Hipparchus or perhaps earlier.

The first Book and the first ten propositions of the second Book of the *Sphaerica* contain the principal properties of great and small circles on the sphere, theorems about tangent planes and the relation between the sizes of the circular sections and their distances from the centre, the position of the poles, circles on a sphere which touch one another (this happens 'when the common section of the planes of the circles touches both circles'), circles on a sphere cutting one another, the 'parallel' circles, and so on. The propositions are developed on lines closely corresponding to those of Euclid's Book III relating to circles.

The second half of Book II and Book III are of astronomical interest only, though the propositions are expressed as propositions in pure geometry without any specific reference to the circles on the heavenly sphere. The object of the propositions is to prove such results as Euclid obtained in his *Phaenomena*, e.g. that different arcs of the ecliptic having the same length take a longer or shorter time to rise according as they are nearer to or farther from the tropic circle.

There is no trigonometry in Theodosius, and he does not discuss spherical triangles as such. The nearest approach is in III. 3 and in III. 11, 12. In the former proposition Theodosius practically proves the congruence-theorem for spherical triangles corresponding to Eucl. I. 4.

In Proposition 11 we have a great circle ACc through the poles A, A'; $CBDc, C'B'D$ are two other great circles, both of which are in planes at right angles to the plane of ACc, but the former is at right angles to AA' while the latter makes with it the acute angle $C'OA$; $AB'BA'$ is any great circle through AA' cutting the arcs CD, $C'D$ in B, B' respectively. Theodosius proceeds to prove that, if R, ρ be the radii of the great circle CBc of the sphere and of the 'parallel' circle through C' respectively,

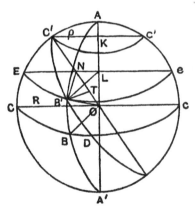

$$2R : 2\rho > (\text{arc } BC) : (\text{arc } B'C') \quad . \quad . \quad . \quad (1)$$

This result is deduced from an intermediate result

obtained in the course of the proof, namely the equivalent of the proportion

$$2R : 2\rho = (\tan ELB') : (\tan C'OB').$$

If a, b, c are the sides of the spherical triangle $AB'C'$ right-angled at C', the latter result is equivalent (since the angle ELB' is equal to the angle A of the spherical triangle) to

$$1 : \sin b = \tan A : \tan a,$$

or $\tan a = \sin b \tan A$;

but Theodosius is unable, for want of trigonometry, to find the actual ratio of a to A, or of the arc $B'C'$ to the arc BC, and only obtains an approximation to it. He is in the same position as Aristarchus, who could only approximate to the values of the trigonometrical ratios which he requires, e.g. sin 1°, cos 1°, sin 3°, by finding upper and lower limits.

The *Sphaerica* of Theodosius has recently (1927) been edited, with Latin translation, by Heiberg, and his other works *On Habitations* and *On Days and Nights*, uniformly with the *Sphaerica*, by Dr. Rudolf Fecht.

So far as we know, the first person to make systematic use of trigonometry was Hipparchus.

HIPPARCHUS, perhaps the greatest astronomer of antiquity, was born at Nicaea, in Bithynia. The period of his activity is indicated by references in Ptolemy to observations made by him, the limits of which are from 161 to 126 B.C. Ptolemy also says that from Hipparchus' time to the beginning of the reign of Antoninus Pius (A.D. 138) was 265 years. Hipparchus' most important observations seem to have been made at Rhodes, though an observation of the vernal equinox at Alexandria on 24 March 146 B.C. recorded by him may have been his own.

The only work of Hipparchus which survives is his *Commentary on the Phaenomena of Eudoxus and Aratus*.

It was an early work written before the date of his great discoveries and in particular that of the precession of the equinoxes. Its subject is mainly the observed times of the risings and settings of various stars, the purpose being the correction of the many inaccuracies contained in the statements of Aratus and Eudoxus.

For our information about Hipparchus' labours generally we are indebted to Ptolemy and other writers.

1. Perhaps his greatest achievement was the discovery of the precession of the equinoxes. He found that, at the time of his observation of it, the bright star Spica was 6° distant from the autumnal equinoctial point, whereas one Timocharis, some 160 years earlier, had made the distance 8°. The motion had therefore amounted to 2° in the period between Timocharis' observations, made in 283 or 295 B.C., and 129/8 B.C., a period, that is, of 154 or 166 years; this gives a rate of about 46·8″ or 43·4″ a year as compared with the true rate of 50·3757″.

2. In a work *On the Length of the Year* Hipparchus compared an observation by himself of the summer solstice with one made by Aristarchus of Samos 145 years earlier, and found that the said solstice occurred one-half of a day-and-night earlier than it should have done on the assumption of $365\frac{1}{4}$ days to the year; hence, he concluded that the *tropical* year contained about $365\frac{1}{4} - \frac{1}{300}$ days. Callippus had taken it to be $365\frac{1}{4}$ days exactly, and the 'cycle' which he constructed, 76 years, contained 27,759 days in which he reckoned 12×76 months *plus* 28 intercalary months, or 940 months in all. As Censorinus (*De die natali*) says that Hipparchus' cycle was 304 years, we may assume that Hipparchus first multiplied Callippus' figures by 4, thus obtaining 3,760 months and 110,136 days respectively, and then subtracted the one day (on

account of the over-estimate of $\frac{1}{300}$th of a day in the assumed length of the year), making the number of days 110,135. This gives as the length of the mean lunar month 29 days 12 hours 44 minutes $2\frac{1}{2}$ seconds, which is less than a second out in comparison with the present accepted figure of 29·530596 days![1]

3. Hipparchus improved on Aristarchus' estimates of the sizes and distances of the sun and moon; he noted the changes in their apparent diameters, and made the mean distances of the sun and moon 1,245 D and $33\frac{2}{3}$ D respectively, and their diameters $12\frac{1}{3}$ D and $\frac{1}{3}$ D respectively, where D is the mean diameter of the earth.

4. Hipparchus made a catalogue of stars to the number of 850 or more, and was apparently the first to state their positions in terms of co-ordinates in relation to the ecliptic (i.e. latitude and longitude); his table also distinguished the apparent sizes of the stars. Ptolemy continued the work, no doubt adding the results of observations of his own, and produced a catalogue of 1,022 stars.

5. Hipparchus made great improvements in the instruments used for observations. Among those which he used were an improved *dioptra*, a 'meridian-instrument' for observations in the meridian only, and a universal instrument (ἀστρολάβον ὄργανον) for more general use.

Hipparchus wrote on geography, mainly by way of criticism and correction of Eratosthenes; he insisted on the necessity of applying astronomy to geography, of fixing the positions of places by latitude and longitude, and of determining longitudes by means of lunar eclipses.

Simplicius (on Aristotle's *De caelo*) quotes two propositions from a lost work by Hipparchus *On Things borne down by their Weight*. Plutarch says that Hipparchus

[1] See also note in Appendix.

considered a certain problem in permutations and combinations. The Arabs attribute to him a work on the art of algebra, known as 'The Rules' (or 'Definitions'); this statement gains in probability now that Babylonian texts of (say) 1900 B.C. recently interpreted are found to contain quadratic equations solved in a way answering, step for step, to what would be our procedure to-day.[1]

The evidence of systematic use of trigonometry by Hipparchus is fairly conclusive. According to Theon of Alexandria in his commentary on Ptolemy's *Syntaxis*, Hipparchus wrote a treatise in twelve Books on straight lines (i.e. chords) in a circle. Now Ptolemy not only gives a table of chords of arcs subtended at the centre of a circle by angles of $\frac{1}{2}°$, $1°$, $1\frac{1}{2}°$, and so on, by halves of a degree, but sets out with great care the minimum number of propositions in geometry (and very elegant they are) which it is necessary to use in order to obtain the measures of all the chords in question. It seems certain that Hipparchus constructed a similar Table, and must have used a like set of propositions for that purpose. Such a Table of Chords is the exact equivalent of a table of trigonometrical sines, as is easily seen from the accompanying figure. Let

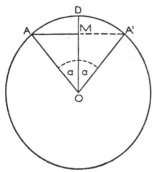

an arc AD of the circle subtend an angle α at the centre O. Draw the radius OD and draw AM from A perpendicular to OD. Let AM be produced to meet the circle again in A'. Then the arcs AD, DA' are equal, and the chord AA' subtends at O an angle of 2α. Now
$\sin \alpha = AM/AO$
$= AA'/\text{(diameter of circle)}$.

[1] Cf. note on Babylonian mathematics in Appendix.

Ptolemy expresses the chord AA' subtended by the angle 2α, which we may for brevity denote by (crd. 2α), as so many 120th parts of the diameter. Thus $\sin \alpha$ is equivalent to (crd. 2α), and Ptolemy's *Table of Chords* is equivalent to a table of sines of angles of $\frac{1}{4}°$, $\frac{1}{2}°$, $\frac{3}{4}°$, and so on by quarters of a degree to $90°$. Cos α is, of course, $\sin(90°-\alpha)$ and therefore equivalent to (crd. $180°-2\alpha$). One of Ptolemy's propositions proves the equivalent of $\sin^2\alpha + \cos^2\alpha = 1$, so that the cosine or tangent of any angle can at once be calculated from a table of sines.

Apart from the evidence of Ptolemy's *Syntaxis*, the use of trigonometry by Hipparchus is indicated (1) by Pappus' mention of numerical calculations (δι' ἀριθμῶν) by Hipparchus, in his book *On the Rising of the Twelve Signs of the Zodiac*, of the time taken by certain arcs of the circle to rise, and (2) by Hipparchus' own statement in the *Commentary on the Phaenomena of Eudoxus and Aratus* that a star lying $27\frac{1}{3}°$ north of the equator describes above the horizon (presumably in the latitude of Rhodes) an arc of 3 minutes less than 15/24ths of the whole circle, a result which he must have obtained by using the equivalent of the formula in the solution of a spherical triangle right-angled at C that $\tan b = \cos A \tan c$.

Much light is thrown on the development of Greek trigonometry by the *Sphaerica* of Menelaus of Alexandria, a work in three Books extant in Arabic. The date of MENELAUS is roughly indicated by an observation of his made in A.D. 98 and referred to by Ptolemy. He was a contemporary of Plutarch, who represents him as being present at the dialogue *De facie in orbe lunae*, and calls him 'the mathematician'. Besides the *Sphaerica*, he wrote six Books on *Chords in a Circle*, which are mentioned by

Theon, while Pappus attributes to him a treatise on the setting or rising of different arcs of the zodiac circle, and an investigation of a certain 'paradoxical' curve, of which nothing more is known. Proclus quotes a demonstration by him of Eucl. I. 25 which is direct instead of by *reductio ad absurdum*. The Arabs attribute to Menelaus yet other works, (1) *Elements of Geometry*, edited by Thābit b. Qurra, in three Books, (2) a book on *Triangles*, and (3) a work with the title (according to Wenrich) of *De cognitione quantitatis discretae corporum permixtorum*, which seems to have been a book on hydrostatics, concerned with such problems as that solved by Archimedes about Hieron's crown.

The *Sphaerica* was edited in 1758 by Halley, who apparently made a free translation from the Hebrew version of the work by Jacob b. Machir (about 1273), but consulted Arabic manuscripts as well. An earlier version in Latin by Gherard of Cremona (1114–87) exists in manuscripts at Paris and elsewhere; and by means of these translations and a Leyden manuscript of an Arabic redaction Björnbo has compiled an adequate reproduction of the contents of the *Sphaerica*. In Book I we have for the first time the conception and definition of a *spherical triangle*, which, according to Pappus, Menelaus called τρίπλευρον, a 'three-side', to distinguish it from τρίγωνον, 'three-angle', a name already appropriated to the plane triangle. The object of Book I was evidently to give the main propositions about spherical triangles in a form corresponding to Euclid's propositions in his Book I about plane triangles. Thus we have all the congruence-theorems (including the 'ambiguous case' and also the case which has no analogue in plane triangles, that in which the three angles of one triangle are respectively equal to the three angles of the other), theorems about isosceles triangles, and theorems corresponding to Eucl. I.

18, 19 (greater side opposite greater angle and vice versa), Euclid I. 24, 25, and so on. Euclid I. 16 and I. 32 are not true of spherical triangles, but Menelaus proves the corresponding propositions, namely, that (with the usual notation for spherical triangles) the exterior angle at C, i.e. $180° - C$, $< =$ or $> A$ according as $(c+a) > =$ or $< 180°$ and conversely, and that the sum of the three angles of a spherical triangle is greater than two right angles.

Book II is of purely astronomical interest; it contains generalizations or extensions of propositions in Theodosius' *Sphaerica*, Book III.

Trigonometry is reserved for Book III, where we find (*a*) 'Menelaus' Theorem' for the sphere, (*b*) deductions from it, furnishing the equivalent of formulae in spherical trigonometry, (*c*) propositions analogous to Eucl. VI. 3. Menelaus in his proof of (*a*) assumes as known the corresponding 'Menelaus' theorem' in plane geometry, while another of his proofs assumes as known the anharmonic property of four great circles drawn through one point on the sphere with reference to any great circle intersecting them all. The latter property can be proved by means of 'Menelaus' theorem' with regard to the sphere, or alternatively by the aid of the corresponding anharmonic property of four straight lines in a plane drawn through one point. It seems probable that both the anharmonic property and 'Menelaus' Theorem' for the sphere were already included in some earlier text-book, and that they were known to Hipparchus. The corresponding plane theorems appear in Pappus among his lemmas to Euclid's *Porisms*, and it is reasonable to infer that they were assumed by Euclid as known. 'Menelaus' Theorem' for the sphere is not enunciated by Menelaus or by Ptolemy (who also proves it) with reference to a spherical *triangle* and a transversal, but with reference to two arcs of great circles ADB, AEC less than a semicircle intersected in D, B, and

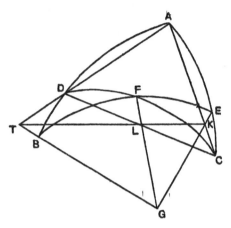

C, E respectively by two other arcs of great circles DFC, BFE less than a semicircle which themselves intersect in F, and the result proved is equivalent to

$$\frac{\sin CE}{\sin EA} = \frac{\sin CF}{\sin FD} \cdot \frac{\sin DB}{\sin BA},$$

which we should rather enunciate with reference, say, to the spherical triangle ADC and a great circle BFE cutting AD in B, DC in F, and CA in E, and in the form of a statement that (apart from sign) the product of three sines is equal to the product of three other sines.

CLAUDIUS PTOLEMY belonged to Alexandria, though according to a tradition he was born at Ptolemaïs ἡ Ἑρμείου. He made observations between A.D. 125 and 141 or perhaps 151, and presumably wrote his great work about the middle of the reign of Antoninus Pius (A.D. 138–61). The original title was Μαθηματικῆς Συντάξεως βιβλία ιγ, or *Mathematical Collection* in thirteen Books. Later, when commentators desired to distinguish it from the lesser introductory treatises on astronomy grouped under the name 'Little Astronomy', μικρὸς ἀστρονομούμενος (sc. τόπος), it became known as the *Great Collection*, μεγάλη σύνταξις, whence it passed into Arabic as Al-majistī (representing ἡ μεγίστη), and thus became *Almagest*, by which name it has ever since been known.

Pappus wrote a commentary, in the form of *scholia*, on at least six Books. Theon of Alexandria, after him, wrote an elaborate commentary (ὑπομνήματα) in eleven Books.

The *Syntaxis* was translated into Arabic first by translators unnamed, then by al-Ḥajjāj, the translator of Euclid, and again by the famous translator Isḥāq b. Ḥunain (d. 910), whose translation, improved by Thābit b. Qurra, still exists in part, as well as the version by Naṣīraddīn aṭ-Ṭūsī (1201–74).

The first edition to be published was the Latin translation from the Arabic by Gherard of Cremona finished in 1175; it was first published in 1515, without the author's name, by Peter Liechtenstein. The *editio princeps* of the Greek text was brought out by Grynaeus at Basel in 1538, after the Latin translation from the Greek made by Georgius of Trebizond and revised by Lucas Gauricus had been published at Venice in 1528. The most complete edition until the end of the last century was that of Halma (1813–16), which included two Books of Theon's commentary in Greek, with Latin translation. The authoritative Greek text of the astronomical works of Ptolemy is now that of Heiberg (1899–1907, Teubner), to which Manitius added a German translation of the *Syntaxis* with notes (1912–13).

The *Syntaxis* represents the definitive achievement of Greek astronomy, and the Ptolemaic system expounded in it remained unchallenged till the age of Copernicus. The

Syntaxis contains very full particulars of the observations and investigations carried out by Hipparchus, as well as of earlier observations recorded by him, e.g. that of a lunar eclipse which occurred in 721 B.C. Ptolemy evidently based himself very largely on Hipparchus, e.g. in the preparation of a Table of Chords, the theories of eccentric circles and epicycles, &c.; it is indeed questionable whether Ptolemy himself contributed anything of great value except a definite theory of the motion of the five planets, for which Hipparchus had only collected material in the shape of observations made by his predecessors and himself.

Book I contains the indispensable preliminaries to the study of the Ptolemaic system, general explanations of the movements of the heavenly bodies in relation to the earth as centre, and what we may call the necessary machinery, the Table of Chords, propositions in spherical geometry leading to trigonometrical calculations of the relations of arcs of the equator, ecliptic, horizon, and meridian, a 'Table of Obliquity' for calculating declinations for each degree-point on the ecliptic, and so on. Book II contains problems on the sphere, with special reference to the differences between various latitudes, the length of the longest day at any latitude, and the like. Book III is on the length of the year and the motion of the sun on the eccentric and epicycle hypotheses, Book IV on the length of the month and the theory of the moon, Book V on the construction of the astrolabe, the theory of the moon continued, the diameters of the sun, the moon, and the earth's shadow, the distance of the sun, and the dimensions of the sun, moon, and earth, Book VI on conjunctions and oppositions of the sun and moon, solar and lunar eclipses and their periods, Books VII and VIII on the fixed stars and precession, and Books IX-XIII on the movements of the planets.

The mathematical interest of Ptolemy's work lies in his use of trigonometry. First come the Table of Chords and his explanation of the steps necessary to its compilation. Ptolemy explains that he will use the division (1) of the circle into 360 equal parts or degrees, and (2) of the diameter into 120 equal parts, and will express fractions of those parts on the sexagesimal system. Accordingly chords are expressed as so many of the 'parts' of the diameter *plus* fractions of such parts; thus the chord subtended at the centre by an angle of 12° is $12^p\ 32'\ 36''$, where 1^p represents one 120th of the diameter.

Ptolemy then gives the geometrical propositions required for the preparation of the Table. These are as follows:

(1) Lemma for the purpose of finding sin 36° and sin 18°.

This takes the form of finding the chords subtending arcs of 72° and 36°, that is to say, the sides of the inscribed regular pentagon and decagon respectively.

Let AB be the diameter of a circle, O its centre, and OC the radius at right angles to AB.

Bisect OB at D, join DC, and with D as centre and DC as radius describe a circle meeting AD in E. Join EC.

Then shall OE be the side of the regular decagon, and EC the side of the regular pentagon, inscribed in the circle.

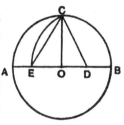

Since OB is bisected at D,

$$BE.EO + OD^2 = DE^2 = DC^2 = CO^2 + OD^2,$$

whence $BE.EO = OC^2 = OB^2,$

and BE is divided at O in extreme and mean ratio.

It follows from Eucl. XIII. 9 that OE is equal to the side of the regular decagon inscribed in the circle.

Again, by Eucl. XIII. 10,

(side of inscr. pentagon)^2 = (side of inscr. hexagon)^2
$$+ \text{(side of inscr. decagon)}^2$$
$$= CO^2 + OE^2 = CE^2;$$

therefore CE is equal to the side of the inscribed pentagon.

The construction in fact gives

$$EO = \tfrac{1}{2}a(\sqrt{5}-1), \quad EC = \tfrac{1}{2}a\sqrt{(10-2\sqrt{5})},$$

where a is the radius of the circle.

Ptolemy does not use these radicals, but calculates the lengths in terms of 'parts' of the diameter thus:

$$DO = 30, \ DO^2 = 900; \ OC = 60, \ OC^2 = 3600;$$

therefore $DE^2 = DC^2 = 4500$, and $DE = 67^p\ 4'\ 55''$ nearly.

Now (side of decagon) or (crd. $36°$) = $DE - DO$
$$= 37^p\ 4'\ 55''.$$

Again $\quad OE^2 = (37^p\ 4'\ 55'')^2 = 1375\ 4'\ 15'',$

while $\quad\quad\quad\quad CO^2 = 3600;$

therefore $\quad\quad\quad CE^2 = 4975\ 4'\ 15'',$

whence CE = (side of pentagon) = (crd. $72°$) = $70^p\ 32'\ 3''$.

The method of extracting square roots is explained by Theon in connexion with $\sqrt{4500}$ (see pp. 33–5 above).

The chords which are the sides of other regular inscribed figures, the hexagon, the square, and the equilateral triangle, are next given, namely

(crd. $60°$) $= 60^p$,
(crd. $90°$) $= \sqrt{(2 \cdot 60^2)} = \sqrt{(7200)} = 84^p\ 51'\ 10''$,
(crd. $120°$) $= \sqrt{(3 \cdot 60^2)} = \sqrt{(10800)} = 103^p\ 55'\ 23''$.

(2) Equivalent of $\sin^2\theta + \cos^2\theta = 1$.

It is obvious that, if x be any arc,

(crd. $x)^2 +$ (crd. $180°-x)^2 =$ (diam.$)^2 = 120^2$,

which is Ptolemy's formula equivalent to the above.

By means of this formula we can deduce (crd. $108°$) from (crd. $72°$), (crd. $144°$) from (crd. $36°$), and so on.

(3) 'Ptolemy's Theorem' leading to the equivalent of

$$\sin(\theta - \phi) = \sin\theta \cos\phi - \cos\theta \sin\phi.$$

The object of this proposition is to enable us to find crd. $(\alpha - \beta)$ when (crd. α) and (crd. β) are given.

Ptolemy proves, as a lemma, the famous theorem known by his name, to the effect that, if AC, BD be the diagonals of a quadrilateral $ABCD$ inscribed in a circle,

$AC \cdot BD = AB \cdot DC + AD \cdot BC$.

The proof is well known and need not be reproduced here.

Now let AB, AC be two arcs of a circle terminating at A, the extremity of the diameter AD. Complete the figure.

Then, by Ptolemy's Theorem,

$AC \cdot BD = AB \cdot DC + AD \cdot BC$.

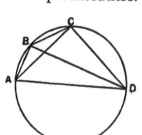

Suppose that the arc $AC(=\alpha)$ is greater than the arc $AB(=\beta)$, and that (crd. α) and (crd. β) are known. Then CD, or crd. $(180°-\alpha)$, and BD, or crd. $(180°-\beta)$, are known.

Since $AB = (\text{crd.}\,\beta)$, $AC = (\text{crd.}\,\alpha)$, and $BC = \text{crd.}\,(\alpha - \beta)$, we have, by substitution in the above formula,

$(\text{crd.}\,\alpha)\,.\,\text{crd.}\,(180° - \beta) = (\text{crd.}\,\beta)\,.\,\text{crd.}\,(180° - \alpha)$
$\qquad\qquad\qquad\qquad + (\text{crd.}\,180°)\,.\,\text{crd.}\,(\alpha - \beta)$

or $\quad \text{crd.}\,180°\,.\,\text{crd.}\,(\alpha - \beta) = \text{crd.}\,\alpha\,.\,\text{crd.}\,(180° - \beta)$
$\qquad\qquad\qquad\qquad - (\text{crd.}\,\beta)\,.\,\text{crd.}\,(180° - \alpha),$

which is equivalent to

$$\sin(\theta - \phi) = \sin\theta\cos\phi - \cos\theta\sin\phi,$$

where $\qquad\qquad \theta = \tfrac{1}{2}\alpha, \qquad \phi = \tfrac{1}{2}\beta.$

By means of this proposition Ptolemy obtains

$$\text{crd.}\,12° = \text{crd.}\,(72° - 60°) = 12^p\,32'\,36''.$$

(4) Equivalent of $\sin^2 \tfrac{1}{2}\theta = \tfrac{1}{2}(1 - \cos\theta)$.

The object of this proposition is to enable us to find (crd. $\tfrac{1}{2}\alpha$) when (crd. α) is given.

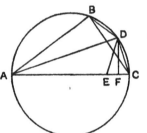

Let AC be the diameter of a circle, and BC an arc equal to α. If D be the middle point of the arc, it is required to find (crd. CD) or (crd. $\tfrac{1}{2}\alpha$).

Join AB, AD, and mark off AE along AC equal to AB. Join DE.

Then, since the angles BAD, DAC are equal, the triangles ABD, AED are equal in all respects. Therefore $DE = DB = DC$; and, if DF be drawn perpendicular to AC, $EF = FC$.

Therefore FC is equal to half the difference between AC and AE, i.e. between AC and AB.

But $\qquad\qquad AC\,.\,CF = CD^2.$

Therefore

$(\text{crd.}\,CD)^2 = \tfrac{1}{2}AC(AC - AB)$
$\qquad\qquad = \tfrac{1}{2}(\text{crd.}\,180°)\,.\,\{\text{crd.}\,180° - \text{crd.}\,(180° - BC)\},$

which is equivalent to

$$\sin^2 \tfrac{1}{2}\theta = \tfrac{1}{2}(1-\cos\theta).$$

By successive applications of his formula Ptolemy obtains (crd. 6°) and (crd. 3°), and finally (crd. $1\tfrac{1}{2}°$) = 1^p 34′ 15″, and (crd. $\tfrac{3}{4}°$ =) 0^p 47′ 8″. But, as a Table progressing by half-degrees is required, we still need (a) a method of calculating (crd. 1°) from (crd. $1\tfrac{1}{2}°$) and (crd. $\tfrac{3}{4}°$), and (b) an addition-formula which will enable us, when crd. α is given, to find crd. ($\alpha + \tfrac{1}{2}°$).

For the latter purpose Ptolemy now proves the
(5) Equivalent of

$$\cos(\theta+\phi) = \cos\theta\cos\phi - \sin\theta\sin\phi.$$

Let AD be the diameter of a circle, and AB, BC two arcs.

Then, given (crd. AB) and (crd. BC), it is required to find (crd. AC).

Draw the diameter BOE and join CE, CD, DE, BD.

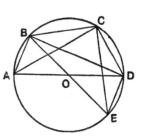

Now

(arc BD) = 180° − (arc AB),
(arc CE) = 180° − (arc BC),
(arc CD) = 180° − (arc AC),

and (arc DE) = (arc AB).

By Ptolemy's Theorem,

$$BD\,.\,CE = BC\,.\,DE + BE\,.\,CD,$$
or $$BE\,.\,CD = BD\,.\,CE - BC\,.\,DE,$$

that is,

(crd. 180°) . crd. (180° − AC)
= crd. (180° − AB) . crd. (180° − BC) − (crd. AB) . (crd. BC),

which is the equivalent of the above formula if $2\theta = AB$, $2\phi = BC$.

This is the addition formula required.

For the purpose (a) of calculating (crd. 1°) from (crd. 1½°) and (crd. ¾°) Ptolemy uses a method of interpolation based on the equivalent of the proposition

(6) If $\tfrac{1}{2}\pi > \alpha > \beta$, then $\sin\alpha/\sin\beta < \alpha/\beta$.

This fact, assumed as known by Aristarchus of Samos (cf. p. 273 above), is proved by Ptolemy in the following proposition.

If AB, BC be unequal chords in a circle, BC being the greater, then shall the ratio of CB to BA be less than the ratio of the arc CB to the arc BA.

Let BD, the bisector of the angle ABC, meet AC in E and the circumference in D. The arcs AD, DC are then equal, as also the chords AD, DC.

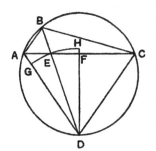

Also (since $CB:BA = CE:EA$) $CE > EA$.

Draw DF perpendicular to AC. Then $AD > DE > DF$, and the circle with centre D and radius DE will meet DA in G and DF produced in H.

Now $\quad FE:EA = \triangle FED : \triangle AED$

$\qquad\qquad\qquad < $ (sector HED) : (sector GED)

$\qquad\qquad\qquad < (\angle FDE) : (\angle EDA)$;

and, *componendo*, $\quad FA:AE < (\angle FDA):(\angle ADE)$.

Doubling the antecedents, we have

$$CA:AE < (\angle CDA):(\angle ADE),$$

and, *separando*,
$$CE:EA < (\angle CDE):(\angle EDA)$$
Therefore $CB:BA < (\angle CDB):(\angle BDA)$
$$< (\text{arc } CB):(\text{arc } BA).$$
It follows that (crd. $1°$) : (crd. $\tfrac{3}{4}°$) $< 1 : \tfrac{3}{4}$,
and again, (crd. $1\tfrac{1}{2}°$) : (crd. $1°$) $< 1\tfrac{1}{2} : 1$.
Therefore $\tfrac{4}{3}$(crd. $\tfrac{3}{4}°$) $>$ (crd. $1°$) $> \tfrac{2}{3}$(crd. $1\tfrac{1}{2}°$).

But (crd. $\tfrac{3}{4}°$) $= 0^p\ 47'\ 8''$, so that $\tfrac{4}{3}$(crd. $\tfrac{3}{4}°$) $= 1^p\ 2'\ 50''$, nearly (actually $1^p\ 2'\ 50\tfrac{2}{3}''$); and (crd. $1\tfrac{1}{2}°$) $= 1^p\ 34'\ 15''$, so that $\tfrac{2}{3}$(crd. $1\tfrac{1}{2}°$) $= 1^p\ 2'\ 50''$.

Since, then, (crd. $1°$) is both less and greater than a length which only differs inappreciably from $1^p\ 2'\ 50''$, we may say that (crd. $1°$) $= 1^p\ 2'\ 50''$, nearly.

From this result Ptolemy deduces that (crd. $\tfrac{1}{2}°$) is very nearly $0^p\ 31'\ 25''$, and so, by the aid of the above propositions, he is able to complete his Table of Chords.

To find approximately the chords of arcs containing a number of minutes intermediate between $0'$ and $30'$, Ptolemy again uses the method of proportionate increase.

There are other cases in Ptolemy in which plane trigonometry is in effect used, and many other calculations of the lengths of arcs are equivalent to spherical trigonometry. In practice Ptolemy obtains what he wants by the use of 'Menelaus' Theorem' for the sphere which (as above noted) does not mention spherical triangles; but the results so arrived at amount to such formulae in the solution of a spherical triangle ABC right-angled at C as the following:
$$\sin a = \sin c\ \sin A,$$
$$\tan a = \sin b\ \tan A,$$
$$\cos c = \cos a\ \cos b,$$
$$\tan b = \tan c\ \cos A.$$

The minor astronomical works of Ptolemy are included in vol. ii of Heiberg's edition. They are the following:

(1) The *Analemma*, lost in Greek except for a few fragments, but translated by William of Moerbeke from the Arabic into Latin and first edited by Commandinus (Rome 1562).

(2) The *Planispherium*, which exists in a Latin translation from the Arabic by 'Hermannus Secundus', published at Basel in 1536 and again edited by Commandinus (Venice 1538);

(3) Φάσεις ἀπλανῶν ἀστέρων, of which only Book II survives;

(4) Ὑποθέσεις τῶν πλανωμένων in two Books, the first of which is extant in Greek, the second in Arabic only;

(5) the inscription in Canobus;

(6) Προχείρων κανόνων διάτασις καὶ ψηφοφορία.

The word *Analemma* means 'taking up' in the sense of making a graphic representation of something, and in particular the representation on a plane of parts of the heavenly sphere. There was another *Analemma* by one Diodorus of Alexandria, on which Pappus wrote a commentary. The problem which Ptolemy sets himself in his *Analemma* is that of determining the position of the sun at any time of the day, and with this view he explains a method of representing on one plane the different points and arcs of the heavenly sphere by means of *orthogonal projection* upon three planes mutually at right angles, the meridian, the horizon, and the 'prime vertical'. Though the method has the appearance of graphic representation only, Zeuthen has shown that the results obtained are equivalent to some by no means easy formulae in spherical trigonometry.

The content of the *Planisphaerium* is somewhat similar, being an explanation of the system of projection known as *stereographic*, by which points on the heavenly sphere are represented on the plane of the equator by projection from one point, a pole. As it is the northern hemisphere which Ptolemy is concerned to represent on a plane, he naturally takes the south pole as the centre of projection. The projections of all circles on the sphere (great circles—other than those through the poles which project into straight lines—and small circles either parallel or not parallel to the equator) are likewise circles. It is easy to prove this generally because, if a cone be described with the pole as vertex and passing through any circle on the sphere, i.e. a circular cone (in general oblique) with the latter circle as base, the section of the cone by the plane of the equator satisfies the criterion for the subcontrary sections found by Apollonius at the beginning of his *Conics*, and is therefore a circle. Ptolemy, however, is content to prove the fact for certain particular circles, the ecliptic, the horizon, &c.

Other works of Ptolemy not dealing with astronomy are the following.

The *Optics*, originally in five Books, was translated from an Arabic version into Latin in the twelfth century by a certain Admiral Eugenius Siculus; Book I, however, and the end of Book V are missing. Books I, II were physical, dealing with generalities; Books III, IV treat of mirrors, but Book V is the most interesting because it contains what seems to be the first attempt at a theory of refraction. Details are given of experiments made with different media, air, glass, and water, and there are tables of angles of refraction (r) corresponding to different angles of incidence (i); these are calculated on the assumption that r and i are connected by an equation of the form
$$r = ai - bi^2,$$
where a, b are constants.

Simplicius alludes to other works of Ptolemy (1) περὶ ῥοπῶν, on balancings or turnings of the scale, in which, among other things, Ptolemy maintained that a bottle full of air is actually lighter than the same bottle empty, instead of heavier as Aristotle held; (2) περὶ διαστάσεως, *On Dimension*, i.e. dimensions, in which he tried to show that the possible number of dimensions is three only. Suidas attributes to him three Books of *Mechanica*.

Lastly, Ptolemy was among those who tried to prove the Parallel-Postulate. The essentials of his argument are given by Proclus; but close examination shows that, in the course of his propositions leading to the supposed proof, he in effect assumes, as Geminus did in his attempted proof, that through a given point only one parallel can be drawn to a given straight line.

XV
MENSURATION: HERON OF ALEXANDRIA

THE question of Heron's date is still in dispute. The possible limits are wide, namely (say) 150 B.C. (since he quotes a work on chords in a circle, and we know of no earlier Table of Chords than that attributed to Hipparchus), and A.D. 250 (since he came before Pappus). He wrote a treatise on Engines of War which in the best manuscript bears the title Ἥρωνος Κτησιβίου Βελοποιϊκά, indicating that it was regarded as a new edition of a similar work by Ctesibius. Some (including a Byzantine writer of the tenth century) inferred that Heron was a pupil of Ctesibius, but this does not follow. There were two persons of the name of Ctesibius, one of whom probably lived under Ptolemy II Philadelphus (285–247 B.C.) or Ptolemy III Euergetes I (247–222) or both, while the other was a barber of the time of Ptolemy Euergetes II, i.e. Ptolemy VII called Physcon (died 117 B.C.). The author of the *Belopoeica* was probably the earlier of the two, since Philon of Byzantium mentions that the first mechanicians had the advantage of living under kings who loved fame and supported the arts, a description which applies much better to Ptolemy II and Ptolemy III. Now Philon, who also wrote on Engines of War, and is thought to have been a generation or so later than Ctesibius, preceded Heron and is referred to by Heron (in his *Automatopoeētica*) as a constructor of 'stationary automata', and in terms which, when closely examined, seem to class Philon among the 'more ancient' rather than among recent or contemporary writers. Vitruvius, writing in 14 B.C., mentions Ctesibius and Philon but not Heron. The tendency, therefore, in recent years has been to regard Heron as belonging to a time as late as the second century A.D. Ingeborg Hammer-Jensen, in her article 'Ptolemäus und Heron' (1913), makes Heron later than Ptolemy (A.D. 100–178) on various grounds; one is that Heron's dioptra, described in his work under that title, is a fine and accurate instrument, and much better than anything that Ptolemy had at his disposal; another is that Heron appears in the *Pneumatica* to criticize certain views attributed to Ptolemy about the weight of a volume of water with water all round it.

The extant works of Heron have now been well edited, by Heiberg and others, in the Teubner series, with the exception of the *Belopoeica* (On Engines of War) and the *Cheiroballistra*, the former of which is included in Thévenot's *Veterum mathematicorum opera* (Paris, 1693), and has been critically edited by Wescher (*Poliorcétique des Grecs*, 1867), while the latter was edited by V. Prou (*Notices et Extraits*, xxvi. 2, 1877).

The works divide themselves into two classes which we may call geometrical and mechanical respectively. The geometrical works are mostly of the nature of mensuration; but in the treatise which, by its style and content, impresses the reader as being the most genuine work of Heron, namely the *Metrica* (discovered as recently as 1896 by R. Schöne in a manuscript at Constantinople), he writes like a competent mathematician having a good grasp of theory as well as practice; this is particularly noticeable in passages referring to the discoveries of his predecessors and especially Archimedes. The geometrical or mensurational handbooks other than the *Metrica* gave more scope for expansion by multiplication of examples, so that it is difficult to say how much of them was Heron's own, and how much was added by others in his name.

The works which have survived in Greek (besides the *Belopoeica* and the *Cheiroballistra*) are the following: (1) the *Metrica*, in three Books; (2) *On the Dioptra*; (3) the *Pneumatica*, in two Books; (4) *On the Art of constructing Automata* (περὶ αὐτοματοποιητικῆς); (5) geometrical works bearing the titles *Definitiones*, *Geometria*, *Geodaesia*, *Stereometrica* I, II, *Mensurae*, and *Liber Geëponicus*.

Only fragments of the *Mechanica* are preserved in Greek, but the three Books are extant in Arabic, though not in their original form, and have been edited with German translation by L. Nix and W. Schmidt (*Heronis Opera*, vol. ii, 1901). The *Catoptrica* included in the same volume is preserved in a Latin translation from the Greek presumed to be by William of Moerbeke.

Other treatises are lost: (1) *On Water-Clocks*, in four Books; (2) a *Commentary on Euclid's Elements*, though a number of excerpts made by Proclus and an-Nairīzī from the latter work enable us to form an idea of its character. Pappus mentions a separate mechanical treatise entitled *Barulcus* ('weight-lifter'), the object of which was 'to move a given weight by means of a given force'; the machine employed consisted of an arrangement of interacting toothed wheels with different diameters. A description of it is given at the end of the Greek text of the *Dioptra* and also appears at the beginning of the *Mechanics* as preserved in the Arabic; this seems to indicate that the beginning of the latter treatise was lost and a chapter from the *Barulcus* inserted instead.

Our account of the geometrical works may properly begin with the extracts made by Proclus and the Arabian commentator from the *Commentary on the Elements*. These contain (*a*) a few general notes, e.g. that Heron would not admit more than three axioms, (*b*) distinctions of *cases*, (c) alternative proofs. Among the alternative proofs are proofs of Eucl. II. 2-10 'without figures', or rather by means of figures consisting of one line only (with divisions marked on it). The effect is to make these propositions consequences of II. 1 corresponding to the algebraical identity

$$a(b+c+d+\ldots) = ab+ac+ad+\ldots$$

This appears to indicate that Heron was the originator of the easy but uninstructive semi-algebraical method which it has been the fashion, in many text-books, to substitute for Euclid's own method of proof. Lastly, Heron gave one or two useful additions to, or extensions of, propositions in Euclid. Thus (1) in a note on I. 37 he proves that, if two triangles have two sides respectively equal and the included angles *supplementary*, the triangles are equal in area. (2) He gives an interesting proof, depending on two preliminary lemmas, that in the figure of Eucl. I. 47 the three straight lines AL, BK, CF meet in a point.

The *Definitiones* is a useful compilation of a great variety of alternative definitions of fundamental concepts. Some we know from other sources to be due to Archimedes (e.g. his 'definition' of a straight line), Apollonius, Posidonius, and others. As the collection contains original definitions of Posidonius, it cannot have been compiled before the first century A.D.; but the content seems to belong in the main to the period before the Christian era. Whether the compilation is actually the work of Heron is disputed; it may in any case have been recast by some later editor.

MENSURATION

Of the reputed works of Heron on mensuration the most important is the *Metrica*, which is more scientific than the others in that it gives the theoretical basis of the formulae used, and is not a mere collection of examples. Of the other collections bearing his name it may be said generally that they are not his work in their present form, although portions may go back to Heron, that some books, e.g. the *Geometrica* and the *Stereometrica*, Book I, seem to have been re-edited, that the *Geodaesia* is not an independent work but only contains extracts from the *Geometria*, as does also the *Liber Geēponicus* to a large extent, and that the *Stereometrica*, Book II, and the *Mensurae* may be of Byzantine origin. The Constantinople MS. of the *Metrica* contains, in addition, some sections which Heiberg has added to the *Geometrica* in his edition; these consist (1) of thirteen indeterminate problems which had been separately published with translation by Heiberg and comments by Zeuthen in the *Bibliotheca Mathematica*, 1907/8, and (2) of some cases of mensuration, mainly of figures inscribed in or circumscribed about others, e.g. squares or circles in triangles, circles in squares, and circles about triangles, and lastly of circles and segments of circles.

The 'Metrica'.

After a short preface in which Archimedes and Eudoxus are mentioned as the pioneers in mensuration, Book I proceeds to the mensuration of squares, rectangles, triangles, parallel-trapezia, and other particular kinds of quadrilaterals (the rhombus, the rhomboid, and the quadrilateral in which one angle is a right angle while all the sides are given in length, but not the quadrilateral inscribed in a circle), regular polygons from the equilateral triangle to the regular dodecagon, circles and segments thereof, the ellipse, a parabolic segment, and the surfaces of a cylinder, an isosceles cone, a sphere, and a segment of a sphere.

Book II gives the mensuration of certain solid figures, finding the solid content of a cone, a cylinder, a parallelepiped, a prism, a pyramid, a frustum of a pyramid and of a cone, a sphere and a segment thereof, a *spire* or *tore* (anchor-ring), the two special solids measured in Archimedes' *Method*, and the five regular solids.

Book III deals with the division of figures into parts having given ratios to one another, first plane figures, then solids, namely a pyramid, a cone and a frustum thereof, a sphere.

Among the cases of triangles in Book I the most interesting is that of the scalene triangle (acute angled or obtuse angled) in which the lengths of the three sides are given. This problem is solved in two ways.

(1) A perpendicular is drawn from a vertex (A) to the opposite side (BC), and the theorems of Eucl. II. 12 and 13 are used in order to find the lengths of the segments into which BC is divided by the perpendicular AD. The length of the perpendicular itself is then deduced, and the area ($=\frac{1}{2} AD \cdot BC$) is thus found.

(2) The second method is to use the formula which we write as

$$\Delta = \sqrt{\{s(s-a)(s-b)(s-c)\}},$$

where s is half the sum of the sides a, b, c; and Heron gives an admirable proof of the formula by pure geometry, as follows.

Let the sides of the triangle ABC be given in length. Inscribe the circle DEF, and let O be its centre.

Then
$$BC \cdot OD = 2\Delta BOC,$$
$$CA \cdot OE = 2\Delta COA,$$
$$AB \cdot OF = 2\Delta AOB;$$

whence, by addition,
$$p \cdot OD = 2\triangle ABC,$$
where p is the perimeter.

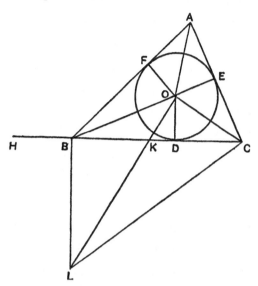

Produce CB to H so that $BH = AF$.

Then, since $AE = AF$, $BF = BD$, and $CE = CD$, we have $CH = BC + AF = \frac{1}{2}p = s$.

Therefore $CH \cdot OD = \triangle ABC$.

But $CH \cdot OD$ is the 'side' of the product $CH^2 \cdot OD^2$, i.e. $\sqrt{(CH^2 \cdot OD^2)}$, so that
$$(\triangle ABC)^2 = CH^2 \cdot OD^2.$$

Draw OL at right angles to OC cutting BC in K, and BL at right angles to BC meeting OL in L. Join CL.

Then, since each of the angles COL, CBL is a right angle, $COBL$ is a quadrilateral in a circle.

Therefore $\angle COB + \angle CLB = 2R$.

But $\angle COB + \angle AOF = 2R$, because AO, BO, CO bisect the angles round O, and the angles COB, AOF are together equal to the angles AOC, BOF, while the sum of all four angles is equal to $4R$.

Therefore $\angle AOF = \angle CLB$.

Accordingly, the right-angled triangles AOF, CLB are similar;

therefore
$$BC : BL = AF : FO$$
$$= BH : OD,$$
and, alternately, $\quad CB : BH = BL : OD$
$$= BK : KD;$$
whence, *componendo*, $CH : HB = BD : DK$.

It follows that
$$CH^2 : CH . HB = BD . DC : CD . DK$$
$$= BD . DC : OD^2, \text{ since } \angle COK \text{ is right.}$$
Therefore $(\triangle ABC)^2 = CH^2 . OD^2$ (from above)
$$= CH . HB . BD . DC$$
$$= s(s-a)(s-b)(s-c).$$

The proposition itself, as we have seen (p. 340), is attributed to Archimedes.

The chapter (8) containing the above proof is otherwise of the highest interest because it explains a method of obtaining approximations to the value of the square roots of numbers which are not squares. This is the only classical method on record, and it enables us to understand how Archimedes may have arrived at the approximations to $\sqrt{3}$ which he merely states without any explanation (cf. pp. 309–10 above).

If A is a non-square number, and a^2 is the nearest

square number to it, so that $A = a^2 \pm b$, Heron's rule amounts to saying that a first approximation (α_1) to $\sqrt{A}$ is

$$\alpha_1 = \tfrac{1}{2}\left(a + \frac{A}{a}\right) \quad \ldots \ldots \quad (1)$$

He says further that a second approximation can be found by substituting for a in the above formula the first approximation α_1; thus

$$\alpha_2 = \tfrac{1}{2}\left(\alpha_1 + \frac{A}{\alpha_1}\right) \quad \ldots \ldots \quad (2)$$

Heron does not himself seem to make any direct use of the formula for a second approximation; but the method is general, and by continuing the process indefinitely we can find any number of successive approximations.

If we substitute in (1) the value $a^2 \pm b$ for A, we obtain the well-known formula

$$\alpha_1 = a \pm \frac{b}{2a}.$$

Examples in the *Metrica* are the following:

(1) $\sqrt{720}$. Since $729\ (= 27^2)$ is the nearest square to 720, the first approximation to $\sqrt{720}$ is

$$\alpha_1 = \tfrac{1}{2}(27 + \tfrac{720}{27}) = \tfrac{1}{2}(27 + 26\tfrac{2}{3}) = 26\tfrac{1}{2}\tfrac{1}{3}.$$

(2) $\sqrt{63}$, says Heron, is nearly $7\tfrac{1}{2}\tfrac{1}{4}\tfrac{1}{8}\tfrac{1}{16}$, which he would clearly obtain thus:

$$\alpha_1 = \tfrac{1}{2}(8 + \tfrac{63}{8}) = \tfrac{1}{2}(8 + 7\tfrac{7}{8}) = 7\tfrac{1}{2}\tfrac{1}{4}\tfrac{1}{8}\tfrac{1}{16}.$$

Heron has occasion to make many such approximations, especially in his mensuration of the regular polygons. In the case of the equilateral triangle he proves that the area Δ is given by $\Delta^2 = 3a^4/16$. In the particular case taken $a = 10$, so that $\Delta = \sqrt{1875} = 43\tfrac{1}{3}$, nearly. Sometimes a well-known approximation to $\sqrt{3}$ is used, as when, in the *Geometrica*, the area of the equilateral triangle of side a is given as $(\frac{1}{3}+\frac{1}{10})a^2$, for in fact $\frac{1}{3}+\frac{1}{10} = \frac{13}{30} = \frac{1}{4}(\frac{26}{15})$.

To find the area of the regular pentagon (c. 18), Heron proceeds in this way. Let ABC be a triangle right angled at C and having the angle at A equal to $\frac{2}{5}R$. Produce AC

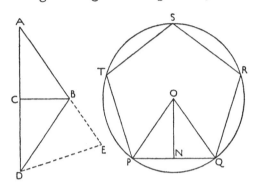

to D, making CD equal to AC, and join BD. Then the angles ABC, DBC are both equal to $\frac{3}{5}R$, so that the angle ABD is the angle of a regular pentagon.

Now, says Heron, if AD be divided in extreme and mean ratio, the greater segment is equal to AB. (In fact, if we produce AB to E so that AE is equal to AD, and join DE, the triangle ADE is the triangle of Eucl. IV. 10).

Therefore, says Heron, $(BA+AC)^2 = 5AC^2$. (Here Heron simply quotes the theorem of Eucl. XIII. 1.)

Now, if PQ be a side of a regular pentagon inscribed in a circle, and ON the perpendicular to it from O, the centre, the triangle ONP is similar to the triangle ACB, and
$$(PO+ON)^2 = 5ON^2.$$

Heron now uses $\frac{9}{4}$ as an approximation to $\sqrt{5}$ and equates $PO+ON$ to $\frac{9}{4}ON$. It follows that $ON = \frac{4}{5}OP$, whence $PN^2 = OP^2 - ON^2 = \frac{9}{25}OP^2$, and $PN = \frac{3}{5}OP = \frac{3}{4}ON$.

The area of the pentagon, being equal to $5ON \cdot NP$, is

therefore equal to $\frac{20}{3}PN^2$ or $\frac{5}{3}PQ^2$, nearly. In the case taken by Heron $PQ = 10$, and the area is $166\frac{2}{3}$.

The regular *hexagon* is, of course, equal in area to six times the equilateral triangle with the same side.

For the regular *heptagon* Heron assumes that, if a be the side, and r the radius of the circumscribing circle, $a = \frac{7}{8}r$, being approximately equal to the perpendicular from the centre of the circle to a side of an inscribed regular hexagon, for $\frac{7}{8}$ is the approximate value of $\frac{1}{2}\sqrt{3}$. This theorem is quoted by Jordanus Nemorarius (died 1237) as an 'Indian rule'. The *Metrica* shows it to be of Greek origin.

In the case of the regular *octagon, decagon*, and *dodecagon* Heron finds the length (p) of the perpendicular on a side from the centre of the circumscribing circle. If OC be the perpendicular from the centre O to the side AB, the angle OAD is made equal to the angle AOD. The angles in the figure are then known.

The case of the *decagon* is, of course, closely connected with that of the pentagon. In this case $\angle ADC = \frac{2}{5}R$, and $AD:DC = 5:4$, nearly. Therefore $AD:AC = 5:3$, and

$$p = \tfrac{1}{2}a(\tfrac{5}{3}+\tfrac{4}{3}) = \tfrac{3}{2}a,$$

where a is the side.

For the regular *enneagon* and *hendecagon* Heron appeals to the Table of Chords (presumably that of Hipparchus). If d be the

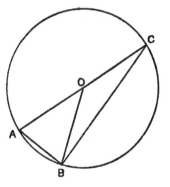

diameter of the circumscribing circle and a the side of the polygon, we are told that for the *enneagon* and *hendecagon* the approximate values of the ratio a/d are $\frac{1}{3}$ and $\frac{7}{25}$ respectively. In the case of the *enneagon*, for example, $AC^2 = 9AB^2$, whence $BC^2 = 8AB^2$, or approximately $\frac{289}{36}AB^2$, and $BC = \frac{17}{6}a$, nearly.

Therefore (area of enneagon) $= \frac{9}{2}\triangle ABC = \frac{9}{4} \cdot \frac{17}{6}a^2 = \frac{51}{8}a^2$.

For the *hendecagon* $AC^2 = \frac{625}{49}AB^2$, and $BC^2 = \frac{576}{49}AB^2$, so that $BC = \frac{24}{7}a$ and

(area of hendecagon) $= \frac{11}{2} \cdot \triangle ABC = \frac{66}{7}a^2$.

As regards the circle Heron gives to π the Archimedean value, $\frac{22}{7}$, while noting that in a work *On Plinthides and Cylinders* Archimedes himself brought the value of π within closer limits than the $3\frac{1}{7}$ and $3\frac{10}{71}$ found in the *Measurement of a Circle* (cf. p. 146 above).

For the area of a *segment* of a circle Heron has only some traditional and inaccurate formulae to give, while explaining that they should only be used for segments less than a semicircle and not even for all of those but only those in which the height is not less than one-third of the base. For smaller segments he knows nothing better than a method exactly modelled on Archimedes' *Quadrature of a Parabola*, which results in making the area of a circular segment equal to that of the parabolic segment with the same base and height.

Coming to the measurement of the surfaces of a cylinder and a right cone, Heron supposes them to be unrolled on a plane, so that the surface becomes that of a parallelogram in the one case, and a sector of a circle in the other. For the surface of a sphere and a segment thereof he uses Archimedes' results.

Book II of the *Metrica* treats of the volume of various solids, beginning with the cone, the cylinder, the 'parallelepiped' (which is interpreted as covering, in addition, what we should call a prism with polygonal base), a pyramid with base of any form, a frustum of a triangular pyramid; these figures are in general oblique. Then follows a wedge-shaped solid, βωμίσκος or σφηνίσκος ('little altar' or 'wedge'); this is a rectilineal figure in which the base is a rectangle and has opposite to it, in a plane parallel to it, a smaller rectangle the sides of which are parallel respectively to the sides of the base but not necessarily in the same proportion as those sides. If a, b be the sides of the rectangular base, c, d the corresponding sides of the upper face of the solid, and h the height of the solid, the solid content is proved to be

$$\{\tfrac{1}{4}(a+c)(b+d)+\tfrac{1}{12}(a-c)(b-d)\}h.$$

A particular case is that of a frustum of a pyramid with square base. If a, a' be the sides of the larger and smaller bases respectively, and h the height, Heron's formula for the volume reduces to

$$[\{\tfrac{1}{2}(a+a')\}^2+\tfrac{1}{3}\{\tfrac{1}{2}(a-a')\}^2]h,$$

which again is easily reduced to $\tfrac{1}{3}h(a^2+aa'+a'^2)$.

A frustum of a cone is then measured in two ways, (1) by the consideration that its volume bears the same ratio to that of the corresponding frustum of the circumscribing pyramid with square base as the area of the circular base of the cone bears to that of the square base of the pyramid, and (2) as the difference between two cones. In either case the volume reduces to $\tfrac{1}{12}\pi h(a^2+aa'+a'^2)$, where h is the height and a, a' the *diameters* of the bases.

In the case of the sphere and segment thereof, and of the two special solids of Archimedes' *Method*, Heron follows Archimedes.

He gives one other interesting case, that of the anchor-ring or *tore*, using a proposition, which he attributes to Dionysodorus, to the effect that, if a be the radius of either circular section of the tore made by a plane through the axis of revolution, and c the distance of the centre of that circle from the axis,

$$\pi a^2 : ac = (\text{volume of tore}) : \pi c^2 . 2a,$$

which gives for the volume of the tore $2\pi^2 a^2 c$. This expression being the product of πa^2 and $2\pi c$, the volume is seen to be the product of the area of the circular section and the circumference of the circle described by its centre in a complete revolution about the axis (cf. p. 385 above). Thus we have a particular case of Guldin's Theorem (as it used to be called, though it is really due to Pappus). Heron is aware of this method of finding the volume, since he speaks of the tore being conceived to be straightened out and made into a cylinder, the section of the cylinder being the section of the tore, and the height of the cylinder equal to the circumference of the circle with c as radius.

Chapters 16–18 contain the measurement of the volume of the five regular solids, and Book II ends with an allusion to the method attributed to Archimedes of measuring the content of irregular bodies by immersing them in water and measuring the amount of fluid displaced.

Book III of the *Metrica* is on divisions of figures, and has much in common with Euclid's book *On Divisions (of figures)* (cf. pp. 258–62 above). The kind of problem dealt with is this: To divide a triangle into two parts having a given ratio by a straight line (*a*) passing through a vertex, (*b*) parallel to a side, or (*c*) through any point on a side; or, To divide a quadrilateral into parts having a given ratio by a straight line passing through a point (*a*) on a side, (*b*) not on a side, or, in the case of a parallel-trapezium, (a) parallel to one of the parallel sides, (b) passing through the point of intersection of the non-parallel sides, (c) passing through a point on one of the parallel sides, or on one of the non-parallel sides, and so on; with similar problems for any polygon. Some of the problems are reduced to problems of cutting off an area solved by Apollonius in his work on that topic (cf. p. 365 above).

We will give one case as an example. If $ABCD$ be any quadrilateral, and E a point on the side AD, it is required to draw through E a straight line EF which shall cut the quadrilateral into two parts in the ratio of AE to ED. (We omit the analysis leading to the solution.)

Draw CG parallel to DA meeting AB produced in G. Join BE, and draw GH parallel to BE, meeting BC in H. Join CE, EH, EG.

Then $\triangle GBE = \triangle HBE$.

Add to each $\triangle ABE$;

therefore

$\triangle AGE =$ (quadrilateral $ABHE$).

Therefore

(quadrilateral $ABHE$) : $\triangle CED = \triangle AGE$: $\triangle CED$
$= AE : ED$.

But the quadrilateral $ABHE$ and the triangle CED are parts of the quadrilateral $ABCD$, and they leave over only the triangle EHC. We have, therefore, only to divide the triangle EHC into parts in the same ratio $AE : ED$ by a straight line EF drawn from E to meet HC. F is found by dividing HC at F in the ratio $AE : ED$, and we have only to join EF, which is therefore the line required.

In cc. 17, 23 Heron divides a sphere into segments such that (a) their surfaces, (b) their volumes, are in a given ratio, by means of Propositions 3, 4 of Archimedes *On the Sphere and Cylinder*, Book II.

Other solid figures divided into parts having a given ratio are the pyramid with base of any form, the cone, and the frustum of a cone, the cutting plane being parallel to the base in each case. These latter problems involve the extraction of the cube root of a number which is not in general an exact cube, and the point of interest is a method by which Heron approximates to the cube root in such a case. The rule is not explained in general terms, but has to be inferred from a particular case in which 100 is the number the cube root of which is required. The numbers do not happen to be such as to settle the rule beyond all doubt, but it appears to be this. Suppose that we require an approximate value for $\sqrt[3]{A}$, where $a^3 < A < (a+1)^3$, a^3 and $(a+1)^3$ being the nearest complete cubes. Suppose, further, that $A - a^3 = d_1$, and $(a+1)^3 - A = d_2$. Then an approximation to the cube root is

$$a + \frac{(a+1)d_1}{(a+1)d_1 + ad_2}.$$

There are several cases of the numerical solution of quadratic equations in Heron. One example may be given from the *Geometrica*. 'Given the sum of the diameter, perimeter, and area of a circle, to find each of them.' If d is the diameter, 212 the given sum, then (assuming $\pi = 3\frac{1}{7}$), the equation becomes

$$\tfrac{11}{14}d^2 + \tfrac{29}{7}d = 212.$$

Heron's solution is like ours except that he does not *divide* the equation throughout by $\tfrac{11}{14}$ in order to make the first term a square; he *multiplies* by such a number as will have this effect, namely, in this case, 154, making the equation

$11^2d^2+58.11d = 212.154$. He then adds, as we do, to each side the number which will make the left side a complete square, namely 29^2. We then obtain

$$(11d+29)^2 = 33489, \text{ and } 11d+29 = 183,$$

whence $d = 14$.

The work *On the Dioptra* is so called from the name of an instrument which served with the ancients the same purpose as a theodolite with us. The instrument is described in cc. 1–5, after which the treatise goes on to deal with (*a*) problems of heights and distances; (*b*) engineering problems, e.g. (1) to bore a tunnel through a mountain, beginning from both ends and fixing the directions so that the working parties will meet in the middle, or (2), given a subterranean canal of any form, to find on the ground above a point from which to sink a perpendicular shaft in order to reach a given point on the canal (e.g. for removing an obstruction); (*c*) problems of mensuration. Heron adds (c. 34) a description of a 'hodometer', or taxameter, consisting in an arrangement of toothed wheels and endless screws on the same axes working on the teeth of the next wheels respectively. C. 35 shows how to find the distance between Rome and Alexandria along great circles of the earth by means of observations of the same eclipse at the two places, the *analemma* for Rome, and a concave hemisphere constructed for Alexandria to show the position of the sun at the time of the said eclipse. The book ends (c. 37) with the problem, With a given force to move a given weight by means of interacting toothed wheels, which really belongs to the *Mechanics*.

The *Mechanics* preserved in the Arabic is evidently far from having kept its original form. The original treatise, no doubt, began with generalities and passed on to the properties of circles, cylinders, &c., with reference to their importance in mechanics. Instead of this, our treatise begins with the problem of 'moving a given weight by a given force' just mentioned, the description being the same as that in the *Dioptra* and that quoted by Pappus from a work of Heron called the 'weight-lifter' (βαρουλκός). Cc. 2–7 discuss the motion of circles or wheels, equal or unequal, revolving on different axes (e.g. interacting toothed wheels) or fixed on one axis, and Heron attempts an explanation of the 'Wheel of Aristotle' (Aristotelian *Problems*, 24), which remained a puzzle till modern times and gave rise to the proverb 'rotam Aristotelis magis torquere quo magis torqueretur'. The treatise further contains the 'parallelogram of velocities' (c. 8), and discusses the force required to keep weights in equilibrium when placed on an inclined plane (I. 20–3), centres of gravity, the five mechanical powers, mechanical questions occurring in ordinary life, and the construction of engines for all sorts of purposes, machines for lifting weights, oil presses, &c. The chapters in Book I on the centre of gravity discuss generally the case of bodies supported by props as well as bodies suspended, the amount of the weight borne by each prop when there are several, and so on; in II. 35–7 Heron finds the centres of gravity of a triangle, a quadrilateral and a pentagon respectively; then, assuming that a triangle of uniform thickness is supported by a prop at each corner, he finds what weight is borne by each prop (*a*) when the props support the triangle only, (*b*) when to the triangle is added a given weight placed at any point on it (cc. 38, 39); lastly, he finds the centre of gravity of a system consisting of a triangle or polygon with known weights placed at each angle (cc. 40, 41).

The *Catoptrica* of Heron has several propositions common with the so-called *Catoptrica* attributed to Euclid (cf. p. 268 above). It also contains problems the purpose of which is to construct mirrors or combinations of mirrors of such shape as to reflect objects in a particular way, e.g. to make the right side appear as the right in the image (instead of the reverse), to enable a person to see his back or to appear head-downwards, and so on. The book contains one proof of interest, namely, that the angles of incidence and reflection in a mirror are equal. This is based on the principle that light travels in a straight line, that is, by the shortest road. Hence, even when the ray is a line broken at a point of the mirror by reflection, it must mark the shortest broken line of the kind connecting the eye and the object. Heron then proves that the broken line connecting the eye and the object is the shortest when the angles made by the two portions, at the point of incidence with the reflecting surface, are equal.

XVI
PAPPUS OF ALEXANDRIA

As we saw, the immediate successors of the great geometers kept up the tradition for a time, but new developments were limited to astronomy with its handmaids sphaeric and trigonometry (Hipparchus, Menelaus, Ptolemy). Meantime the study of higher geometry seems to have languished or even been in abeyance until Pappus arose to kindle fresh interest in it. From the way in which Pappus thinks it necessary to describe the contents of the classical works comprised in what he calls the *Treasury of Analysis* (τόπος ἀναλυόμενος) we may infer that many of them, if not wholly or partially lost, were already forgotten. Presumably such interest as Pappus was able to arouse soon died out. But our obligation to him cannot be over-estimated, for his *Collection* constitutes, after the works of the great mathematicians which have actually survived, the most important of all our sources.

Pappus lived at the end of the third century A.D. This we infer from a scholium to a Leyden MS. of chronological tables by Theon of Alexandria which says, opposite to the name of Diocletian (who reigned from 284 to 305), 'In his time Pappus wrote.' Suidas, it is true, makes Pappus a contemporary of Theon of Alexandria (fourth century), but this is probably due to some confusion.

Besides the *Collection* (συναγωγή) with which we are mainly concerned, Pappus wrote commentaries on the *Elements* and *Data* of Euclid, Ptolemy's *Syntaxis* and *Planispherium*, and an *Analemma* by one Diodorus. A great deal of the commentary on the *Syntaxis* was incorporated by Theon of Alexandria in his own commentary on the same work; that on the *Planispherium* is said to have been translated into Arabic by Thābit b. Qurra. The commentary on Euclid's *Elements* has left traces in Proclus and Eutocius, while fragments of the portion relating to Book X survive in Arabic (cf. pp. 241, 376 above). Among Proclus' references we may mention the following. Pappus said that the converse of Euclid's Postulate 4 (that all right angles are equal) is not true, for you may have an angle equal to a right angle which is not a right angle, e.g. the curvilinear angle between the coterminous arcs of two semicircles the diameters of which are two equal straight lines forming a right angle, as in the figure annexed. Pappus also said that, besides the genuine Axioms of Euclid, there were others on record about unequals added to equals and equals added to unequals; other axioms given by him are, says Proclus, involved by the definitions, e.g. that 'all 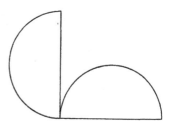 parts of the plane and of the straight line coincide with one another'. Pappus again gave a pretty proof of Eucl. I. 5 which modern editors have utterly spoiled when introducing it into text-books. If AB, AC are the equal sides of an isosceles triangle, Pappus compares the triangles ABC and ACB (as if he were comparing the triangle ABC seen from the front with the same triangle seen from the back), and shows that they satisfy the conditions of I. 4 and are therefore equal in all respects, whence the desired result follows.

THE 'SYNAGOGE' OR *COLLECTION*

While it covers the whole range of Greek geometry, the *Collection* is a handbook or guide rather than an encyclopaedia; it was intended, that is, to be read with the original works (where extant) rather than independently. Where, however, the history of a subject is given, e.g. that of the duplication of the cube, or the finding of the two mean proportionals, the solutions themselves are reproduced, presumably because they were not readily accessible but had to be collected from scattered sources. Even when some accessible classic is being described, the opportunity is taken to give alternative methods or to make improvements in proofs, extensions, and so on. Without pretending to great originality, the whole work shows, on the part of its author, a complete grasp over the subjects treated, independence of judgement, and mastery of technique; the style is concise and clear; in short, Pappus stands out as an accomplished and versatile mathematician and a worthy representative of the classical Greek geometry.

The first published edition of the *Collection* was the Latin translation by Commandinus (Venice 1589, reissued 'Pisauri ... 1602'). Up to 1876 portions only of the Greek text had been published, e.g. the remains of Book II by John Wallis (*Opera Mathematica*, iii, Oxford 1699), and Books VII, VIII in Greek and German by C. J. Gerhardt, 1871. Editors of ancient works, and restorers of some of those of Euclid and Apollonius which are lost, naturally give extracts from the Greek text of Pappus relating to the particular works. In 1876-8 appeared the fine edition by Friedrich Hultsch in three volumes, containing the complete Greek text, with apparatus, Latin translation, commentary, appendices, and indices; this first monument of the revived study of Greek mathematics in the last half of the nineteenth century has properly been the model for the definitive editions of the original texts of Greek mathematicians by Heiberg and others which have lately appeared.

THE *COLLECTION*. BOOKS I-III 437

Book I and the first thirteen propositions of Book II are lost. The lost propositions of Book II, like the rest of the Book, treated of Apollonius' method of writing and working with large numbers expressed in terms of the successive powers of the myriad 10000 (cf. p. 19 above).

Book III is in four sections. Section (1) is a sort of historical account of the problem of *finding two mean proportionals in continued proportion between two given straight lines*. It begins with some general remarks about the distinction between theorems and problems, and lays special stress on the fact that it is the duty of the solver of a problem to determine under what conditions the solution is possible or impossible and, 'if possible, when, how, and in how many ways it is possible'. This brings Pappus to the case of a person unnamed who, though he appeared to be an able geometer, claimed to know how to solve the problem of the two mean proportionals by a 'plane' method, and asked Pappus for his judgement on a particular construction for that purpose. Pappus says that the author gave no proof; he therefore describes the construction and examines its consequences, for the purpose of proving that it does not solve the problem as the author claims. Pappus does not, however, seem to have noticed that the method, though not actually solving the problem, does enable us to make a series of any number of successive approximations to the correct solution.

Pappus passes on to the distinction between three classes of problems, (1) *plane* problems, by which are meant problems which can be solved by means of straight lines and circles only; (2) *solid* problems, which require for their solution the use of one or more conic sections, these curves being sections of the surface of a solid figure, namely a cone; (3) *linear* problems, being those which necessitate recourse to higher curves still, namely curves such as spirals, *quadratrices*, cochloids, or cissoids, the generation of which is more complicated, nay, sometimes even 'forced and unnatural'. The problem of the two mean proportionals is in its nature 'solid'; hence for its solution the ancients used either conic sections or some equivalent which they found in various mechanical devices, e.g. the *mean-finder* of Eratosthenes, those described in the *Mechanics* of Philon and Heron, and that of Nicomedes (the 'cochloidal' curve, or conchoid). Pappus then gives the solutions by Eratosthenes, Nicomedes, and Heron, and adds a fourth which he claims as his own, though it is practically the same as the solution attributed by Eutocius to Sporus (see pp. 169-70 above).

Section (2) of Book III is on the theory of Means. It begins with the statement of a problem enunciated by some other geometer, namely, to exhibit in a semicircle all three means, namely, the arithmetic, geometric, and harmonic means. Let ADC be a semicircle on the diameter AC, O its centre, and B any point on OC. Draw BD at right angles to AC meeting the semicircle in D. Join OD, and draw BF perpendicular to OD.

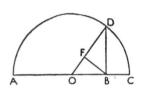

Then clearly OC is the arithmetic mean, and BD the geometric mean, between AB, BC.

The 'other geometer' seems to have said that DF is the harmonic mean. Very strangely, Pappus complains that he did not indicate how DF is the harmonic mean or between what straight lines. But DF is actually the harmonic mean between AB, BC, as is easily proved.

For, since ODB is a right-angled triangle, and BF is perpendicular to OD,

$$DF : BD = BD : DO,$$
or $\qquad DF \cdot DO = BD^2 = AB \cdot BC.$

But $\qquad DO = \tfrac{1}{2}(AB + BC);$

therefore $\qquad DF \cdot (AB + BC) = 2AB \cdot BC.$

Therefore $AB \cdot (DF - BC) = BC \cdot (AB - DF),$

or $\qquad AB : BC = (AB - DF) : (DF - BC),$

that is, DF is the harmonic mean between AB, BC.

Pappus gives other constructions to show how, given any two of three terms a, b, c in arithmetic, geometric, or harmonic progression, the third can be found. He then proceeds to discuss three other means treated by Nicomachus and others, and lastly four further means added by 'more recent writers' to the first six (cf. pp. 52-3 above).

Section (3) contains a series of propositions, curious rather than geometrically important, which seems to have been taken direct from a collection of *Paradoxes* by one Erycinus. The first set connect themselves with Eucl. I. 21, which proves that, if from the extremities of the base of any triangle two straight lines be drawn meeting at any point within the triangle, the straight lines are together less than the two sides of the triangle other than the base, but contain a greater angle. It is proved that, if the straight lines are allowed to be drawn from points on the base other than its extremities, their sum may be equal to or greater than the remaining two sides of the triangle. Pappus begins with the following simple case. Given a triangle ABC right angled at B, it is required to find an internal point F and to draw from it (1) a straight line to C, the extremity of the base, and (2) a straight line meeting the base in a point D

such that the sum of FD, FC may be greater than the sum of AB, AC.

Draw any straight line AD from A to meet BC in D. Measure DE along DA equal to BA. Bisect AE at F and join FC. Then shall $(DF+FC)$ be $> (BA+AC)$.

This is obvious, for $EF+FC = AF+FC > FC$, and, adding the equals DE, AB respectively, we have

$$DF+FC > BA+AC.$$

The propositions which follow are more elaborate, e.g.:

In any triangle, except an equilateral triangle or an isosceles triangle with base less than either of the equal sides, it is possible to draw from points on the base (1) two straight lines meeting within the triangle such that their sum is *equal* to the sum of the other two sides of the triangle, and (2) two other such lines the sum of which is *greater* than that of the same two sides (Props. 29-31).

Other propositions (35-9) have a similar relation to the Postulate of Archimedes (*On the Sphere and Cylinder*, I) about continuous broken lines one of which wholly encloses the other.

Section (4) shows how *to inscribe each of the five regular solids in a given sphere* (Props. 54-8, following preliminary lemmas forming Props. 43-53). The procedure in each case is by analysis followed by synthesis. The constructions are different from those of Euclid in XIII. 13-17, where the problem is, not to inscribe the solid in a given sphere, but first to construct the solid and then to 'comprehend' it in a sphere, i.e. to determine the circumscribing sphere.

At the beginning of Book IV the title and preface are missing, and the first section begins with the famous extension or generalization (as Pappus himself calls it) of the Pythagorean theorem of the square on the hypotenuse (Eucl. I. 47).

If *ABC* be *any* triangle and on *AB*, *AC* any parallelograms whatever be described, as *ABDE*, *ACFG*, and if *DE*, *FG* produced meet in *H* and *HA* be joined, then the parallelograms *ABDE*, *ACFG* are together equal to the parallelogram contained by *BC*, *HA* in an angle which is equal to the sum of the angles *ABC*, *DHA*.

Produce *HA* to meet *BC* in *K*, draw *BL*, *CM* parallel to *KH*, meeting *DE* in *L* and *FG* in *M*, and join *LNM*.

Then *BLHA* is a parallelogram, and *HA*, *BL* are equal and parallel.

Similarly, *CMHA* is a parallelogram, and *HA*, *MC* are equal and parallel.

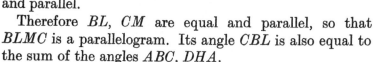

Therefore *BL*, *CM* are equal and parallel, so that *BLMC* is a parallelogram. Its angle *CBL* is also equal to the sum of the angles *ABC*, *DHA*.

Now, of the parallelograms,

(*ABDE*) =(*BLHA*), in the same parallels,
=(*BLNK*), for the same reason.

Similarly, (*ACFG*) =(*ACMH*) =(*NKCM*).

Therefore, by addition,

(*ABDE*)+(*ACFG*) =(*BLMC*).

There follow some propositions which are quite interesting in themselves, though their bearing is not clear. In two of them we meet with two of the irrationals classified in Eucl. Book X, and in two others we have cases of analysis followed by synthesis as an aid to the proof of theorems. The most interesting propositions belong to the subject of *tangencies*, and lead up to the proposition that 'Given three unequal circles touching one another externally two and two, the diameter of the circle including them and touching all three is also given'. The enunciation is first given by itself between Propositions 6 and 7, and the presumption is that the succeeding propositions (7–9) were interposed as facilitating the final proof, which purports to be given in Proposition 10; but the method is not clear. One of the propositions is to the effect that, if there are two equal circles and a point external to both, the diameter of the circle passing through the point and touching both circles is given.

Section (2) of Book IV is highly interesting and ingenious. It relates to circles inscribed in the figure called ἄρβηλος ('shoemaker's knife') owing to its shape. If AB be a straight line divided into two parts, generally unequal, at C, and if semicircles be described on one and the same side of AB with AB, AC, CB as diameters respectively, the

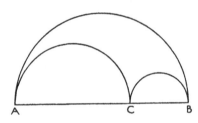

figure included between the three circumferences is the ἄρβηλος. We have seen (p. 339) that some propositions about the ἄρβηλος and circles inscribed in it are found in the Book of Lemmas which has come down to us in the Arabic under the name of Archimedes. Pappus' propositions form a more elaborate series and lead up to what, according to Pappus, was an ancient proposition to the effect that, if a series of circles be inscribed in the ἄρβηλος touching the semicircles and one another in the manner shown in the subjoined figure and proceeding from the first circle with centre P towards A, one end of the ἄρβηλος, and if the centres of the circles be P, Q, R ..., their diameters d_1, d_2, d_3 ... respectively, and if p_1, p_2, p_3 ... be the lengths of the respective perpendiculars from P, Q, R ... on AB, then

$p_1 = d_1, p_2 = 2d_2, p_3 = 3d_3 ...$

A number of lemmas precede the final proof which comes in Proposition 6.

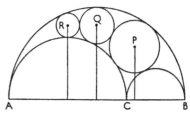

The same proposition holds if the series of circles, instead of being placed between the large and one of the small semicircles, come down between the two small semicircles.

Pappus proceeds to deal with other cases, e.g. (1) where, instead of the two smaller semicircles, we take straight lines drawn at right angles to AB from its extremities; (2) where, instead of one of the smaller semicircles, we have the straight line through C at right angles to AB; and (3) where we have no semicircle on CB as diameter but make the first circle touch CB itself and the two other semicircles.

Pappus' propositions include, as particular cases, those which are attributed to Archimedes in the Book of Lemmas.

The remaining sections of Book IV are not very clearly delimited, but they deal with special curves used for one or other of the two problems of squaring the circle and of trisecting any angle (or dividing it into two parts in any ratio).

First we have a description of the genesis and property of the *Spiral of Archimedes*. When Pappus speaks of the theorem about the spiral 'propounded by Conon' and the wonderful proof of it by Archimedes, he seems to be in error as regards Conon, since it is clear from Archimedes' preface to his own treatise *On Spirals* that it was he who first propounded the theorem to Conon rather than the reverse.

Pappus proceeds to find the area of the spiral (1) included between the first complete turn and the initial line, (2) included between the curve and the radius vector to any point on the first turn. His method is exceedingly interesting, being quite different from that of Archimedes, though both are equivalent to integrations. Pappus proves that the area of the spiral cut off by any radius vector OB to a point B on the first turn is one-third of the area of the sector of the circle described with centre O and radius OB which is cut off between the initial line and the radius OB. In the next propositions he proves that, if r, r' be two radii vectores, the areas cut off by them respectively are in the ratio of r^3 to r'^3; and it follows, as a consequence, that, if we take the radii vectores to the points reached when the revolving line has passed through angles of $\frac{1}{2}\pi$, π, $\frac{3}{2}\pi$, and 2π, the areas cut off by the four radii vectores are in the ratio of $(\frac{1}{4})^3$, $(\frac{1}{2})^3$, $(\frac{3}{4})^3$, 1^3 or of 1, 8, 27, 64, so that the areas of the spiral included in the four quadrants are in the ratio of 1, 7, 19, 37 (Prop. 22).

Pappus next describes the *conchoid* or *cochloid* of Nicomedes, its construction, and its use for the purpose of doubling the cube by finding two mean proportionals between two given straight lines. He mentions that the curve is the '*first* cochloid', while there have also been discovered a second and a third and a fourth, which are of use for other theorems; but he gives no details regarding these latter curves.

The *quadratrix* next claims attention; as to this Pappus observes that it was used by Dinostratus, Nicomedes, and other later geometers for squaring the circle, a use from which it acquired its name. The discussion of the curve

THE *COLLECTION*. BOOK IV

of Hippias includes (1) a description of the generation of the curve and its property, (2) an account of Sporus' objection to the construction as involving, practically, a *petitio principii*, an objection in which Pappus concurs (cf. p. 144 above), (3) two interesting alternative 'geometrical' constructions for the *quadratrix*, (a) one which uses 'loci on surfaces', the locus actually used being the cylindrical helix, (b) another which uses a right cylinder erected on an Archimedean spiral as base.

Before passing to the discussion of the problem of trisecting any angle, Pappus observes, by way of digression, that, just as we have an Archimedean spiral described in a plane by a point moving uniformly along a straight line while the straight line itself revolves uniformly round one extremity, and a cylindrical helix described on a cylinder by a fixed point on a straight line having two uniform motions, so we may have a *spiral described on a sphere* in a certain way.

Take a hemisphere with O as centre and H as pole

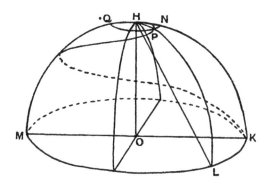

bounded by the great circle MLK. Suppose that the quadrant HNK of a great circle in a plane perpendicular to the plane of the circle MLK revolves uniformly about HO as axis, returning to its original position, so that K describes the circle KLM, and suppose that, concurrently, a point moves at uniform speed along the arc of the moving quadrant from H to K at such speed that the point arrives at K at the same moment at which HK assumes its original position. If, when the quadrant has reached the position HPL, the point has arrived at P, P describes a *spiral on a sphere*.

Pappus then finds, by an application of the method of exhaustion equivalent to an integration, the area of the portion of the surface of the hemisphere cut off towards the pole between the spiral and the arc HNK. Drawing in a separate diagram a quadrant ABC equal to the quadrant of a great circle in the sphere, he proves that the area of the portion of the surface of the hemisphere cut off by the spiral as described is to the surface of the hemisphere itself

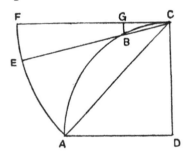

as (in the auxiliary figure) the area of the segment ABC is to that of the sector CAF (which is equal in area to the quadrant $ABCD$). He refers elements in the three-dimensional figure to corresponding elements in the auxiliary plane figure, and it would be difficult to imagine a more elegant piece of pure geometry. The result can be verified by using 'spherical co-ordinates' and integrating.

The final section of Book IV is mainly devoted to the problem of trisecting any given angle or dividing it into two parts in a given ratio. Pappus begins by referring again to the distinction between the three classes of problem called (1) plane, (2) solid, (3) linear, according as they can be solved by means of (1) the straight line and circle only, (2) conics or their equivalent, or (3) higher curves, i.e. curves 'which have a more complicated and forced or unnatural origin, being produced from more irregular surfaces and involved motions'. He mentions as such higher curves 'loci on surfaces', other curves discovered by one Demetrius of Alexandria in his *Linear Considerations*, and by Philon of Tyana, who interlaced 'plectoids' and other surfaces of all sorts, a certain curve called by Menelaus παράδοξος, besides of course spirals, *quadratrices*, cochloids, cissoids, &c. Pappus adds the often-quoted reflection on the error committed by geometers when they solve a problem by resort to an 'inappropriate' class (of curve or its equivalent), illustrating this by the use by Apollonius, in his Book V, of a rectangular hyperbola as a means of finding the feet of normals to a parabola passing through a given point (where a circle would serve the purpose), and by the assumption by Archimedes (*On Spirals*) of a solution of a certain νεῦσις which is 'solid'.

Pappus then solves the problem of trisecting any angle (which is a 'solid' problem), first by means of a certain νεῦσις (Prop. 33, chs. 41-2), and then directly (in two ways) by means of a hyperbola (Props. 34, 35). The νεῦσις itself is solved by means of the intersection of a hyperbola and a circle (cf. pp. 148-50 above). In one of the direct solutions by means of a hyperbola the focus-directrix property is used (cf. pp. 152-3 above). Solutions follow of the problem of dividing any angle in a given ratio by means (1) of the *quadratrix*, (2) of Archimedes' spiral.

Pappus appends some problems depending on the results obtained in the trisection of any angle or the division of it in any ratio, e.g. To construct an isosceles triangle in which either of the base angles has a given ratio to the vertical angle, To inscribe in a circle a regular polygon with any given number of sides, and To rectify a circle or a segment thereof (Props. 37-9).

The Book concludes with the solution, by means of the intersection of a parabola and a hyperbola, of the νεῦσις which Pappus considered that Archimedes had unnecessarily assumed in *On Spirals*, Proposition 8.

Book V is mainly devoted to the subject of *isoperimetry*, or the comparison of the content of figures which have different shapes but the same contour, or (in the case of solids) equal surfaces. The Book begins with an attractive passage on the sagacity of bees shown in their choice of the shape of the cells in which they store their honey; the passage offers one instance among many which show how the Greek mathematicians, when free from the restraint of mathematical phraseology, could write in language of the highest literary quality and charm. After generalities and a reference to the admirable orderliness of bees, their submission to the queens who rule in their state, and so on, Pappus continues:

'Presumably because they know themselves to be entrusted with the task of bringing from heaven to the cultured portion of mankind a share of ambrosia in the form of honey, they do not think it proper to pour it carelessly on ground or wood or any other ugly and irregular material; but, first collecting the sweets of the fairest flowers which grow on the earth, they make from them, for the reception of the honey, the vessels which we call honey-combs, with cells all equal, similar, and contiguous to one another, and hexagonal in form. And that they have contrived this by virtue of a certain geometrical forethought we may infer in this way. They would necessarily think that the figures must be such as to be contiguous to one another, that is to say, to have their sides common, in order that no foreign matter could enter the interstices between them and so defile the purity of their produce. Now there were three rectilinear figures which were capable of fulfilling this

THE 'COLLECTION'. BOOKS V, VI 449

condition—I mean ordered figures which are equilateral and equiangular—for the bees would have none of the figures which are not uniform.... There being then three figures capable by themselves of exactly filling up the space about the same point, namely, the (equilateral) triangle, the square, and the (regular) hexagon, the bees by reason of their instinctive wisdom chose for their construction the figure which has the most angles, because they conceived that it would hold more honey than either of the two others....

'We,' Pappus goes on, 'claiming as we do a greater share in wisdom than bees, will investigate a problem of still wider import, namely that, of all equilateral and equiangular plane figures having an equal perimeter, that which has the greater number of angles is always greater (in area), and the greatest plane figure of all those which have a perimeter equal to that of the polygons is the circle.'

Section (1) of Book V dealing with plane figures (cc. 1–10) seems to have followed very closely the exposition of Zenodorus in his work *On Isometric Figures* (cf. pp. 382–3 above), but Pappus adds an interesting proposition (preceded by some lemmas) that, of all segments of circles having the same length of arc, the semicircle is the greatest in area (Prop. 17).

Section (2) on the volume of the sphere in comparison with that of other solids having an equal surface with it, begins with a digression on the *semi-regular* solids of Archimedes (cf. pp. 337–8 above), and the number of edges and solid angles which they contain respectively (c. 19). Pappus then proves that the content of the sphere is greater than that of any of the regular solids which have the same surface (Prop. 18), and also greater than any cone or cylinder of equal surface (Prop. 19).

Section (3) proves a series of propositions on the same lines as those of Archimedes' *On the Sphere and Cylinder* and leading to the same results (cc. 21–43).

The last section (4) returns to the main problem of the Book, and Pappus, after a number of preliminary lemmas, finally proves that, of the five regular solids assumed to have their surfaces equal, that is greater in volume which has the more faces (cc. 43–65). The preliminary lemmas are contained in cc. 44–59; they include propositions about the perpendiculars from the centre of the circumscribing sphere to a face of the octahedron and the icosahedron respectively (Props. 39, 43), the proposition (48) that, if a dodecahedron and an icosahedron are inscribed in the same sphere, the same small circle of the sphere circumscribes both the pentagon of the dodecahedron and the triangle of the icosahedron (cf. Prop. 2 of the so-called 'Book XIV' of Euclid's *Elements* written by Hypsicles), and the proposition (49) that twelve of the regular pentagons inscribed in a circle are greater than twenty of the equilateral triangles inscribed in the same circle.

Book VI is mostly astronomical, dealing with the treatises included in the collection known as the 'Little Astronomy', that is, the smaller astronomical treatises which were studied as an introduction to the great *Syntaxis* of Ptolemy. The occasion for the exposition seems to be the sins of omission and commission on the part of many who taught the books in question, which seemed to Pappus to call for notice and correction. The *Sphaerica* of Theodosius, Autolycus' *On the Moving Sphere*, Theodosius' *On Days and Nights*, Aristarchus' *On the Sizes and Distances of the Sun and Moon*, Euclid's *Optics* and *Phaenomena* are all treated at some length. Allusions also occur to Menelaus' *Sphaerica*; in particular we learn that Menelaus in his *Sphaerica* used the word τρίπλευρον, *three-side*, to denote a spherical triangle.

There are some propositions of mathematical interest

in the section of the Book relating to Euclid's *Optics*. Two relate to straight lines drawn to a plane from an external point. (1) If from a point A above a plane a straight line AB be drawn perpendicular to the plane, and if from B, the foot of the perpendicular, a straight line BD be drawn perpendicular to EF, any straight line in the plane, then will AD also be perpendicular to EF. (2) If, from a point A above a plane, AB be drawn meeting the plane obliquely at B, and if BM be the orthogonal projection of BA on the plane, then, if we have any number of other lines BP drawn through B and lying in the plane, the angle ABM is the least of all the angles ABP, and the angle ABP increases as BP moves away from BM on either side of it; further, given one such angle ABP, there is only one other straight line through B and lying in the plane which makes with BA an angle equal to ABP, namely, a straight line BP' on the other side of BM making with BM an angle equal to the angle MBP. Other propositions follow (some of which recall propositions in Euclid's *Optics*) leading up to the main problem, 'Given a circle $ABDE$, and any point C within it, to find outside the plane of the circle a point from which the circle will have the appearance of an ellipse with C as centre.'

Book VII is historically the most important of all the Books, since it gives an account of the works constituting the 'Treasury of Analysis', as it was called, a collection of treatises which, after the *Elements* of Euclid, provided the body of doctrine necessary for the professional mathematician to know if he was to be regarded as fully equipped for the solution of problems arising in geometry. All but two of the works in question were by Euclid and Apollonius, and, as regards such of them as are now lost, Pappus' description, with such further indications of content as are to be gathered from his lemmas to each book, is our only source of information.

The treatises forming the Treasury of Analysis are enumerated by Pappus as follows: Euclid's *Data*, one Book, Apollonius' *Cutting-off of a Ratio*, two Books, *Cutting-off of an Area*, two Books, *Determinate Section*, two Books, and *Contacts*, two Books, Euclid's *Porisms*, three Books, Apollonius' *Inclinationes* or *Vergings* (νεύσεις), two Books, *Plane Loci*, two Books, and *Conics*, eight Books, Aristaeus' *Solid Loci*, five Books, Euclid's *Surface-Loci*, two Books, and Eratosthenes' *On Means*, two Books.

Pappus begins with a definition of *analysis* and *synthesis* which, as being perhaps the most formal statement we possess of the Greek view of the subject, shall be quoted in full:

Analysis takes that which is sought as if it were admitted and passes from it through its successive consequences to something which is admitted as the result of synthesis; for in analysis we assume that which is sought as if it were already done (γεγονός), and we inquire what it is from which this results, and again what is the antecedent cause of the latter, and so on, until, by so retracing our steps, we come upon something already known or belonging to the class of first principles, and such a method we call analysis as being solution backwards (ἀνάπαλιν λύσιν).

But in *synthesis*, reversing the process, we take as already done that which was last arrived at in the analysis and, by arranging in their natural order as consequences what before were antecedents, and successively connecting them one with another, we arrive finally at the construction of that which was sought; and this we call *synthesis*.

For completeness it ought to be added that each step in the chain of inference in the *analysis* must be *unconditionally convertible*, that is to say that, when in the

THE 'COLLECTION'. BOOK VII 453

analysis we say that, if A is true, B is true, we must be sure that the truth of A equally follows from the truth of B.

Pappus goes on to speak of the application of analysis in the case of theorems and problems respectively, observing that (1) where we begin by assuming a certain *theorem* to be true, if (*a*) we finally arrive at something admittedly true, the theorem is true, but if (*b*) we arrive at something admittedly false, the supposed theorem is also false; (2) where we assume a certain problem solved, if (*a*) we finally arrive at something admittedly possible or, as mathematicians say, *given*, the problem is possible, but if (*b*) we arrive at something admittedly impossible, the problem is impossible.

Pappus gives a short description of the contents of the above-mentioned works down to Apollonius' *Conics*, but not of Aristaeus' *Solid Loci*, or of Euclid's *Surface-Loci* or of Eratosthenes' *On Means*; nor has he any lemmas in the last three cases except two on the *Surface-Loci* of Euclid.

It is in his account of Apollonius' *Conics* that Pappus speaks of Euclid's *Conics* and Aristaeus' *Solid Loci*, with special reference to the 'loci with respect to three or four lines'. If p_1, p_2, p_3 or p_1, p_2, p_3, p_4 be the lengths of straight lines drawn from a point to meet three or four given straight lines at given angles respectively, and if in the one case $p_1 p_2 = \lambda p_3^2$, or in the second case $p_1 p_2 = \lambda p_3 p_4$ (where λ is a constant ratio), then the locus of the point is a conic section. Pappus goes on to say that a similar proposition holds for five or six lines; if, that is, p_1, p_2, p_3, p_4, p_5 or $p_1, p_2 \ldots p_6$ be the lengths of straight lines drawn from a point to meet each of the five or six given straight lines at a given angle, and if in the first case $p_1 p_2 p_3 = \lambda p_4 p_5 a$, (where a is a given length), or in the second case $p_1 p_2 p_3 = \lambda p_4 p_5 p_6$, the locus of the point is, in each case, a certain curve given in position. A similar theorem holds with regard to any odd or even number of lines, though, as there are not more than three dimensions in geometry, it cannot be expressed in the same form (of equating products); we have therefore to put it in the form of equating a product of ratios to a certain constant. If, then,

$$\frac{p_1}{p_2} \cdot \frac{p_3}{p_4} \ldots \frac{p_n}{a} \text{ or } \frac{p_1}{p_2} \cdot \frac{p_3}{p_4} \ldots \frac{p_n}{p_{n+1}} = \lambda,$$

the locus in either case is still a curve. The locus is, of course, a curve of a higher order than conics, and Pappus says that none of these loci had been investigated in his time except one, which, however, he does not describe. The general proposition, known as Pappus' Problem, is important historically because it was with reference to this particular problem that Descartes came to formulate, in his *Géométrie*, the method of co-ordinates.

By way of further emphasizing that new theorems remain to be discovered and established, Pappus adds that he will present his readers with another. This proves to be nothing less than the equivalent of the theorem commonly attributed to R. Guldin (1577-1643) to the effect that, if a plane figure revolve about an axis, the solid content of the figure produced by the revolution is equal to the product of the area of the revolving figure and the length of the path of its centre of gravity. Pappus' own statement is a little more complicated, as it speaks of the ratios between the contents of two different figures so revolving, and also includes uncompleted revolutions; but, as he says, his propositions are practically one, and enable all sorts of theorems about curves, surfaces, and solids to be proved by one demonstration.

THE 'COLLECTION'. BOOK VII

After the descriptions of the treatises come the lemmas supplied for use with them respectively. It is difficult to give any general idea of their nature, as they are so various; many are quite difficult and require full command of all the resources of pure geometry. Their number is greatly increased by the addition of alternative proofs, often requiring lemmas of their own. Where the treatise to which they relate is lost, it is often impossible to see the connexion of the lemmas with one another and the problems to which they relate.

The lemmas to the *Sectio rationis* and *Sectio spatii* of Apollonius are mostly by way of filling in details of the theory of proportion, e.g. by showing how to transform greater and less ratios *componendo*, *convertendo*, &c., in the same way as Euclid transforms equal ratios.

The *Determinate Section* of Apollonius seems, as we have seen (p. 366), to have amounted to a sort of Theory of Involution. In the lemmas we have at all events many remarkable applications of what Zeuthen has called geometrical algebra. The results are for our purposes best expressed in algebraical notation. The following are characteristic examples.

(1) If $ax = by$, then

(α) $\dfrac{b}{y} = \dfrac{(a+b)(b \pm x)}{(a \pm y)(x+y)}$ (Props. 22, 25, 29, 30, 32, 34);

(β) $\dfrac{a}{x} = \dfrac{(a-y)(a-b)}{(b-x)(y-x)}$ and $\dfrac{b}{y} = \dfrac{(a-b)(b-x)}{(a-y)(y-x)}$.

(Props. 35, 36.)

(2) If $\dfrac{a(a-d)}{c(c-d)} = \dfrac{(a-b)^2}{(b-c)^2}$, then $\dfrac{ac}{(a-d)(c-d)} = \dfrac{b^2}{(b-d)^2}$.

(Prop. 39.)

(3) If $(d-a)(d-c) = (b-d)(e-d)$ and $k = (e-a)+(b-c)$, then $(x-a)(x-c)+(x-e)(b-x) = k(x-d)$. (Props. 45-56.)

The results of Propositions 41-3 are obtained by giving x the values of a, b, c, e respectively in this equation.

In Propositions 61, 62, 64 (to which Propositions 59, 60, and 63 are lemmas) Pappus investigates the 'singular and minimum' (or maximum) values of the ratio

$$AP.PD : BP.PC,$$

where (A, D), (B, C) are point-pairs on a straight line and P is another point on the straight line. He finds the values for three different positions of P in relation to the four given points.

On the Νεύσεις of Apollonius Pappus gives the solution of the νεῦσις with reference to a rhombus (cf. pp. 373-5 above), and after that the solution by one Heraclitus of the same problem with regard to a square (Props. 71, 72). The problem is, *Given a square ABCD, to draw through B a straight line meeting CD in H and AD produced in E such that HE has a given length.*

Pappus first gives a lemma to the effect that, if (as in the annexed figure) any straight line BHE cut CD in H and AD produced in E, and if EF be drawn at right angles to BE meeting BC produced in F, then $CF^2 = BC^2 + HE^2$.

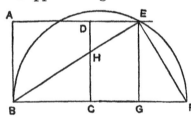

Draw EG perpendicular to BF.

Then the triangles BCH, EGF are similar, and $BC = EG$; therefore $EF = BH$.

Now $BF^2 = BE^2 + EF^2$,

or $BC.BF + BF.FC = BH.BE + BE.EH + EF^2$.

But, the angles HCF, HEF being right angles, the points H, C, F, E lie on a circle; therefore
$$BC.BF = BE.BH.$$
Hence, by subtraction,
$$\begin{aligned}BF.FC &= BE.EH+EF^2\\&= BE.EH+BH^2, \text{from above,}\\&= BH.HE+EH^2+BH^2\\&= EB.BH+EH^2\\&= FB.BC+EH^2.\end{aligned}$$
Taking away the common part $BC.CF$, we have
$$CF^2 = BC^2+EH^2.$$

Suppose now that we have to draw BHE through B so as to make HE equal to a given length k. Since BC, EH are both given, we have only to determine a length x such that $x^2 = BC^2+k^2$, produce BC to F so that $CF = x$, draw the semicircle on BF as diameter, produce AD to meet the semicircle in E and join BE. BE is then the straight line required.

We have already noticed the important lemmas to Apollonius' book on *Tangencies* (pp. 367–9). The lemmas to the *Plane Loci* of Apollonius are mostly propositions in geometrical algebra worked out by the methods of Euclid, Books II and VI. They include (Props. 122, 125)

(1) the well-known theorem that, if D be the middle point of the side BC of a triangle ABC
$$BA^2+AC^2 = 2(AD^2+DC^2);$$

(2) the very remarkable proposition that, if C, D be two points on a straight line AB,
$$AD^2+\frac{AC}{BC}.DB^2 = AC^2+AC.CB+\frac{AB}{BC}.CD^2.$$

This is equivalent to
$$AD^2.BC+BD^2.CA+CD^2.AB+BC.CA.AB = 0,$$
which is the general relation between four points on a straight line discovered by R. Simson, and therefore wrongly known as 'Stewart's Theorem' (Simson discovered this theorem for the more general case where D is a point outside the line ABC).

The 38 Lemmas to the *Porisms* of Euclid form an important collection, of which full use is made in the 'restorations' by Chasles and others of the original treatise. Six of them (Lemmas 1, 2, 4, 5, 6, 7) deal with a quadrilateral cut by any transversal, and one of them is equivalent to one of the equations by which we express the involution of six points. If A, A'; B, B'; C, C' be the points in which the transversal meets the pairs of opposite sides and the two diagonals respectively, Pappus' result is equivalent to
$$\frac{AB.B'C}{A'B'.BC'} = \frac{CA}{C'A'}.$$
Six lemmas (3, 10, 11, 14, 16, 19) deal with the equality of the anharmonic ratios which four straight lines issuing from a point determine on two transversals. Four (12, 13, 15, 17) may be regarded as expressing a property of the hexagon inscribed in two straight lines, namely that, if the vertices of a hexagon are situate, three and three, on two straight lines, the points of concourse of opposite sides are in a straight line.

Lemmas 20, 21 prove that, if one angle of one triangle be equal or supplementary to one angle of another, the areas of the triangles are in the ratio of the rectangles contained by the sides containing the equal or supplementary angles.

Seven other lemmas are of the nature of geometrical

algebra. Lemmas relating to the circle include the harmonic properties of the pole and polar, whether the pole is external (Prop. 154) or internal (Prop. 161), and also the proposition (156) that the straight lines drawn from the extremities of a chord to any point on the circumference divide harmonically the diameter perpendicular to the chord. Proposition 157 is remarkable in that (without any mention of a conic) it is practically identical with Apollonius' *Conics*, III. 45, about the foci of a central conic.

Among the lemmas to the *Conics* of Apollonius there are a large number containing geometrical algebra of the usual kind. There are also propositions about the construction of the hyperbola, and another proving that two hyperbolas having the same asymptotes do not intersect.

Two propositions (221, 222) give the equivalent of an obvious trigonometrical formula. If $ABCD$ be a rectangle, and AF any straight line meeting BC, CD (produced if necessary) in F, E, it is proved that

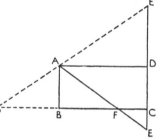

$$EA \cdot AF = ED \cdot DC + CB \cdot BF.$$

Dividing by $DE \cdot BF$, and denoting by θ the angle DAE, we have

$$\sec \theta \operatorname{cosec} \theta = \tan \theta + \cot \theta.$$

Of the lemmas to the *Surface-Loci* of Euclid, the second is important in that it is the first statement on record of the focus-directrix property of the three conic sections. Pappus gives the proof in full, and it is a fair inference that Euclid was aware of the property and assumed it without proof (cf. p. 266 above).

Book VIII begins with an interesting historical preface on the claim of theoretical, as distinct from practical or industrial, mechanics to rank as a mathematical subject. Archimedes, Philon of Byzantium, and Heron are referred to as the principal authorities on the subject, while Carpus of Antioch is also mentioned as having applied geometry to 'certain (practical) arts'. (Carpus, whose date is uncertain, is otherwise known as having squared the circle by means of a curve generated by a double motion, and as having had views on certain geometrical questions. Proclus says that he placed the 'angle' in the category of *quantity* as representing a sort of 'distance', 'extending one way', by which he no doubt meant *divergence* measured by rotation in one sense.)

According to Pappus, Heron regarded the theoretical part of mechanics as made up of geometry, arithmetic, astronomy, and physics, and the practical part ($\chi\epsilon\iota\rho\text{oup-}$ $\gamma\iota\kappa\acute{o}\nu$) as comprising work in metal, architecture, carpentering, and painting. But other varieties of mechanical work included by the ancients under the general term mechanics were (1) the use of the mechanical powers or devices for moving great weights by a small force, (2) the construction of engines of war, (3) the use of machines for pumping water from great depths, (4) the devices of 'wonder-workers' ($\theta\alpha\upsilon\mu\alpha\sigma\iota\text{oup}\gamma\text{o}\acute{\iota}$), some depending on pneumatics (as in Heron's *Pneumatica*), some using strings, &c., to produce movements like those of living things (cf. Heron's *Automatopoeëtice*), water-clocks (on which again Heron wrote), and spheres constructed to imitate the movements of the heavenly bodies with the 'uniform circular motion of water' as the moving power.

Pappus adds that he will give ancient propositions proved by geometrical methods, and others of his own, with a view to treating of such questions as (1) a comparison between the forces required to move a given weight along a horizontal plane and up an inclined plane respectively, (2) the finding of two mean proportionals between two unequal straight lines, and (3) the construction of a toothed wheel with a given number of teeth which will work on a given toothed wheel with any number of teeth.

Pappus defines the centre of gravity as 'the point within a body which is such that, if the weight be conceived to be suspended from the point, it will remain at rest in any position in which it is put'. The practical method of determining the point by means of the intersection, first of planes, then of straight lines, meeting within the body is next described, after which Pappus proves (Prop. 2) a theorem of some difficulty, namely that, if D, E, F be points on the sides of a triangle ABC dividing those sides in the same proportion, i.e. if

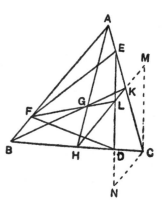

$$BD : DC = CE : EA = AF : FB,$$

then the triangles ABC, DEF have the same centre of gravity. In the proof Pappus in effect uses 'Menelaus' Theorem' twice, though he does not quote it, but proves the particular cases in lemmas *ad hoc*.

Another geometrical proposition is also of interest (Prop. 7).

Given two straight lines AB, AC, and B, a fixed point on AB, if CD be drawn with its extremities on AC, AB respectively and such that $AC : BD$ is equal to a given ratio, then the centre of gravity of the triangle ADC will lie on a certain straight line.

Bisect AC at E, and take F on DE such that $DF = 2FE$.

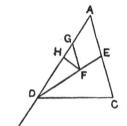

Also let H be a point on BA such that $BH = 2HA$. Draw FG parallel to CA.

Then $AG = \frac{1}{3}AD$, and $AH = \frac{1}{3}AB$; therefore $HG = \frac{1}{3}BD$.

Again $FG = \frac{2}{3}AE = \frac{1}{3}AC$; therefore, since the ratio AC to BD is given, the ratio of FG to GH is given.

And the angle FGH ($= A$) is given; therefore the triangle FGH is given in species, and hence the angle GHF is given.

But H is a given point; therefore HF is a given straight line; and it contains the centre of gravity of the triangle ADC.

The attempt to find the force required to haul a weight in the shape of a sphere up an inclined plane is curious as being apparently the first or only attempt in ancient times to investigate motion on an inclined plane; but it is crude and of no value (c. 10).

Pappus next gives, from Heron's *Barulcus*, a description of the machine consisting of a pulley, interacting toothed wheels, and an endless screw working on the last wheel and turned by a handle. At the end of the chapter he repeats his construction for finding two mean proportionals (cf. pp. 169–70 above).

Chapters 13–17 are interesting because they contain the solution of the problem of *constructing a conic through five given points*. The problem arises in this way. Suppose that we are given a broken piece of the surface of a cylindrical column such that no portion of the circumference of either of its bases is left intact, and let it be required to find the diameter of the circular section of the cylinder. The method adopted is this. We take any two points A, B on the surface of the fragment, and with five different radii we draw five pairs of circles having A, B respectively as centres. The pairs of circles with equal radii, intersecting at points on the surface, determine five points in one plane section (in general oblique) made by the plane bisecting AB at right angles, and all lying on the surface. The five points are then represented on any plane by triangulation, and the construction proceeds.

When the conic section is drawn through the five points, we still have to find the principal axes, in order to determine the diameter of the circular section of the cylinder, for the minor axis of the section will give that diameter. A pair of conjugate diameters arises out of the construction of the conic, as given by Pappus, and Pappus now shows, lastly, how, given any pair of conjugate diameters, we can find the axes (c. 17), though the proof is omitted.

Chapter 23 contains an interesting problem:

To inscribe in a circle seven equal regular hexagons in such a way that one is about the centre of the circle, while the six others stand on its sides and have their opposite sides in each case placed as chords in the circle.

Suppose $GHKLNM$ to be the hexagon so described on a side HK of the inner hexagon; OKL is then a straight line. Produce OL to meet the circle in P.

Then $OK = KL = LN$. Therefore, in the triangle OLN, $OL = 2LN$. And the included angle OLN is given ($= 120°$).

Therefore the triangle OLN is given in species, so that the ratio $ON:NL$ is given; and since one side, ON, is given, NL is given.

The auxiliary construction for NL is effected thus.

Take AF equal to the radius OP, and let $AC = \frac{1}{3}AF$. On AC as base describe a segment of a circle containing

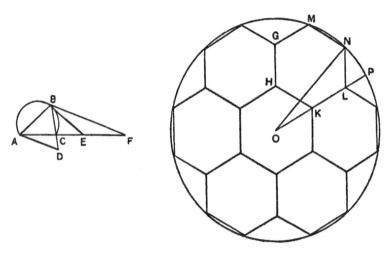

an angle of $60°$. Take CE along CF equal to $\frac{4}{5}AC$. Draw from E the tangent EB to the circle, and join AB.

Then shall AB be equal in length to the side of the hexagon required.

Produce BC to D so that $BD = BA$, and join AD. The triangle ABD is then equilateral.

Now, since EB is a tangent to the circle, $AE \cdot EC = EB^2$, or $AE : EB = EB : EC$, and the triangles EAB, EBC are therefore similar.

Therefore $\quad BA^2 : BC^2 = AE^2 : EB^2$
$$= AE : EC$$
$$= 9 : 4.$$

Hence $BC = \frac{2}{3}BA = \frac{2}{3}BD$, so that $BC = 2CD$.

Now $CF = 2CA$; therefore $AC : CF = DC : CB$, and AD, BF are parallel.

Therefore $BF:AD = BC:CD = 2:1$,
and $BF = 2AD = 2AB$.

Also $\angle FBC = \angle BDA = 60°$, so that $\angle ABF = 120°$, and the triangle ABF is therefore equal and similar to the required triangle NLO, which can therefore be drawn.

The rest of the Book is devoted to the construction (1) of toothed wheels with a given number of teeth, each of which is equal to those of a given toothed wheel, and (2) of a cylindrical helix, the *cochlias*, indented so as to work on a toothed wheel. The text is evidently defective at the end, and an interpolator has inserted extracts from Heron's *Mechanics* on the mechanical powers, the simple and compound pulley, the lever, the wedge, and the screw.

XVII

ALGEBRA: DIOPHANTUS OF ALEXANDRIA

GREEK tradition traces the beginnings of algebra no less than geometry to the Egyptians. The Greeks were familiar with the Egyptian methods of calculation. Thus the scholiast to Plato's *Charmides* (165 E) speaks of λογιστική, the science of calculation, as including 'the so-called Greek and Egyptian methods in multiplications and divisions, and the additions and subtractions of fractions'. Plato himself (*Laws*, 819 A–C) says that free-born boys should, as is the practice in Egypt, learn, along with their reading, simple mathematical calculations adapted to their age and presented in a form combining amusement with instruction, e.g. problems of distributing apples, garlands, and the like, such as occur in ordinary life.

Of just this character are the Egyptian *hau*-calculations found in the Papyrus Rhind. *Hau*, meaning a *heap*, is best translated as 'quantity'; the calculations in terms of it are equivalent to the solution of simple equations with one unknown, though the form is different. Problems are enunciated thus: 'A quantity to which its two-thirds, its half, and its seventh are added becomes 33'; 'two-thirds added and one-third taken away, 10 remains': the equivalents are

$$x + \tfrac{2}{3}x + \tfrac{1}{2}x + \tfrac{1}{7}x = 33,$$

$$x + \tfrac{2}{3}x - \tfrac{1}{3}(x + \tfrac{2}{3}x) = 10.$$

The solutions in different cases can be variously regarded, namely as examples (1) of direct division, (2) of the use of the rule of three or proportion, or (3) of 'false hypothesis' or *regula falsi*. Two instances of (3) may be given. The first is from the Rhind Papyrus, and is a case of an arithmetical progression. 'A hundred loaves to five men, one-seventh of the three first men to the two last. What is the difference of share?' This is an elliptical way of saying that 'the shares are in arithmetical progression, and the sum of the last two is 1/7th of the sum of the first three; what is the common difference?' Ahmes says, 'The doing as it occurs, the difference of share is $5\frac{1}{2}$'; then, assuming 1 as the smallest share, he writes down the terms 23, $17\frac{1}{2}$, 12, $6\frac{1}{2}$, 1 (with $5\frac{1}{2}$ as the common difference).

Now the sum of the five terms is 60, and 100 is $1\frac{2}{3}$ times 60. Ahmes says simply 'you are to count with $1\frac{2}{3}$', meaning 'multiply the terms by $1\frac{2}{3}$', and thus obtains $38\frac{1}{3}$, $29\frac{1}{6}$, 20, $10\frac{2}{3}\frac{1}{6}$, $1\frac{2}{3}$, numbers which, when added together, produce 100.

To obtain $5\frac{1}{2}$ as the common difference the solver must have tacitly solved the equivalent of an auxiliary equation thus. If we assume 1 as the smallest share, and d as the common difference, the five terms (beginning from the smallest) will be 1, $1+d$, $1+2d$, $1+3d$, $1+4d$, and the conditions of the problem give

$$(1+2d)+(1+3d)+(1+4d) = 7\{1+(1+d)\},$$
that is, $\qquad 9d+3 = 7d+14,$
whence $\qquad d = 5\frac{1}{2}.$

The second example (from the Berlin Papyrus 6619) is the problem 'To divide 100 square cubits into two squares the sides of which are in the ratio $1 : \frac{3}{4}$', which is equivalent to the simultaneous equations

$$x^2+y^2 = 100,$$
$$x:y = 1:\tfrac{3}{4}, \qquad \text{or } y = \tfrac{3}{4}x.$$

468 ALGEBRA

It is first assumed that $x = 1$, and x^2+y^2 is thus found to be $\frac{25}{16}$. In order to make 100, $\frac{25}{16}$ has to be multiplied by 64. The true value of x is therefore 8 times 1, or 8.

Many other examples of the type of problem alluded to by Plato are to be found in the arithmetical epigrams of the Greek Anthology. Most of these appear under the name of Metrodorus, a grammarian who lived, probably, in the time of the Emperors Anastasius I (A.D. 491–518) and Justin I (A.D. 518–27). They were obviously only collected by Metrodorus, from ancient as well as more recent sources. Among the epigrams (46 in number) we find problems of dividing, say, a number of apples or nuts among a certain number of persons, which lead to simple equations. For example, a number of apples has to be found such that, if four persons out of six receive one-third, one-eighth, one-fourth, and one-fifth respectively of the total number, while the fifth receives 10 apples, one apple remains over for the sixth person. This is equivalent to the equation

$$\tfrac{1}{3}x+\tfrac{1}{8}x+\tfrac{1}{4}x+\tfrac{1}{5}x+10+1 = x.$$

Plato's remark shows that the origin of such problems dates back, at least, to the fifth century B.C.

The problems in the Anthology may be thus classified.

(1) Twenty-three lead to simple equations of the above type; one is an epigram about the age of Diophantus (xiv. 126).

(2) Twelve lead to easy simultaneous equations with two unknowns, like Dioph. I. 6, one other to simultaneous equations between three, and one other to simultaneous equations between four, unknowns, namely (xiv. 51 and 49),

$$x = y+\tfrac{1}{3}z, \qquad y = z+\tfrac{1}{3}x, \qquad z = 10+\tfrac{1}{3}y,$$

and $x+y = 40$, $x+z = 45$, $x+u = 36$, $x+y+z+u = 60$.

With these equations may be compared Dioph. I. 16-21, as well as the general solution of simultaneous equations of the latter type with any number of unknowns given by Thymaridas, which was known as the ἐπάνθημα, 'flower' or 'bloom', of Thymaridas (cf. pp. 57-9 above).

(3) Six are problems of the usual sort about the filling and emptying of vessels by pipes, and the like; e.g. (xiv. 130), if one pipe fills the vessel in one day, a second in two days, and a third in three; how long will all three running together take to fill it?

(4) The Anthology also contains two cases of *indeterminate* equations of the first degree. One is a distribution of apples, $3x$ in number, into parts satisfying the equation $x-3y=y$, where y is not less than 2; the other leads to the equivalent of three equations connecting four unknowns,

$$x+y = x_1+y_1,$$
$$x = 2y_1,$$
$$x_1 = 3y,$$

the general solution of which is $x = 4k$, $y = k$, $x_1 = 3k$, $y_1 = 2k$. Diophantus solves these very equations (I. 12), but makes them determinate by assuming

$$x+y = x_1+y_1 = 100.$$

Enough has been said to show that Diophantus was not the inventor of algebra. Nor was he the first to solve indeterminate problems of the second degree. We have already seen (p. 47) that Pythagoras and Plato were both credited with the discovery of formulae for finding any number of square numbers which are the sum of two square numbers, the formulae being

$$n^2 + \{\tfrac{1}{2}(n^2-1)\}^2 = \{\tfrac{1}{2}(n^2+1)\}^2 \qquad (n \text{ odd})$$
and
$$(2n)^2 + (n^2-1)^2 = (n^2+1)^2.$$

470   ALGEBRA

Euclid's more general formula (Lemma following X. 28) is
$$m^2 n^2 p^2 q^2 = \{\tfrac{1}{2}(mnp^2 + mnq^2)\}^2 - \{\tfrac{1}{2}(mnp^2 - mnq^2)\}^2.$$
The Pythagoreans, too, as we have seen (pp. 55–7), discovered the series of 'side-' and 'diameter-' numbers furnishing any number of successive solutions of the equations
$$2x^2 - y^2 = \pm 1.$$
Before we come to Diophantus, it is necessary to notice some further examples of indeterminate problems discovered by Heiberg in the Constantinople manuscript from which Schöne edited the *Metrica* of Heron; they are now included in vol. iv of Heiberg's edition of Heron. The problems are as follows.

1. To find two rectangles such that the perimeter of the second is three times that of the first, and the area of the first is three times that of the second. If we substitute n times for 3 times, the problem is equivalent to the solution of the equations
$$u + v = n(x+y), \quad \ldots \quad (1)$$
$$xy = n \cdot uv, \quad \ldots \quad (2)$$
and the solution in the text is equivalent to
$$x = 2n^3 - 1, \qquad y = 2n^3,$$
$$u = n(4n^3 - 2), \qquad v = n.$$
It is not shown how the solution was arrived at; but Zeuthen suggested the following as a possible method.

As the problem is indeterminate, it is natural to start with some hypothesis, e.g. to put $v = n$. It follows from (1) that u must be a multiple of n, say nz. Thus
$$x + y = 1 + z,$$
while, by (2), $xy = n^3 z;$
hence $xy = n^3(x+y) - n^3$, or $(x - n^3)(y - n^3) = n^3(n^3 - 1).$

An obvious solution is $x-n^3 = n^3-1$, $y-n^3 = n^3$, which gives $z = 2n^3-1+2n^3-1 = 4n^3-2$, so that
$$u = nz = n(4n^3-2).$$

2. A similar problem about two rectangles is equivalent to the solution of the equations
$$x+y = u+v,$$
$$xy = n.uv;$$
and the solution given in the text, without further explanation, is
$$x+y = u+v = n^3-1,$$
$$u = n-1, \quad v = n(n^2-1),$$
$$x = n^2-1, \quad y = n^2(n-1).$$

3. To find a right-angled triangle with sides in the ratios of rational numbers and having an area of 5 square feet. We are told to multiply 5 by some square number containing 6 as a factor, e.g. 36. This gives 180, and this is the area of the right-angled triangle (9, 40, 41). Divide each side by 6, and we have the triangle required. The author was, then, aware that *the area of a right-angled triangle with sides in whole numbers is divisible by* 6.

4. The last four problems are particular cases of one problem, To find a rational right-angled triangle such that the arithmetical sum of its area and its perimeter is a given number. The solution depends on the following formulae (easily proved by means of a figure), where a, b are the perpendicular sides and c the hypotenuse of a right-angled triangle, Δ the area, r the radius of the inscribed circle, and $s = \tfrac{1}{2}(a+b+c)$:
$$\Delta = rs = \tfrac{1}{2}ab, \quad r+s = a+b, \quad c = s-r.$$
Solving the first two equations to find a, b, we have
$$\left.\begin{matrix}a\\b\end{matrix}\right\} = \tfrac{1}{2}[r+s \mp \sqrt{\{(r+s)^2-8rs\}}].$$

472   ALGEBRA

Bearing in mind that $\Delta+2s$ or $s(r+2)$ is equal to a given number (A, say), we have to find suitable values for s and r by separating A into suitable factors and equating $s, r+2$ to these factors respectively; the factors must be so chosen as to make sr, the area, divisible by 6. In one case taken $A = 280$, and the author chooses 8, 35 as factors, making $s = 35$, $r+2 = 8$, because sr is then 6.35. The writer then says that

$a = \frac{1}{2}[6+35-\sqrt{\{(6+35)^2-8.6.35\}}] = \frac{1}{2}(41-1) = 20$,
$b = \frac{1}{2}(41+1) = 21$,
$c = 35-6 = 29$.

The above problems about right-angled triangles in rational numbers would be suitable for Diophantus' Book VI. He does not, however, include them, although he finds rational right-angled triangles such that (1) the area plus or minus *one* of the perpendicular sides, (2) the area plus or minus the sum of *two* sides, is a given number.

The geometrical algebra of the Greeks included, from the time of the Pythagoreans onwards, the geometrical solution of any quadratic equations having real roots. When the method was first applied arithmetically is uncertain, but the numerical solution of quadratic equations is well established in Heron's writings, so that Diophantus was not the first to treat quadratic equations algebraically.[1] What Diophantus did was to take the first step towards an algebraical *notation*.

DIOPHANTUS of Alexandria flourished probably about A.D. 250 or not much later. This is gathered from the statement in a letter of Psellus, published by Tannery in his edition of Diophantus, that Anatolius (Bishop of Laodicea about A.D. 280) dedicated to Diophantus a concise treatise on the Egyptian method of calculation.

[1] See also note in Appendix on Babylonian mathematics, pp. 524–8.

DIOPHANTUS OF ALEXANDRIA

An epigram in the Anthology purports to give some details of the life of Diophantus: his boyhood lasted $\frac{1}{6}$th of his life; his beard grew after $\frac{1}{12}$th more; after $\frac{1}{7}$th more he married, and his son was born 5 years later; the son lived to half his father's age, and the father died 4 years after the son. If x was his age at death,

$$\tfrac{1}{6}x + \tfrac{1}{12}x + \tfrac{1}{7}x + 5 + \tfrac{1}{2}x + 4 = x,$$

which gives $x = 84$.

The great work of Diophantus is the *Arithmetica*, of which six Books (out of the original thirteen) survive. In addition, we have a fragment of a tract on *Polygonal Numbers*. Allusions in the *Arithmetica* imply the existence of a collection of propositions under the title of *Porisms*; for in three propositions of Book V Diophantus quotes as known certain facts in the Theory of Numbers, introducing them with the words 'We have it in the *Porisms* that . . .'

THE 'ARITHMETICA'

None of the manuscripts which we possess contain more than six Books, although some few of them divide the content of the six Books into seven, and one or two give the fragment on Polygonal Numbers as Book VIII. The missing Books were apparently lost before the tenth century, for there is no sign that the Arabs ever possessed them. It is possible that Hypatia's commentary extended to six Books only, and that, partly in consequence, the remaining Books were first forgotten and then lost (cf. the case of Apollonius' *Conics*, where the only Books which have survived in Greek, Books I–IV, are those on which Eutocius wrote a commentary).

Opinions have differed as to the probable relation of the lost to the surviving Books. Nesselmann thought that the last portion bore to the whole a ratio much less than would be implied by the proportion of 7 to 13, and that the missing portion came in the middle rather than at the end, possibly containing, besides the *Porisms*, the treatment of determinate equations of the second degree (which Diophantus promises in his preface); these topics would, however, be far from sufficient to fill seven Books. Tannery, on the other hand, held that it is the last and most difficult Books which are lost, pointing to the immense distance that separates even the most difficult of Diophantus' problems from, say, the Cattle-Problem attributed to Archimedes. The question remains an open one.

Tannery thought that the *Porisms* were not a separate collection of propositions, but were scattered here and there in the *Arithmetica* as *corollaries*. But it would be strange if Diophantus quoted such propositions under a separate title 'the *Porisms*' when it would be easier to refer to particular places in the various Books where they occurred, or to say generally 'for this has been proved'; hence I think it preferable to suppose, with Hultsch, that the *Porisms* formed a separate work.

The name of Diophantus came to be used, as were the names of Euclid, Archimedes, and Heron, for the purpose of palming off the compilations of much later authors. Three fragments are included in Tannery's edition under the heading 'Diophantus Pseudepigraphus'; but not one of them has anything to do with Diophantus.

The early commentators on Diophantus include (1) Hypatia, the daughter of Theon of Alexandria, who was murdered by Christian fanatics in A.D. 415, (2) the Arabians Abū 'l Wafā al-Būzjānī (940–98), Qusṭā b. Lūqā (d. about 912), and probably Ibn al-Haitham (about 965–1039), (3) Georgius Pachymeres (1240 to about 1310), who wrote

DIOPHANTUS OF ALEXANDRIA 475

in Greek a paraphrase of some portion of Diophantus, sections 25–44 of which, relating to Book I, Def. 1–Prop. 11, survive, (4) Maximus Planudes (about 1260–1310), who wrote a systematic commentary on Books I, II.

To Regiomontanus belongs the credit of being the first to call attention to the work of Diophantus as being extant in Greek and calling for a translator: 'No one', he said in an *Oratio* delivered in 1463, 'has yet translated from the Greek into Latin the fine thirteen Books of Diophantus, in which the very flower of the whole of arithmetic lies hidden, the *ars rei et census* which to-day they call by the Arabic name of Algebra'. Rafael Bombelli found a manuscript of the work in the Vatican and, towards 1570, thought of publishing it. With Antonio Maria Pazzi, he translated five Books out of the seven into which the manuscript was divided, but did not publish them; instead, he took all the problems of the first four Books and some from Book V and incorporated them in his *Algebra* (1572) among problems of his own.

Next Xylander (Wilhelm Holzmann), by dint of extraordinary industry and care, produced a most meritorious Latin translation, with commentary (1575). This was used, but without sufficient appreciation, by Bachet, who in 1621 published the *editio princeps* of the Greek text with Latin translation and notes. A second edition, carelessly printed and untrustworthy as regards the text, which was brought out in 1670, is precious for the reason that it contained the epoch-making notes of Fermat; it was edited by S. Fermat, his son. Unfortunately neither Xylander nor Bachet was able to use the best manuscripts. The best of all is that of Madrid (Matritensis 48 of the thirteenth century), which, however, was unfortunately spoiled by corrections made, especially in Books I, II, from a manuscript of an inferior class; but here Vat. Gr. 191, which was copied from the Madrid manuscript before the general alterations referred to, comes to our assistance. These two are the first of the manuscripts used by Paul Tannery in preparing his definitive Greek text (Teubner, 1893, 1895).

Simon Stevin published in 1585 a French version of Books I–IV, to which Albert Girard added Books V–VI, the complete edition appearing in 1625. German translations were brought out by Otto Schulz in 1822 and by G. Wertheim in 1890. A reproduction of Diophantus in modern notation, with introduction and notes, by the present writer (second edition, 1910) is based on the text of Tannery, and includes a supplement on the work of Fermat and Euler in relation to Diophantine theorems and problems.

The *Arithmetica* can best be dealt with under three main headings, (1) the notation and definitions, (2) the principal methods employed, so far as they can be classified, (3) the nature of the contents, including the assumed Porisms, with some typical examples of the devices by which the problems are solved.

It is in Diophantus that we find the first systematic use of a notation to express certain constantly recurring quantities, although the signs used are of the nature of mere abbreviations rather than algebraical symbols in our sense. There is also, as we shall see, one sign of operation.

We will take first the sign for the unknown quantity ($=x$). Diophantus defines the unknown as 'an indeterminate or undefined multitude of units' (πλῆθος μονάδων ἀόριστον), adding that it is called ἀριθμός, i.e. *number*, simply, and is denoted by a certain sign. In the earliest (the Madrid) manuscript the sign takes the form Ч, in Marcianus

308 it is ς. In the printed editions before Tannery's it was represented by a final sigma with an accent, ς'. For the plural the sign was doubled, and for the different cases of ἀριθμός the case-endings were added above and to the right, like an exponent, e.g. ς`` for ἀριθμόν, as τ`` for τόν, ς^{οῦ} for ἀριθμοῦ, ςς^{ούς} for ἀριθμούς, &c.; ςς^{οἱ} ια (with the numeral ια added) means 11 ἀριθμοί ($=11x$), and so on. The sign is often used for ἀριθμός in its ordinary sense of 'number', and not the unknown quantity, which shows that it was merely an abbreviation for the word. It appears also in slightly different forms in the manuscripts of other Greek mathematicians, e.g. (1) in the Bodleian MS. of Euclid (D'Orville, 301), in the forms ϛ̣, ϛ̈, or as a curved line similar to the abbreviation for καί, (2) in the manuscripts of Archimedes' *Sand-reckoner* in a form approximating to ς, (3) in a manuscript of the *Geodaesia* included in the Heronian collections edited by Hultsch, where it sometimes resembles ζ, sometimes ρ, sometimes ο, and once ξ.

In the Bodleian manuscript of Diophantus the sign appears in the form 'ϛ̄, quite unlike the final sigma. This form, combined with the fact that Xylander's manuscript in one place read αρ for the full word, suggested to me years ago that the sign might be merely a contraction for the letters αρ, especially as I found a note in Gardthausen of a contraction for αρ in the form ῠρ, occurring in a papyrus of A.D. 154. This might have become ϛ̓, and the loss of the downward stroke or of the first half of the loop would give a fair approximation to the forms which we know. I am not aware that this hypothesis as to the origin of the sign has since been improved upon. It has the immense advantage that it brings the sign for ἀριθμός into the same category as the signs for the powers of the unknown, e.g. $Δ^Υ$ for δύναμις (square), $Κ^Υ$ for κύβος (cube), and the sign for the unit, $\overset{\circ}{\mathsf{M}}$ (μονάς), the sole difference being that the two letters coalesce into one instead of being separate.

The powers of the unknown, corresponding to our $x^2, x^3, \ldots, x^6$, are represented as follows:

Δ$^\Upsilon$ (for δύναμις, square) represents x^2,
Κ$^\Upsilon$ (for κύβος, cube) ,, x^3,
Δ$^\Upsilon$Δ (for δυναμοδύναμις, square-square) ,, x^4,
ΔΚ$^\Upsilon$ (for δυναμόκυβος, square-cube) ,, x^5,
Κ$^\Upsilon$Κ (for κυβόκυβος, cube-cube) ... ,, x^6.

Diophantus does not go beyond the sixth power. Each of the above terms (except δύναμις) may be used for the particular power of any ordinary number as well as of the unknown; but a particular *square* number is spoken of as τετράγωνος ἀριθμός, not δύναμις.

The reciprocals of the various powers of the unknown, i.e. our $1/x$, $1/x^2$ ... are denoted by the same signs with a special mark attached, just as an aliquot part is ordinarily denoted by the sign for the corresponding number with an accent, e.g. $\gamma' = \frac{1}{3}$, $\iota\alpha' = \frac{1}{11}$. Tannery prints the mark in the case of the unknown and its powers thus, $^\chi$, that is,

ς$^\chi$ (for ἀριθμοστόν) represents $1/x$,
Δ$^{\Upsilon\chi}$ (for δυναμοστόν) represents $1/x^2$,

and so on.

The coefficient of any term is represented by the ordinary numeral following the sign: thus ΔΚ$^\Upsilon$ κς $= 26x^5$, Δ$^{\Upsilon\chi}$ σν $= 250/x^2$.

Diophantus does not need any signs for the operations of multiplication and division. Addition is indicated by mere juxtaposition: thus Κ$^\Upsilon$αΔ$^\Upsilon$ιγ ς ε corresponds to $x^3 + 13x^2 + 5x$. When there are units in addition, they are expressed by $\overset{\circ}{\mathsf{M}}$ (for μονάδες) with the appropriate numeral.

There is a sign for subtraction or *minus*. The full term is λεῖψις, a *wanting*, as opposed to ὕπαρξις, a *forthcoming*, which denotes a positive term. The symbol indicating a *wanting*, corresponding to our sign for *minus*, is ⋏, and the text explains that 'the sign for a *wanting* is a Ψ turned downwards and truncated'. The description is evidently interpolated, and it is certain that the sign has nothing to do with Ψ. Nor is it confined to Diophantus, for it appears in practically the same form in Heron's *Metrica*, where we find the expression μονάδων οδ ⋏ ι'δ' in the sense of 74 units minus $\frac{1}{14}$. In Diophantus the sign is generally resolved into λείψει, the dative of λεῖψις, but in other places the sign is used instead of parts of the verb λείπειν, namely λιπών or λείψας (meaning 'having dropped' or 'lost', i.e. having had something subtracted from it) and once even λίπωσι. The proper construction was no doubt λιπών or λείψας followed by the accusative of the quantity subtracted; in one place λείψει is actually followed by the accusative, showing that the abbreviation was there wrongly resolved into λείψει instead of λείψας. Heron's phrase written in full would therefore probably have been μονάδων οδ λιπουσῶν (or λειψασῶν) τεσσαρακαιδέκατον. It thus appears probable that ⋏ is a mere compendium for the root of the verb λείπειν, and is made up of Λ and Ι placed in the middle (cf. Τ̄ for ΤΛ as an abbreviation for τάλαντον).

Lastly, there is a sign for ἴσος, equal, connecting the two sides of an equation. The archetype seems to have had ι^σ for ἴσος, equal; but copyists introduced a sign which was sometimes confused with ϥ, the sign for ἀριθμός.

Since Diophantus has no sign for *plus*, it is necessary for him, where an expression contains several terms with different signs, to write all the positive terms together, and to place all the negative terms together after the sign for *minus*. Thus for the equivalent of x^3-5x^2+8x-1 he would have to write $K^Y \alpha \varsigma \eta \pitchfork \Delta^Y \epsilon \overset{o}{M} \alpha$.

Diophantus' method of writing fractions as well as large numbers has been explained above (pp. 19–22). It is only necessary to add that, where the numerator and denominator are complicated expressions, he writes the numerator first, then ἐν μορίῳ (or μορίου), meaning 'divided by', followed by the denominator, e.g.

$\Delta^Y \xi \overset{o}{M} \,\beta\phi\kappa \; \grave{\epsilon}\nu \; \mu o\rho\acute{\iota}\omega \; \Delta^Y\Delta \; \alpha \; \overset{o}{M} \; \gamma \pitchfork \Delta^Y \xi$
$= (60x^2+2520)/(x^4+900-60x^2).$

For a *term* in a particular expression Diophantus uses εἶδος, 'species', which primarily means the 'denomination' according to the particular power of the unknown that it contains. He has directions for simplifying equations by getting rid of negative terms, and so on. The object, he says, is to reduce the equation until we have one term equal to one term; 'but', he adds, 'I will show you later how, in the case also where two terms are left equal to one term, such a problem is solved'. Wherever he can, Diophantus endeavours, by suitable assumptions, to reduce any equation either to a simple equation or to a *pure* quadratic. The solution of a mixed quadratic is clearly assumed in several places, but Diophantus never gives the promised general explanation with regard to it.

Diophantus was under a serious disability in that he possessed notation for one unknown only, whereas we can use, at the outset, as many as we please. Diophantus gets over the difficulties in the most ingenious way by an infinite variety of expedients. First, he shows great adroitness in choosing his unknown in such a way that he can at once express most, if not all, of the quantities

DIOPHANTUS OF ALEXANDRIA

occurring in the problem in terms of the one unknown; this is equivalent to an elimination of unknown quantities. Secondly, when he arrives at an indeterminate equation containing two or more unknowns, he assumes for one or other of them some particular number arbitrarily chosen, while making it clear that we may equally choose any other number, so that, as a rule, there is no real loss of generality. In two cases (II. 28, 29), where it is imperatively necessary to use two unknowns, Diophantus calls one of them ς as usual; the second, for want of a term, he agrees to call, in the first instance, 'one unit'. Then later, having completed the part of the solution necessary for finding the x, he substitutes its value and uses x over again for what he had originally called '1'. He has, therefore, to put his finger on the place to which, in the working, the 1 has passed, which happens, in the particular cases, not to be too difficult.

It should be observed that Diophantus will only admit solutions in rational numbers, that is, integers, or fractions which are ratios of integers; he excludes not only surds and imaginary quantities, but also negative quantities. Of a negative quantity *per se*, i.e. without some greater quantity from which it is subtracted, he had apparently no conception, notwithstanding that he says generally that a *wanting* (i.e. *minus*) multiplied by a *wanting* makes a *forthcoming* (i.e. *plus*), and a *wanting* multiplied by a *forthcoming* makes a *wanting*. Such equations then as lead to irrational or negative roots he regards as useless for his purpose; the solutions are then 'impossible' ($\dot{a}\delta\acute{v}\nu a\tau o\varsigma$). When he arrives in the course of his working at such an equation, he retraces his steps, to see how it arose, and how he may, by making different assumptions, substitute for it an equation which will give a rational result. This gives rise, in general, to a subsidiary problem, the solution of which ensures a rational solution of the original problem.

The problems in Diophantus and his devices for solving them are so various that, to appreciate them properly, the whole book must be read. Some general methods, however, emerge, which can best be exhibited in modern notation. They may be classified as follows. It must be understood that, where I have used the letters $a, b \ldots p, q \ldots$ to denote determinate numbers, Diophantus always has specific numbers.

I. DIOPHANTUS' TREATMENT OF EQUATIONS

(A) *Determinate equations*

Diophantus solved without difficulty equations of the first and second degrees; there is only one example of a cubic equation in the *Arithmetica*, and that is a very special case.

(1) *Pure determinate equations.*

These arise when, after all possible simplifications, by combining like terms, cancelling out, &c., we can make one term containing a power of the unknown equal to one term (a certain number), e.g. $Ax^m = B$. The equation is then considered solved. Diophantus uses one root only, and that 'rational', i.e. positive and integral or a ratio of integers.

(2) *Mixed quadratic equations.*

Diophantus requires these to be stated in the form of two positive terms equated to another positive term. With him, therefore, they take one of the forms

(a) $mx^2 + px = q$,   (b) $mx^2 = px + q$,   (c) $mx^2 + q = px$.

It is clear from his explanations in various cases that

DIOPHANTUS OF ALEXANDRIA

he solved the equations practically in the same way as we do (except that, when making the term in x^2 a square, he multiplies, rather than divides, by m), and that he stated the roots in a form equivalent to

(a) $\dfrac{\sqrt{(\frac{1}{4}p^2+mq)}-\frac{1}{2}p}{m}$,   (b) $\dfrac{\sqrt{(\frac{1}{4}p^2+mq)}+\frac{1}{2}p}{m}$,

(c) $\dfrac{\sqrt{(\frac{1}{4}p^2-mq)}+\frac{1}{2}p}{m}$.

Diophantus uses only the positive sign with the radical and thus obtains one root only. Whether he knew that a quadratic equation has two roots is a moot point. But, having regard to the procedure in Eucl. VI. 27-9 for the geometrical solution of the equivalent of a mixed quadratic equation, it seems hardly credible that he did not; he may have ignored it for the reason that his only object throughout is to obtain one solution.

A few cases of quadratic equations which actually occur may be quoted, with the results:

$$325x^2 = 3x+18;\ x = \tfrac{78}{325} = \tfrac{6}{25}.$$
$$84x^2+7x = 7;\ x = \tfrac{1}{4}.$$
$$17x^2+17 < 72x < 19x^2+19;\ x \text{ not} > \tfrac{66}{17} \text{ and not} < \tfrac{66}{19}.$$
$$22x < x^2+60 < 24x;\ x \text{ not} < 19 \text{ but} < 21.$$

(The limits in the last two cases are *a fortiori* limits, and in the first of the two $\tfrac{66}{19}$ should have been $\tfrac{67}{19}$.)

(3) *Simultaneous equations involving quadratics.*

(a) $\left.\begin{array}{l}x+y = 2a\\ xy = B\end{array}\right\}$,   (b) $\left.\begin{array}{l}x+y = 2a\\ x^2+y^2 = B\end{array}\right\}$,   (c) $\left.\begin{array}{l}x-y = 2a\\ xy = B\end{array}\right\}$.

If we denote Diophantus' one unknown by ξ, his solutions are as follows:

(a) Let $x-y = 2\xi\ (x > y)$.

Adding and subtracting, we have $x = a+\xi$, $y = a-\xi$, so that $a^2 - \xi^2 = xy = B$, a pure quadratic in ξ.

(b) and (c). Diophantus similarly puts $2\xi = x-y$ and $x+y$ respectively, and obtains a pure quadratic in ξ.

(4) *Cubic equation.*

Only one case occurs, namely in VI. 17, where Diophantus obtains the equivalent of
$$x^2 + 2x + 3 = x^3 + 3x - 3x^2 - 1,$$
and says simply 'whence x is found to be 4'. In fact the equation reduces to $x^3 + x = 4x^2 + 4$, and we can divide out by the factor $x^2 + 1$.

(B) *Indeterminate equations*

Nothing is said of indeterminate equations of the first degree. The first class to be dealt with is therefore

(a) *Indeterminate equations of the second degree.*

The form in which these occur is invariably this: one or two (but never more) functions of x of the form $Ax^2 + Bx + C$ or simpler forms have to be made rational square numbers by finding a suitable value for x. That is, we have in the general case to solve one or two equations of the form $Ax^2 + Bx + C = y^2$.

(1) *Single equation.*

In most of the cases treated the equation takes a simpler form in consequence of one or other of the terms on the left-hand side being absent.

1. When A or C or both $=0$, the equation can always be solved rationally. We may have $Bx = y^2$ or $Bx + C = y^2$, and in either case Diophantus puts for y^2 any determinate square and finds x at once. Or we may have $Ax^2 + Bx = y^2$,

DIOPHANTUS OF ALEXANDRIA

in which case Diophantus obtains a simple equation in x by substituting for y any multiple of x, as $\frac{m}{n}x$.

2. When $B = 0$, we have the form $Ax^2 + C = y^2$.

According to Diophantus this equation can be solved rationally

(α) when A is positive and a square, say a^2; in this case we assume $a^2x^2 + C = (ax \pm m)^2$, and

$$x = \pm \frac{C - m^2}{2ma},$$

m and the sign being so chosen as to make x positive;

(β) when C is positive and a square, as c^2; in this case we assume $Ax^2 + c^2 = (mx \pm c)^2$, and

$$x = \pm \frac{2mc}{A - m^2};$$

(γ) when one solution is known, any number of other solutions can be found. This is only stated by Diophantus of the case $Ax^2 - C = y^2$, but it is true in the general case $Ax^2 + Bx + C = y^2$.

Given that x_0 satisfies the equation $Ax^2 - C = y^2$, and that $Ax_0^2 - C = q^2$, Diophantus finds a greater value for x by substituting $x_0 + x$ for x, and $(q - kx)$ for y, in the original equation.

We have then $\quad A(x_0 + x)^2 - C = (q - kx)^2$.

But $\quad\quad\quad\quad\quad Ax_0^2 - C = q^2$;

therefore $\quad\quad 2x(Ax_0 + kq) = x^2(k^2 - A)$,

whence $\quad\quad x = 2(Ax_0 + kq)/(k^2 - A)$,

and the new value of x is $x_0 + \{2(Ax_0 + kq)/(k^2 - A)\}$.

The equation $Ax^2 - c^2 = y^2$, the particular case where C is a square, can (according to Diophantus) only be solved when A is the sum of two squares.

In fact, if $x = p/q$ satisfies the equation, so that
$$A\left(\frac{p}{q}\right)^2 - c^2 = k^2,$$
we have $Ap^2 = c^2q^2 + k^2q^2$, and
$$A = \left(\frac{cq}{p}\right)^2 + \left(\frac{kq}{p}\right)^2.$$

As a particular case of (γ), the equation $Ax^2 + C = y^2$ has an infinite number of solutions if $A + C$ is a square, i.e. if $x = 1$ is a solution. For we have only to substitute $1 + x$ for x and $q - kx$ for y, where $q^2 = A + C$ and k is some integer.

3. Form $Ax^2 + Bx + C = y^2$.

This could be reduced to the form in which the second term is wanting by replacing x by $z - \dfrac{B}{2A}$. Diophantus, however, treats it separately.

The equation can readily be solved if A is a square, say a^2, or C is a square, say c^2, for in the first case we put $y = ax - m$, and in the second $y = mx - c$, which has the effect of producing a simple equation in x.

It can also be solved if $\frac{1}{4}B^2 - AC$ is positive and a square. Diophantus does not enunciate this generally, but a case occurs in IV. 31, where the equation to be solved is $3x + 18 - x^2 = y^2$. Diophantus' solution comes to this. We put $y = px$, say, whence $(p^2 + 1)x^2 - 3x - 18 = 0$, and we have to find a value for p which will make this equation solvable. It will be so if $18(p^2 + 1) + \frac{9}{4}$ can be made a square, i.e. if $72p^2 + 81$ can be made a square. This is possible, by the usual method, namely by equating the expression to $(kp + 9)^2$; Diophantus equates to $(8p + 9)^2$ and obtains $p = 18$. Thus the original equation becomes
$$325x^2 - 3x - 18 = 0, \text{ and } x = \tfrac{78}{325} \text{ or } \tfrac{6}{25}.$$

(2) *Double equation.*

The Greek term is διπλοϊσότης, διπλῆ ἰσότης, or διπλῆ ἴσωσις. Here two functions of the unknown have to be made squares simultaneously. The general case is, To solve in rational numbers the equations

$$mx^2 + \alpha x + a = u^2,$$
$$nx^2 + \beta x + b = v^2.$$

A necessary preliminary condition is that each of the two expressions can be made a square. There are only a few types of cases in Diophantus where one or both of the expressions to be made squares are quadratic; in most of the cases which Diophantus solves the term in x^2 is absent and the expressions are of the first degree in x.

1. *Double equations of the first degree.*

The equations are in general

$$\alpha x + a = u^2, \quad \ldots \ldots (1)$$
$$\beta x + b = v^2. \quad \ldots \ldots (2)$$

(a) Diophantus' general method of solution depends on the identity

$$\{\tfrac{1}{2}(p+q)\}^2 - \{\tfrac{1}{2}(p-q)\}^2 = pq.$$

If the difference between the two expressions can be separated into two factors p, q, the expressions themselves can be equated to $\{\tfrac{1}{2}(p+q)\}^2$ and $\{\tfrac{1}{2}(p-q)\}^2$ respectively. As Diophantus himself says (II. 11), 'we equate either the square of half the difference of the two factors to the lesser (of the expressions) or the square of half the sum to the greater'.

The cases solved by Diophantus are such that the final equation in x reduces to a simple equation. This happens in the following cases:

(1) when α, β are in the ratio of a square to a square,

e.g. of m^2 to n^2, since, if $\alpha x+a$ is a square, so is $(\alpha x+a)n^2$, and if $\beta x+b$ is a square, so is $(\beta x+b)m^2$.

Now if $\alpha/\beta = m^2/n^2$, $\alpha n^2 = \beta m^2$. We therefore multiply (1) by n^2 and (2) by m^2, and we have

$$\alpha n^2 x + an^2 = u^2 n^2 = u'^2, \text{ say,}$$
$$\beta m^2 x + bm^2 = v^2 m^2 = v'^2, \text{ say.}$$

Subtracting, we obtain (since $\alpha n^2 = \beta m^2$)

$$an^2 - bm^2 = u'^2 - v'^2.$$

Since $an^2 - bm^2$ is a number, we can resolve it into two numerical factors p, q, and put

$$u' + v' = p,$$
$$u' - v' = q,$$

whence $u'^2 = \{\tfrac{1}{2}(p+q)\}^2$ and $v'^2 = \{\tfrac{1}{2}(p-q)\}^2$.

Thus $\alpha n^2 x + an^2 = \{\tfrac{1}{2}(p+q)\}^2$,
and $\beta m^2 x + bm^2 = \{\tfrac{1}{2}(p-q)\}^2$;

from either of which equations we obtain one and the same value of x satisfying the given conditions.

(2) when a, b are both square numbers.

If the equations are

$$\alpha x + c^2 = u^2,$$
$$\beta x + d^2 = v^2,$$

we have, by subtraction,

$$(\alpha - \beta)x + c^2 - d^2 = u^2 - v^2.$$

Of the factors of the expression on the left hand to be taken, one, as p, must be either $c+d$ or $c-d$ in order that the final equation in x may reduce to a simple equation. The factors are then

$$c \pm d, \quad \frac{\alpha - \beta}{c \pm d}x + c \mp d,$$

and, equating these respectively to $u+v$, $u-v$, we find x from the equation

$$\alpha x + c^2 = \tfrac{1}{4}\left\{c \pm d + \frac{\alpha-\beta}{c \pm d}x + c \mp d\right\}^2$$
$$= \tfrac{1}{4}\left(2c + \frac{\alpha-\beta}{c \pm d}x\right)^2.$$

Ex. from Diophantus:
$$10x+9 = u^2,$$
$$5x+4 = v^2.$$

The difference is $5x+5 = 5(x+1)$, and we have $10x+9 = \tfrac{1}{4}(x+6)^2 = \tfrac{1}{4}x^2 + 3x + 9$, and $x = 28$.

Or, again, we may multiply the equations respectively by d^2, c^2, making them

$$\alpha d^2 x + c^2 d^2 = u'^2,$$
$$\beta c^2 x + c^2 d^2 = v'^2.$$

The difference is $(\alpha d^2 - \beta c^2)x$, and, if px, q are the factors taken, we obtain the solution from

$$\alpha d^2 x + c^2 d^2 = \tfrac{1}{4}(px+q)^2.$$

In order that this may reduce to a simple equation, we must have $q = 2cd$, so that

$$p^2 x = 4\alpha d^2 - 2pq = 4\alpha d^2 - 2(\alpha d^2 - \beta c^2) = 2(\alpha d^2 + \beta c^2),$$

and, since $p = (\alpha d^2 - \beta c^2)/2cd$,

$$x = 2(\alpha d^2 + \beta c^2)/p^2 = 8c^2 d^2 (\alpha d^2 + \beta c^2)/(\alpha d^2 - \beta c^2)^2.$$

An alternative solution is found in one case, where the equations are

$$hx + n^2 = u^2,$$
$$(h+f)x + n^2 = v^2.$$

Suppose that $hx + n^2 = (y+n)^2$; therefore $hx = y^2 + 2ny$.

Substituting in the second equation, we have

$$(y+n)^2 + \frac{f}{h}(y^2 + 2ny) = v^2.$$

To make this expression a square, assume that it is equal to $(py-n)^2$, and the equation reduces to a simple equation in y.

2. *Double equations of the second degree.*

The double equations in which one or both of the expressions to be made squares are of the second degree are of three types only. Two are in forms so special that the usual method of solution explained above can be applied.

The third is as follows:

$$\alpha x^2 + ax = u^2,$$
$$\beta x^2 + bx = v^2,$$

and requires different treatment. It could have been transformed into a case of a double-equation of the first degree by substituting $1/y$ for x, but Diophantus makes a different substitution.

He assumes that $u^2 = m^2 x^2$, and substitutes the resulting value of x in the second equation. The value of x is $a/(m^2-\alpha)$; therefore $\beta\left(\dfrac{a}{m^2-\alpha}\right)^2 + \dfrac{ba}{m^2-\alpha}$ has to be made a square, that is, $a^2\beta + ab(m^2-\alpha)$ must be made a square.

We have, therefore, to solve the equation in m

$$abm^2 + a(a\beta - \alpha b) = y^2,$$

and the above methods have to be applied. The equation can be solved in Diophantus' manner if either the ratio a/b or the ratio $(a\beta - \alpha b)/a$ is that of a square to a square. Examples are VI. 12, 14.

(b) *Indeterminate equations of degree higher than the second.*

(1) *Single equations.*

There are two classes, namely those in which expressions in x have to be made squares or cubes respectively. In none of the cases is the given expression of a degree higher than the sixth. Diophantus' efforts are generally directed towards making the ultimate expression in x reduce to a simple equation, and his solutions by this means depend on some of the coefficients of the powers of x in the expressions to be made squares or cubes being squares or cubes respectively. Where this is the case, I shall indicate it by using $a^2, b^2 \ldots a^3, b^3 \ldots$ for the coefficients.

(i) The types of cases in which expressions are to be made *squares* are these.

1. $Ax^3 + Bx^2 + Cx + d^2 = y^2$.

Here Diophantus assumes $y = \dfrac{C}{2d}x + d$, and so effects the reduction to a simple equation.

2. $a^2x^4 + Bx^3 + Cx^2 + Dx + E = y^2$,
or $Ax^4 + Bx^3 + Cx^2 + Dx + e^2 = y^2$.

Diophantus assumes
$$y = ax^2 + \frac{B}{2a}x + m \text{ or } y = mx^2 + \frac{D}{2e}x + e$$
in the two cases respectively, and then determines m in such a way that the term in x^2 may vanish, leaving a simple equation in x.

3. $a^2x^4 + c^2x^2 + e^2 = y^2$.
4. $Ax^4 + E = y^2$.

The case occurring in Diophantus is $x^4 + 97 = y^2$ (V. 29).

Diophantus tries one assumption, $y = x^2 - 10$, and finds that this gives $x^2 = \frac{3}{20}$, which leads to no 'rational' result.

He therefore retraces his steps, and sets himself to find out how he may alter his original assumptions in order to replace the refractory equation by one which he can solve rationally. He ultimately obtains the equation $x^4+337 = y^2$, and also a suitable assumption for y, namely $y = x^2-25$, which produces a rational result, $x = \frac{12}{5}$.

5. $x^6 - Ax^3 + Bx + c^2 = y^2$.

Assuming $y = x^3 + c$, we have $-Ax^2 + B = 2cx^2$, which gives a rational solution if $B/(A+2c)$ is a square. Where this is not the case (in IV. 18), Diophantus harks back and replaces the equation $x^6 - 16x^3 + x + 64 = y^2$ by another, namely $x^6 - 128x^3 + x + 4096 = y^2$, which satisfies the condition.

(ii) The expressions made into *cubes* are as follows.

1. Two particular cases of the form $Ax^2 + Bx + C = y^3$.

The first is $x^2 - 4x + 4 = y^3$; and, as $x^2 - 4x + 4$ is already a square, Diophantus naturally makes $(x-2)$ a cube.

The second occurs in VI. 17, where a cube has to be found exceeding a square number by 2. Diophantus assumes $(x-1)^3$ for the cube and $(x+1)^2$ for the square, which gives

$$x^3 - 3x^2 + 3x - 1 = x^2 + 2x + 3,$$
or $$x^3 + x = 4x^2 + 4.$$

We divide out by $x^2 + 1$, and obtain $x = 4$. The assumptions were evidently made with knowledge and intention, and were meant to lead up to the known result $5^2 + 2 = 3^3$.

2. $a^3x^3 + Bx^2 + Cx + D = y^3$,

or $Ax^3 + Bx^2 + Cx + d^3 = y^3$.

Here Diophantus assumes $y = ax + \dfrac{B}{3a^2}$ or $y = \dfrac{C}{3d^2}x + d$ as the case may be, and so obtains a simple equation in x.

(2) *Double equations.*

These are cases in which one expression has to be made a square and the other a cube. They are mostly very simple, and only one need be noticed here, namely
$$2x^2+2x = y^2,$$
$$x^3+2x^2+x = z^3.$$
Diophantus assumes $y = mx$, whence $x = 2/(m^2-2)$ and, by substitution in the second equation, we have
$$\left(\frac{2}{m^2-2}\right)^3 + 2\left(\frac{2}{m^2-2}\right)^2 + \frac{2}{m^2-2} = z^3,$$
which reduces to
$$\frac{2m^4}{(m^2-2)^3} = z^3,$$
and we have only to make $2m^4$, or $2m$, a cube.

II. METHOD OF APPROXIMATION TO LIMITS

This is a distinctive and remarkable method which calls for a short description; the Greek expression for it is παρισότης or παρισότητος ἀγωγή ('inducement of approximation'). The object is to solve such problems as that of finding two or three square numbers the sum of which is a given number, while each of them either approximates to one and the same number or is subject to particular limits. Examples are:

(1) to divide 13 into two square numbers each of which > 6 (V. 9);

(2) to divide 10 into three squares such that the first > 2, the second > 3, and the third > 4 (V. 12).

(3) to divide 30 into four squares each of which < 10 (V. 14).

We will illustrate by giving the working in case (1). Take one-half of 13, i.e. $6\frac{1}{2}$, and find what *small* fraction

$\frac{1}{x^2}$ added to $6\frac{1}{2}$ will give a square; thus $6\frac{1}{2}+\frac{1}{x^2}$, and therefore $26+\frac{1}{y^2}$, must be a square.

Seeing that 5^2 is near to 26, Diophantus assumes that
$$26+\frac{1}{y^2}=\left(5+\frac{1}{y}\right)^2.$$
Therefore $1/y = \frac{1}{10}$; accordingly $1/x^2 = \frac{1}{400}$, and
$$6\tfrac{1}{2}+\tfrac{1}{400}=(\tfrac{51}{20})^2.$$
We have now to divide 13 into two squares each of which is as nearly as possible equal to $(\tfrac{51}{20})^2$.

Now $13 = 3^2 + 2^2$ [it is necessary that the original number shall be capable of being expressed as the sum of two squares, not necessarily integral], and
$$3 > \tfrac{51}{20} \text{ by } \tfrac{9}{20},$$
$$2 < \tfrac{51}{20} \text{ by } \tfrac{11}{20}.$$
But if we took $3-\tfrac{9}{20}$, $2+\tfrac{11}{20}$ as the sides of two squares, the sum of the squares would be $2(\tfrac{51}{20})^2 = \tfrac{5202}{400}$, which is > 13.

Accordingly we assume $3-9x$ and $2+11x$ as the sides of the squares respectively (so that x is not exactly $\tfrac{1}{20}$ but near it).

Therefore $(3-9x)^2+(2+11x)^2 = 13$,

and we find $x = \tfrac{5}{101}$.

The sides of the required squares are therefore $\tfrac{258}{101}$, $\tfrac{257}{101}$.

III. PORISMS AND PROPOSITIONS IN THE THEORY OF NUMBERS

(i) Three propositions stated without proof are introduced with the words 'We have it in the Porisms that...'.

One is general and states that 'the difference of two

cubes is also the sum of two cubes'. Diophantus enunciates this (V. 16), but neither proves it nor shows how to make the transformation. [The subject of the transformation of sums and differences of cubes was investigated by Vieta, Bachet, and Fermat.]

The other two are particular propositions which, when once established, enable two characteristic problems of Diophantus to be solved almost instantaneously.

1. If a is a given number, and x, y numbers such that $x+a = m^2$, $y+a = n^2$, then, if $xy+a$ is also a square, m and n differ by unity (V. 3).

This should rather have been stated the other way. 'If m, n differ by unity, $xy+a$ is also a square', and this is easily proved; but $m = n \pm 1$ is not the only possible condition under which $xy+a$ might be a square.

2. If $X(=m^2)$, $Y(=\overline{m+1}^2)$ be consecutive squares, and a third number Z be taken equal to $2\{m^2+(m+1)^2\}+2$, or $4(m^2+m+1)$, then the three numbers have the property that $XY+X+Y$, $YZ+Y+Z$, $ZX+Z+X$, and also $XY+Z$, $YZ+X$, $ZX+Y$ are all squares (V. 5). The truth of this is easily verified.

Of precisely the same kind are three other propositions stated without proof.

1. If $X = a^2x+2a$, $Y = (a+1)^2x+2(a+1)$, or, in other words, if $xX+1 = (ax+1)^2$ and $xY+1 = \{(a+1)x+1\}^2$, then (IV. 20) $XY+1$ is also a square.

2. If $X \pm a = m^2$, $Y \pm a = (m+1)^2$, and $Z = 2(X+Y)-1$, then $YZ \pm a$, $ZX \pm a$, $XY \pm a$ are all squares (V. 3, 4).

3. If $X = m^2+2$, $Y = (m+1)^2+2$,
and $Z = 2\{m^2+(m+1)^2+1\}+2$,

then the six expressions
$$YZ-(Y+Z),\ ZX-(Z+X),\ XY-(X+Y)$$
$$YZ-X,\ ZX-Y,\ XY-Z$$
are all squares (V. 6).

(ii) *Theorems on the composition of numbers as the sum of two, three, or four squares.*

These are propositions in the Theory of Numbers which are for the first time stated or assumed in Diophantus, though how far he possessed scientific proofs of them is a matter of doubt.

(α) *On numbers which are the sum of two squares.*

(1) Any square number can be resolved into two squares in any number of ways (II. 8).

(2) Any number which is the sum of two squares can be resolved into two squares in any number of ways (II. 9). (The squares may everywhere be either integral or fractional.)

(3) If each of two whole numbers is the sum of two squares, their product can be resolved into two squares in two ways.

In fact $(a^2+b^2)(c^2+d^2) = (ac \pm bd)^2 + (ad \mp bc)^2$.

This theorem is used for the purpose of finding four right-angled triangles in rational numbers which have the same hypotenuse. Diophantus speaks of a rational right-angled triangle being 'formed from two numbers a, b'; by this he means that we take as the sides a^2+b^2, a^2-b^2, $2ab$. In this case we first form right-angled triangles from (a, b) and (c, d) respectively, namely $(a^2+b^2, a^2-b^2, 2ab)$ and $(c^2+d^2, c^2-d^2, 2cd)$.

Then, if we multiply the sides of the former by c^2+d^2, and the sides of the latter by a^2+b^2, we have two different right-angled triangles with the same hypotenuse $(a^2+b^2)(c^2+d^2)$. The other two are obtained from the above formula, which only fails if the relations between a, b, c, d are such as to cause one or other of the perpendicular sides of the resulting triangles to vanish, e.g. if either $ac-bd$ or $ad-bc = 0$.

It is on this proposition of Diophantus (III. 19) that Fermat has a famous note about the number of ways in which a prime number of the form $4n+1$ and its powers will be (a) the hypotenuse of a right-angled triangle, (b) the sum of two squares, and the number of ways in which the product of one such prime number and the various powers of another such prime number will be the sum of two squares. The hypotenuses of the two rational right-angled triangles used by Diophantus are 5 and 13, the smallest prime numbers of the form $4n+1$. It was Fermat who first proved that all such prime numbers are the sum of two squares, and Euler who first proved that prime numbers of the form $4n+1$ and numbers arising from the multiplication of such numbers are the only classes of numbers that are always the sum of two squares.

(4) More remarkable is a condition of possibility of solution prefixed to V. 9, 'To divide 1 into two parts such that, if a given number be added to each part, the result will be a square'. The condition is stated by Diophantus in two parts. There is no doubt about the first part, 'The given number must not be odd' [i.e. no number of the form $4n+3$ or $4n-1$ can be the sum of two squares]; the text of the second part is corrupt, but it is quite likely that corrections suggested by Hankel and Tannery give the true intention of the original, 'nor must the double of the given number *plus* 1 be measured by any prime number which is less by 1 than a multiple of 4'. This is not far from the condition stated by Fermat, 'The given number

must not be odd, and the double of it increased by 1, when divided by the greatest square which measures it, must not be divisible by a prime number of the form $4n-1$'.

(β) *On numbers which are the sum of three squares.*

In V. 11 the number $3a+1$ has to be divisible into three squares. Diophantus says that a must not be 2 or any multiple of 8 increased by 2. That is, a number of the form $24n+7$ cannot be the sum of three squares. As a matter of fact, the factor 3 in the 24 is irrelevant, and Diophantus might have said that a number of the form $8n+7$ cannot be the sum of three squares. This is true, but the form $8n+7$ does not include *all* the numbers which cannot be the sum of three squares. Fermat gives the conditions to which a must be subject, proving that $3a+1$ cannot be of the form $4^n(24k+7)$ or $4^n(8k+7)$, where $k = 0$ or any integer.

(γ) *Numbers which are the sum of four squares.*

Diophantus has three problems in which a number has to be divided into four squares. In this case he states no necessary condition, as he does when numbers have to be divided into *two* or *three* squares. Now every number is either a square or the sum of two, three, or four squares (a theorem enunciated by Fermat and proved by Lagrange, who followed up results obtained by Euler). This shows that any number can be divided into four squares (integral or fractional), since any square number can be divided into two other squares integral or fractional.

It is difficult to give an adequate idea of the ingenuity and the variety of the devices used by Diophantus in solving his problems without transcribing the whole book. Short of that, the best that can be done within the limits permissible in this work is to make a selection of a few characteristic problems and show how they are solved. This can be done most concisely by using modern notation. In general, Diophantus asks us to find two or more numbers such that certain functions of them are squares. We will designate the required numbers as x, y, z .... As a rule Diophantus does not choose any of these as his unknown (s); he gives his unknown in each case the signification that will be most convenient in respect that it enables as many of the required numbers as possible to be expressed in terms of the one unknown, thereby avoiding the necessity for eliminations. I shall as a rule denote Diophantus' unknown by ξ. When in the equations expressions are said to be equal to $u^2, v^2, w^2, t^2 \ldots$, this will mean simply that they are to be made squares respectively. Given numbers will be indicated by $a, b, c \ldots m, n \ldots$; these letters will take the place of the numbers used by Diophantus, which are always specific numbers.

We need not consider determinate systems of equations of the first degree or reducible to the first degree, nor easy determinate systems leading to equations of the second degree.

We will begin with two cases in which Diophantus uses the method of 'false hypothesis' (*regula falsi*) (IV. 15). To solve the equations $(y+z)x = a$, $(z+x)y = b$, $(x+y)z = c$.

Diophantus takes the third number z as his unknown; thus $x+y = c/z$.

He then assumes $x = p/z$, $y = q/z$;

it follows that $\dfrac{pq}{z^2} + p = a$, $\dfrac{pq}{z^2} + q = b$.

These equations are inconsistent unless $q-p = b-a$.

With the specific numbers which Diophantus assumes

this condition is not fulfilled; he therefore seeks to replace them by others which will satisfy the condition, and this he does by solving the auxiliary problem of finding two numbers the sum of which is c and the difference $b-a$.

Then, having determined p, q, we have $\frac{pq}{z^2} = a-p$ or $b-q$.

For a rational solution we must have $pq/(a-p)$ a square, and the numbers originally taken must be such as to secure this. In Diophantus' case a, b, c are $27, 32, 35$ respectively, and p, q are found to be $15, 20$ respectively. Thus $pq/(a-p) = 300/12 = 25$, a square. Therefore $z = 5$, whence $x = 3$, $y = 4$.

(IV. 37) $yz = a(x+y+z)$, $zx = b(x+y+z)$, $xy = c(x+y+z)$.

Diophantus begins by giving $x+y+z$ an *arbitrary* value; then, finding that the result produced by this assumption is not 'rational', he seeks to find a value for $x+y+z$ which will lead to a rational result.

Now, if $w = x+y+z$, we have $x = cw/y$, $z = aw/y$, so that $zx = acw^2/y^2$. But $zx = bw$, by hypothesis; therefore

$$y^2 = \frac{ac}{b}w.$$

In order to make the last expression a square, Diophantus substitutes $\frac{ac}{b}\xi^2$ for w, so that $y^2 = \frac{a^2c^2}{b^2}\xi^2$.

Then $x+y+z = \frac{ac}{b}\xi^2$, $y = \frac{ac}{b}\xi$, $x = c\xi$, $z = a\xi$.

Eliminating x, y, z, we obtain $\xi = (bc+ca+ab)/ac$, and $x = (bc+ca+ab)/a$, &c.

We pass to indeterminate analysis of the second degree, which is, as it were, the staple of Diophantus' trade.

(II. 30) $xy+(x+y) = u^2$, $xy-(x+y) = v^2$.

Since $m^2+n^2\pm 2mn$ is a square, Diophantus assumes that $xy = (m^2+n^2)\xi^2$, $x+y = 2mn\xi^2$.

Now put $x = p\xi$, $y = q\xi$, where p, q are such numbers that $pq = m^2+n^2$;
therefore $(p+q)\xi = 2mn\xi^2$, and ξ is found.

(II. 34) $x^2+(x+y+z) = u^2$, $y^2+(x+y+z) = v^2$,
$$z^2+(x+y+z) = w^2.$$

Since $\{\tfrac{1}{2}(m-n)\}^2+mn$ is a square, take any number separable into two factors (m, n) in three ways. This gives three values, say p, q, r, for $\tfrac{1}{2}(m-n)$.

Assume that $x = p\xi$, $y = q\xi$, $z = r\xi$, and $x+y+z = mn\xi^2$; then $(p+q+r)\xi = mn\xi^2$, and ξ is found.

Diophantus has for mn the number 12, which has three pairs of factors, $(1, 12)$, $(2, 6)$, and $(3, 4)$; p, q, r, are then $5\tfrac{1}{2}$, 2, $\tfrac{1}{2}$. Therefore $8\xi = 12\xi^2$, whence $\xi = \tfrac{2}{3}$, and the numbers are $\tfrac{11}{3}$, $\tfrac{4}{3}$, $\tfrac{1}{3}$.

(III. 5) $x+y+z = t^2$; $y+z-x = u^2$, $z+x-y = v^2$,
$$x+y-z = w^2.$$

Diophantus gives two solutions, the second of which is the more elegant. The sum of the last three expressions is $x+y+z$. Seek, therefore, three square numbers (u^2, v^2, w^2), the sum of which is a square (t^2), and we have a solution. Diophantus says, 'If I add two [square] numbers, e.g. 4 and 9, and if I seek what square with 13 added makes a square, I shall find 36'. Then, solving the last three equations with 9, 36, 4 in place of u^2, v^2, w^2, he finds for x, y, z the values 20, $6\tfrac{1}{2}$, $22\tfrac{1}{2}$ respectively, the sum of which is a square (49).

(IV. 16) $x+y+z = t^2$, $x^2+y = u^2$, $y^2+z = v^2$, $z^2+x = w^2$.

Put $4m\xi$ for y, and, by means of the factors $2m\xi$, 2, we can make the second expression a square, namely by making x equal to half the difference, or $m\xi-1$.

We can satisfy the third condition by subtracting $(4m\xi)^2$ from some square, say $(4m\xi+1)^2$; thus $z = 8m\xi+1$.

By the first condition, $x+y+z = 13m\xi$ must be a square. Let it be $169\eta^2$, so that $m\xi = 13\eta^2$. The numbers are therefore $13\eta^2-1$, $52\eta^2$, $104\eta^2+1$, and the last condition gives $10816\eta^4+221\eta^2 =$ a square, or

$$10816\eta^2+221 = \text{a square} = (104\eta+1)^2, \text{ say.}$$

Therefore $\eta = \frac{220}{208} = \frac{55}{52}$, and the problem is solved.

(IV. 29) $x^2+y^2+z^2+w^2+x+y+z+w = a$.

Since $x^2+x+\frac{1}{4}$ is a square,

$(x^2+x)+(y^2+y)+(z^2+z)+(w^2+w)+1$ is the sum of four squares.

We have, therefore, only to separate $(a+1)$ into four squares.

In Diophantus' case $a = 12$, and

$$13 = 4+9 = (\tfrac{64}{25}+\tfrac{36}{25})+(\tfrac{144}{25}+\tfrac{81}{25});$$

the values of x, y, z, w are $(\tfrac{8}{5}-\tfrac{1}{2})$, &c., or $\tfrac{11}{10}, \tfrac{7}{10}, \tfrac{19}{10}, \tfrac{13}{10}$.

(Lemma to V. 7) $xy+x^2+y^2 = u^2$.

Assume that $y =$ some specific number, say 1.

Then $x^2+x+1 =$ a square $= (x-2)^2$, say;

therefore $x = \tfrac{3}{5}$, and $(\tfrac{3}{5}, 1)$ or $(3, 5)$ is a solution.

In another lemma (To find three rational right-angled triangles with equal area) Diophantus verifies that, if we take numbers such as those just found, i.e. such that $mn+m^2+n^2 = p^2$, then, if we 'form' rational right-angled triangles from (p, m), (p, n), and $(p, m+n)$, the triangles will have equal areas. The sides being $(p^2-m^2, 2pm, (p^2+m^2)$, &c., the areas are $(p^2-m^2)pm$, $(p^2-n^2)pn$, and $\{(m+n)^2-p^2\}p(m+n)$, and, by means of the relation

$mn+m^2+n^2 = p^2$, we easily prove that each of these three expressions is equal to $mnp(m+n)$.

By means of this lemma Diophantus is able to solve such a problem as the following.

(V. 7) To find x, y, z such that $x^2 \pm (x+y+z)$, &c., are all squares.

Since, in any right-angled triangle, (hypotenuse)$^2 \pm$ (four times the area) is a square, we have only to find three right-angled triangles with equal areas, make $x+y+z$ equal to four times the common area multiplied by ξ^2, and x, y, z equal respectively to the three hypotenuses multiplied by ξ, and then equate the two values of $x+y+z$. Diophantus uses the three triangles (40, 42, 58), (24, 70, 74), (15, 112, 113), with the common area 840. The three hypotenuses are then $x = 58\xi$, $y = 74\xi$, $z = 113\xi$, while $x+y+z = 4 \cdot 840\,\xi^2$.

Therefore $\qquad 245\,\xi = 3360\,\xi^2$,

whence $\xi = \frac{245}{3360} = \frac{49}{672} = \frac{7}{96}$, and x, y, z are found.

Indeterminate analysis of degrees higher than the second.

(IV. 38) $(x+y+z)x =$ a triangular number $= \frac{1}{2}u(u+1)$,

$\qquad (x+y+z)y =$ a square (v^2),

$\qquad (x+y+z)z =$ a cube (w^3).

Diophantus assumes, in addition, that $x+y+z =$ a square, say ξ^2.

Then $\qquad x = \dfrac{u(u+1)}{2\xi^2}, \qquad y = \dfrac{v^2}{\xi^2}, \qquad z = \dfrac{w^3}{\xi^2}$.

Therefore $\qquad \xi^2 = \dfrac{\frac{1}{2}u(u+1)+v^2+w^3}{\xi^2}$,

or $\qquad \xi^4 = \frac{1}{2}u(u+1)+v^2+w^3$.

Diophantus has at first assumed 6, 4, 8 as the triangular

number, square, and cube respectively, and this assumption leads to $\xi^4 = 18$, which cannot be solved, since 18 is not a fourth power.

He therefore seeks a triangular number, a square, and a cube such that their sum is a fourth power. If ξ^4 be the fourth power, assume $v^2 = \xi^4 - 2\xi^2 + 1$, so that $\frac{1}{2}u(u+1) + w^3 = 2\xi^2 - 1$. If the cube is, say, 8, we have to make $2\xi^2 - 9$ a triangular number.

Now 8 times a triangular number *plus* 1 = a square.

Therefore $16\xi^2 - 71 =$ a square $= (4\xi - 1)^2$, say; thus $\xi = 9$, the triangular number is 153, the square 6400, and the cube 8.

Assume then as the numbers $x = 153/\eta^2$, $y = 6400/\eta^2$, $z = 8/\eta^2$, and $x + y + z = \eta^2$.

Therefore $\eta^4 = 6561$, so that $\eta^2 = 81$, and $(\frac{153}{81}, \frac{6400}{81}, \frac{8}{81})$ is a solution.

(V. 15) $(x+y+z)^3 + x = u^3$, $(x+y+z)^3 + y = v^3$,
$$(x+y+z)^3 + z = w^3.$$

Assume $x + y + z = \xi$, $u^3 = m^3 \xi^3$, $v^3 = n^3 \xi^3$, $w^3 = p^3 \xi^3$.

Therefore $\xi = \{(m^3 - 1) + (n^3 - 1) + (p^3 - 1)\}\xi^3$, and we have to find three cubes m^3, n^3, p^3 such that $m^3 + n^3 + p^3 - 3$ is a square.

Diophantus assumes as the sides of the cubes $k+1$, $2-k$, 2; this gives $9k^2 - 9k + 14 =$ a square $= (3k-4)^2$, say.
Therefore $k = \frac{2}{15}$, and the three cubes are $(\frac{17}{15})^3$, $(\frac{28}{15})^3$, 2^3.
Substituting these values in the above equation, we have
$$\xi = \tfrac{43740}{3375}\xi^3 = \tfrac{2916}{225}\xi^3.$$
Therefore $\xi = \frac{15}{54}$, and $x = (m^3 - 1)\xi^3$, &c., are found.

(V. 29) $x^4 + y^4 + z^4 = u^2$.

Suppose, says Diophantus, $x^2 = \xi^2$, $y^2 = p^2$, $z^2 = q^2$.
Therefore $\xi^4 + p^4 + q^4 =$ a square $= (\xi^2 - r)^2$, say.

DIOPHANTUS OF ALEXANDRIA 505

Therefore $\xi^2 = \dfrac{r^2-(p^4+q^4)}{2r}$, and we have to make this expression a square.

Diophantus assumes $r = p^2+4$, $q^2 = 4$, so that the expression reduces to $8p^2/(2p^2+8)$ or $4p^2/(p^2+4)$. To make this a square, let $p^2+4 = (p+1)^2$, say; therefore $p = 1\frac{1}{2}$, and $p^2 = 2\frac{1}{4}$, $q^2 = 4$, $r = 6\frac{1}{4}$, or, multiplying by 4, $p^2 = 9$, $q^2 = 16$, $r = 25$, which solves the problem.

RATIONAL RIGHT-ANGLED TRIANGLES

Book VI deals with problems of constructing right-angled triangles with sides in rational numbers and fulfilling certain conditions. If we call the perpendicular sides x, y, and the hypotenuse z, we have always $z^2 = x^2+y^2$ as a condition applying in all cases, in addition to the other conditions specified.

(VI. 2) $z+x = u^3$, $z+y = v^3$. $\qquad [z^2 = x^2+y^2]$

Form a right-angled triangle from m, ξ, so that its sides are $m^2+\xi^2$, $2m\xi$, $m^2-\xi^2$. Thus $z+y = 2m^2$, and as this must be a cube, we put $m = 2$.

Form the triangle, then, from 2, ξ; the sides are then $4+\xi^2$, 4ξ, $4-\xi^2$, and $z+x = \xi^2+4x+4$ must be a cube. Therefore $\xi+2$ must be a cube, while ξ^2 must be less than 4, so that ξ is less than 2. The cube must, therefore, lie between 2 and 4. And $\frac{27}{8}$ is such a cube. We therefore put

$$\xi+2 = \tfrac{27}{8}, \text{ whence } \xi = \tfrac{11}{8}.$$

The triangle is therefore $(\frac{377}{64}, 5\frac{1}{2}, \frac{135}{64})$, or, if we multiply by the common denominator 64, (377, 352, 135).

(VI. 6) $\frac{1}{2}xy+x = a$.

Assume that the triangle is $h\xi$, $p\xi$, $b\xi$ ($h\xi$ being the hypotenuse). Then we must have $\frac{1}{2}pb\xi^2+p\xi = a$; and, for a rational solution of this equation, $(\frac{1}{2}p)^2+a(\frac{1}{2}pb)$ must be a square.

We seek, therefore, a rational right-angled triangle (h, p, b), such that $(\frac{1}{2}p)^2+a(\frac{1}{2}pb)$ is a square. Suppose $p=1, b=m$; then $\frac{1}{2}am+\frac{1}{4}$, or $2am+1$, must be a square. But, since h is rational, m^2+1 must also be a square. We have, therefore, the double equation

$$2am+1 = u^2,$$
$$m^2+1 = v^2.$$

The difference $m^2-2am = m(m-2a)$.

Equating $2am+1$ to $[\frac{1}{2}\{m-(m-2a)\}]^2$, or a^2, we have $m = (a^2-1)/2a$, and $m^2+1 = (a^2+1)^2/4a^2$.

Substituting the values $(a^2+1)/2a$, 1, $(a^2-1)/2a$ for h, p, b respectively in the expressions for the sides of our original right-angled triangle, we must have

$$\frac{a^2-1}{4a}\xi^2+\xi = a,$$

or $$(a^2-1)\xi^2+4a\xi = 4a^2,$$

whence $\xi = \dfrac{2a}{a+1}$, and the required triangle is

$$\left(\frac{a^2+1}{a+1}, \frac{2a}{a+1}, a-1\right).$$

(VI. 17) $\frac{1}{2}xy+z = u^2$, $x+y+z = v^3$. $\qquad [z^2 = x^2+y^2]$

Let $\xi = \frac{1}{2}xy$, the area, and let $z = k^2-\xi$.

Since $2\xi = xy$, suppose $x=2$, $y=\xi$. Therefore $2+k^2$ must be a cube. As we have seen (p. 492), Diophantus takes $(m-1)^3$ for the cube and $(m+1)^2$ for the square, thus obtaining $m^3-3m^2+3m-1 = m^2+2m+3$, whence $m^3+m = 4m^2+4$, and, dividing out by m^2+1, we have

$m=4$. Therefore $k=5$, and we assume $\frac{1}{2}xy=\xi$, $z=25-\xi$, $x=2$, $y=\xi$ (see above).

Now $z^2=x^2+y^2$; therefore $(25-\xi)^2=4+\xi^2$, and $\xi=\frac{621}{50}$, so that the triangle is $(2, \frac{621}{50}, \frac{629}{50})$.

(VI. 19) $\frac{1}{2}xy+x=w^2$, $x+y+z=v^3$.

Here Diophantus forms a rational right-angled triangle according to the *Pythagorean* formula

$$m^2+\{\tfrac{1}{2}(m^2-1)\}^2 = \{\tfrac{1}{2}(m^2+1)\}^2,$$

where m is any odd number, say $2\xi+1$. The sides are therefore $2\xi+1$, $2\xi^2+2\xi$, $2\xi^2+2\xi+1$.

Since therefore $x+y+z=v^3$,

$$4\xi^2+6\xi+2 = (4\xi+2)(\xi+1) = \text{a cube},$$

or, if we divide the sides by $\xi+1$, $4\xi+2$ has to be made a cube.

Also $\frac{1}{2}xy+x = \dfrac{2\xi^3+3\xi^2+\xi}{(\xi+1)^2} + \dfrac{2\xi+1}{\xi+1} = $ a square.

The latter expression reduces to $2\xi+1=$ a square.

But $4\xi+2$ is a cube. We therefore put 8 for the cube, and $\xi=1\frac{1}{2}$.

THE TREATISE ON POLYGONAL NUMBERS

We know that the subject of Polygonal Numbers goes back to the Pythagoreans. After them Philippus of Opus and Speusippus are said to have written upon it. Hypsicles (about 170 B.C.) is cited by Diophantus as the author of a definition of a polygonal number (see pp. 383-4 above), which amounts to saying generally that, if there be an a-gonal number of side n, the number is equal to

$$1+\{1+(a-2)\}+\{1+2(a-2)\}+ \ldots +\{1+(n-1)(a-2)\}$$
$$= \tfrac{1}{2}n\{2+(n-1)(a-2)\}.$$

508   ALGEBRA

Theon of Smyrna, Nicomachus, and Iamblichus all devote considerable space to polygonal numbers. Nicomachus in particular has rules for obtaining a polygonal number of $(a+1)$ sides from a polygonal number of a sides. Thus he says:

1. If we put two consecutive triangular numbers together, we get a square.

In fact, $\frac{1}{2}(n-1)n + \frac{1}{2}n(n+1) = n^2$.

2. If we add to a square a triangular number with side less by 1 than that of the square, we obtain a pentagon; if we add to a pentagon a triangular number with side less by 1 than that of the pentagon, we have a hexagon; and so on.

In fact, $\frac{1}{2}n\{2+(n-1)(a-2)\} + \frac{1}{2}(n-1)n$
$= \frac{1}{2}n[2+(n-1)\{(a+1)-2\}]$.

Plutarch, a contemporary of Nicomachus, mentions, and Diophantus uses, the theorem that, if we take eight times a triangular number and then add 1, we have a square.

In fact, $8 \cdot \frac{1}{2}n(n+1) + 1 = (2n+1)^2$.

The treatise of Diophantus, as we have it, is only a fragment. It is quite different in character from the *Arithmetica*, in that the proofs used in it are all geometrical. The main results obtained are as follows.

1. Diophantus generalizes the above theorem that eight times a triangular number *plus* 1 is a square, by proving that, if P be any polygonal number with a angles,

$8P(a-2)+(a-4)^2 = $ a square.

2. Diophantus further proves that this square is equal to $\{2+(2n-1)(a-2)\}^2$, and deduces rules as follows:

(a) To find the number from its side:
$$P = \frac{\{2+(2n-1)(a-2)\}^2-(a-4)^2}{8(a-2)}.$$
(b) To find the side from the number:
$$n = \tfrac{1}{2}\left(\frac{\sqrt{\{8P(a-2)+(a-4)^2\}}-2}{a-2}+1\right).$$

The last proposition, which breaks off in the middle, and is therefore not worked out, is, *Given a number, to find in how many ways it can be a polygonal number.*

XVIII
COMMENTATORS AND MINOR WRITERS

WITH Pappus and Diophantus the story of Greek mathematics as a living study comes to an end. It only remains, by way of epilogue, to say a word about the commentators and others who, by reason of the historical allusions and the quotations from original treatises now lost which are found here and there in their works, have contributed scraps of information throwing light on dark places in the history of our subject, for which we cannot be too grateful.

CLEOMEDES wrote a small work, still extant, in two Books with the title κυκλικὴ θεωρία μετεώρων, or *De motu circulari corporum caelestium*, which has little or no mathematical interest, but is of some historical value because it is, even more than the works attributed to Geminus, based upon Posidonius. As Cleomedes mentions no author later than Posidonius, it is permissible to suppose that he wrote about the middle of the first century B.C. or not much later. The first Book deals with general topics of astronomy and mathematical geography, the finite universe surrounded by infinite void, the circles in the heaven, the zones, habitable and uninhabitable, the motion of the fixed stars, and the independent movements of the sun, moon, and planets, the inclination of the axis of the universe and its effects on the lengths of days and nights at different places and their rates of variation, the different lengths of the seasons, the habitable regions of the globe; Cleomedes mentions Britain and the island of Thule, said to have been visited by Pytheas, where, when the sun is in Cancer and visible, the day is a month long. The last chapters purport to prove that the earth is a sphere and is in the centre of the universe, and that the earth is in the relation of a point to, i.e. is negligible in size in comparison with, the universe and the sun's circle (but not the moon's circle). A chapter (10) on the size of the earth gives details of the measurements of the earth by Eratosthenes and Posidonius respectively.

Book II begins with what is evidently the main purpose of the book, an elaborate refutation of the view of Epicurus and his followers that the sun is just as large as it *looks*, and that the stars are lit up as they rise and extinguished as they set. The long chapter (1) on this subject seems to be almost wholly taken from Posidonius. In the rest of the Book Cleomedes seems to have deserted his guide, and the quality of the content falls off. After dealing with the sizes of the sun and moon, the phases of the moon, and eclipses, Cleomedes gives (c. 7) estimates of the maximum elongations of Mercury and Venus from the sun (20° and 50° respectively), the maximum deviations in latitude of the five planets (Venus 5°, Mercury 4°, Mars and Jupiter $2\frac{1}{2}$°, Saturn 1°) and of their synodic periods (Mercury 116 days, Venus 584 days, Mars 780, Jupiter 398, and Saturn 378 days). One passage relating to eclipses is worth notice. In c. 6 Cleomedes mentions stories of extraordinary eclipses which the more ancient of the mathematicians had vainly tried to explain; the supposed 'paradoxical' case was that in which, while the sun seems to be still above the horizon, the *eclipsed* moon rises in the east. It seems clear, however, from Cleomedes' remarks that the true explanation had in fact been found, namely that the phenomenon is due to atmospheric refraction, which makes the sun visible to us though it is actually below the horizon; for he speaks of cases of atmospheric refraction which had been observed in the neighbourhood of Pontus, and compares the case of the ring at the bottom of a jug where the ring, just out of sight when the jug is empty, is brought into view when water is poured in.

THEON of Smyrna (second century A.D.) was the author of a book which has come down with the title *Expositio rerum mathematicarum ad legendum Platonem utilium*. The book is a curious medley, but contains many important historical notices. In c. 1 Theon promises a short and summary account of the mathematical theorems which are specially important for the student of Plato to know, in arithmetic, music, and geometry, with its application to stereometry and astronomy. The promise is, however, by no means kept as regards geometry and stereometry; there are geometrical definitions of point, line, straight line, the three dimensions, rectilinear plane and solid figures, especially parallelograms and parallelepipedal figures including cubes, *plinthides* (square bricks), and δοκίδες (beams), and lastly *scalene* figures with sides unequal every way; but this is all, except in so far as the treatment of figured numbers, plane and solid, borders on geometry. The arithmetical section has already been dealt with (pp. 70–1); the most important part of it for us is the account of the Pythagorean system of 'side-' and 'diameter-' numbers.

The section on Music, in which Theon distinguishes two kinds of music, the audible or instrumental, and the theoretical subsisting in numbers, contains substantial extracts from Thrasylus and Adrastus, besides references to views of Aristoxenus, Hippasus, Archytas, Eudoxus, and Plato. We have next a general discussion of ratios, proportions, and means, more than once interrupted by other matters such as the mystic properties of the numbers 2 to 10.

The section on Astronomy is the longest, as it is the most important. Here again Theon is mainly dependent upon Adrastus, from whom he makes long quotations. He begins with the sphericity of the earth, explaining that the unevennesses represented by mountains, &c., are negligible in comparison with the size of the whole; he mentions that Eratosthenes and Dicaearchus claimed to have discovered the highest mountain to be no more than 10 stades higher than the normal level of the land. Then, after describing the principal circles in the heaven, he gives the assumed maximum deviations in latitude of the sun, moon, and planets (the sun 1°, the moon and Venus 12°, Mercury 8°, Mars 5°, Jupiter 5°, and Saturn 3°). The order of the orbits (on the geocentric system) is next discussed; two orders are mentioned; the first, that assumed by Plato and the early Pythagoreans, was (reckoning outwards from the earth) moon, sun, Venus, Mercury, Mars, Jupiter, Saturn; the other, adopted by some later Pythagoreans, was the Chaldaean order, and only differed from the other in putting Mercury and Venus between the sun and the moon. Theon then explains two assignments of musical notes to the heavenly bodies constituting the 'harmony in the universe'. The first gave one note to each of nine bodies, the earth (though stationary), the sun, the moon, the five planets, and the sphere of the fixed stars, the notes being so arranged as to bring the nine into an octave; the other was that of Eratosthenes and Plato, who excluded the earth, and give the eight notes of the octachord to the other eight. The next chapters deal with the forward movements, the stationary points, and the retrogradations, as they seem to us, of the five planets and the 'saving of the phenomena' by the alternative hypotheses of eccentric circles and epicycles; the identity of the motions produced by the two is shown in the case of the sun after Adrastus (cc. 26, 27). We have next an allusion to the famous system of concentric spheres devised by Eudoxus for the purpose of explaining the apparent irregularities in the motions of the planets, with the modifications made by Callippus and Aristotle, and a description of a further system in which the 'carrying' spheres (called 'hollow') have between them 'solid spheres which by their own motion will roll the carrying spheres in the opposite direction, being in contact with them'. These solid spheres (carrying the planet at a point on their surface) act in practically the same way as epicycles. It is in connexion with this description that Theon (i.e. Adrastus) speaks of two alternative hypotheses which, by comparison with Chalcidius' commentary on the *Timaeus*, we can recognize as the hypotheses of Plato and Heraclides of Pontus respectively, so that we can definitely conclude that Heraclides made Mercury and Venus revolve in circles about the sun like satellites, while the sun itself revolved about the earth as centre. Theon gives the maximum arcs separating Venus and Mercury respectively from the sun as 20° and 50°, which are also Cleomedes' figures. The last chapters quoted from Adrastus deal with conjunctions, transits, occultations, and eclipses, and in conclusion Theon gives a considerable extract from Dercyllides, at the beginning of which comes the well-known passage based on Eudemus' *History of Astronomy*:

 Eudemus relates in his Astronomy that it was Oenopides who first discovered the girdling of the zodiac and the revolution (or cycle) of the Great Year, that Thales was the first to discover the eclipse of the sun and the fact that the sun's period with respect to the solstices is not always the same, that Anaximander discovered that the earth is (suspended) on high and lies [substituting κεῖται for the reading of the manuscripts κινεῖται, 'moves'] about the centre of the universe, and that Anaximenes said that the moon has its light from the sun and (explained) how its eclipses come about ['Anaximenes' is here apparently a mistake for 'Anaxagoras'].

SERENUS, of Antinoeia or Antinoupolis, a city in Egypt founded by Hadrian, most probably belonged to the fourth century A.D. He wrote a commentary on the *Conics* of Apollonius. It was evidently in connexion with this that he wrote also two small treatises, which are extant in Greek, *On the Section of a Cylinder* and *On the Section of a Cone*. The occasion and the object of the former are stated in the preface. Serenus observes that many students of geometry were under the erroneous impression that the oblique section of a cylinder was different in character from the oblique section of a cone, whereas they are, of course, the same curve; he thought it necessary, therefore, to prove in a systematic treatise that the said oblique sections are equally ellipses whether they are sections of a cylinder or of a cone.

The treatise *On the Section of a Cone* is even less important, since it deals mainly with the various *triangular* sections made by planes passing through the vertex and either through the axis or not through the axis. Serenus investigates the areas of such classes of triangles, showing when the area of a certain triangle of a particular class is a maximum, under what conditions two triangles of one class may be equal in area, and so on. The last twelve propositions constitute a separate section dealing with the volumes of right cones in relation to their heights, their bases, and the areas of the triangular sections through the axis.

COMMENTATORS AND MINOR WRITERS

THEON of Alexandria lived towards the end of the fourth century A.D. He was the author of a commentary on Ptolemy's *Syntaxis* in eleven Books; this is valuable for certain historical allusions, and for a useful account of the method of operating with sexagesimal fractions, which is illustrated by examples of multiplication and division and the extraction of the square root of a non-square number by way of approximation. One allusion is of capital importance; here Theon observes that the portion of Eucl. VI. 33 relating to *sectors* (of circles) was inserted by himself in his edition of the *Elements*; this is the guide to the distinction between the Theonine and the ante-Theonine texts of Euclid (cf. pp. 207-8 above).

HYPATIA, the daughter of Theon, who, according to tradition, was mistress of the whole of pagan science, and especially of philosophy and medicine, is said to have written commentaries on Diophantus' *Arithmetica*, the *Astronomical Canon* (of Ptolemy), and the *Conics* of Apollonius. She was murdered by a mob of Christians in March 415.

PROCLUS (A.D. 410-85), the Neo-Platonist, was more a philosopher than a mathematician, but he has placed the historian of mathematics under a great obligation by his *Commentary on Euclid, Book I*, which is one of the main sources of our information on the history of elementary geometry. The great value of this Commentary arises from the fact that Proclus had access to a number of historical and critical works which are now lost except for fragments preserved by himself and others. Of these works the great *History of Geometry* in four Books by Eudemus was by far the most important, and its loss can never be sufficiently deplored. Another work, to which Proclus was even more indebted, was that of Geminus on the *Doctrine* or *Theory* of the *Mathematics*; this also seems to have been a very comprehensive work. Again, the commentary of Proclus on the *Republic* of Plato contains some passages of great interest to the historian of mathematics; one relates to Plato's 'Geometrical Number', the other, of more importance, indicates that the theorems of Eucl. II. 9, 10 are Pythagorean propositions formulated for the purpose of proving geometrically the law of formation of the 'side-' and 'diameter-' numbers which give any number of successive approximations to the value of $\sqrt{2}$ (cf. pp. 55–7 above).

Another extant work of Proclus is the *Hypotyposis of Astronomical Hypotheses*, a sort of easy and readable introduction to the astronomical system of Hipparchus and Ptolemy.

DOMNINUS of Larissa, a pupil of Syrianus at the same time as Proclus, wrote a *Manual of Introductory Arithmetic*, a concise and well-arranged sketch of the elements of the Theory of Numbers. It represents a serious attempt at a reaction against the *Introductio Arithmetica* of Nicomachus, and a return to the doctrine of Euclid. It was edited by Boissonade, and Paul Tannery wrote two articles upon it and left a translation, with prolegomena, which has since been published in vol. iii of his *Mémoires scientifiques* edited by Heiberg and Zeuthen.

To SIMPLICIUS, the famous commentator on Aristotle, we owe two long extracts of great importance for the history of mathematics and astronomy. The first, contained in his commentary on Aristotle's *Physics*, is his account, based upon, and to a large extent quoted word for word from, Eudemus, of the attempt of Antiphon to square the circle, and of the quadratures of *lunes* by Hippocrates of Chios (cf. pp. 121–31 above). The other is contained in Simplicius' commentary on the *De Caelo*, and gives the well-known detailed account of the system of concentric spheres as first invented by Eudoxus for the purpose of explaining the apparent motions of the sun, moon, and planets, and of the modifications made in this system by Callippus and Aristotle (cf. pp. 188-9 above).

Simplicius also wrote a commentary on Euclid's *Elements*, Book I, from which an-Nairīzī, the Arabian commentator, made valuable extracts.

Simplicius lived in the first half of the sixth century and was a pupil, first of Ammonius of Alexandria, and then of Damascius, the last head of the Platonic school at Athens.

Contemporary with Simplicius, or nearly so, was EUTOCIUS of Ascalon, the commentator on Archimedes and Apollonius. He dedicated the commentary on Book I *On the Sphere and Cylinder* to Ammonius, a pupil of Proclus and teacher of Simplicius, while the commentary on the first four Books of Apollonius' *Conics* was dedicated to Anthemius of Tralles, the architect of Saint Sophia at Constantinople in 532. Eutocius' commentaries on Archimedes seem to have extended only to the three works *On the Sphere and Cylinder*, *Measurement of a Circle*, and *Plane Equilibriums*. Among the valuable historical notes contained in these commentaries are three of special importance: (1) the account of the various solutions of the problem of Delos, the doubling of the cube or the finding of the two mean proportionals; (2) the fragment, discovered by Eutocius himself, containing the solution, promised by Archimedes in *On the Sphere and Cylinder*, II. 4, but not there given, of the auxiliary problem equivalent to the cubic equation $(a-x)x^2 = bc^2$, which is solved by means of conics; (3) the solutions (a) by Dionysodorus of the cubic equation actually arising out of *On the Sphere and Cylinder*,

II. 4, (b) by Diocles of the original problem in that proposition.

The last person claiming our attention is ANTHEMIUS of Tralles, the architect already mentioned. That he was a competent mathematician is evident from a fragment of a work of his *On burning-mirrors*. This document is of some importance for the history of conic sections; originally edited by L. Dupuy in 1777, it was reprinted in Westermann's Παραδοξογράφοι (*Scriptores rerum mirabilium Graeci*), 1839, pp. 149-58. The first portion of the fragment gives a solution of the problem, 'To contrive that a ray of the sun (admitted through a small hole or window) shall fall at a given point without moving away at any hour or season.' This is effected by constructing an ellipsoidal mirror, one focus of which is at the point where the ray of the sun is admitted, and the other at the point to which the ray is required to be reflected at all times. An interesting feature of the construction and proof is that Anthemius uses the well-known method of drawing an ellipse by means of a string stretched tight round two pins fixed at the foci and a pencil moving round inside the string in such a way as to keep the string always taut. In the third portion of the fragment Anthemius proves that any number of parallel rays can be reflected to one single point from a mirror of parabolic section in which the point is the focus. The *directrix* of the parabola is used in the construction, though not under that name; and it is interesting to note that this passage is the only place except the passages of Pappus referred to above (pp. 153, 266, 459) in which there is any mention of the line in question.

APPENDIX
ADDITIONAL NOTES
1. *Egyptian Mathematics* (cf. pp. 76–80)

UNTIL the Moscow Papyrus is published in full and can be studied as a whole, any estimate of Egyptian geometry must be subject to a certain reserve. The papyrus belongs definitely to the XIIth Dynasty and is understood to contain at least two things of importance. The first is a determination (in Prob. 14) of the volume of a truncated pyramid with a square base. It is apparently based on the well-known formula $V = \frac{1}{3} h (a^2 + ab + b^2)$, where a, b are the sides of the two square faces (top and bottom) and h the height. If the height taken in the papyrus is the vertical height, the solution is correct and must be regarded as a remarkable achievement. The formula is relatively complicated and not such as we should expect to be discovered empirically. On the other hand, if it had a theoretical basis, it is hard to see how it could be got except by deduction from the formula for the volume of a pyramid with square base, namely $\frac{1}{3}ha^2$. But this is a particular case of the theorem which, according to Archimedes, Democritus first discovered, namely, that the volume of any pyramid is one-third of that of the prism with equal base and height. Now the scientific proof of this is in two parts. The second part proves that any prism with triangular bases can be divided into three pyramids equal in volume; but the proof of this depends on the fact proved in the first part, namely, that pyramids with equal bases and heights are equal in content. This first part cannot be theoretically proved except by using the method of infinitesimals in some form, e.g. by dividing the pyramids by planes parallel to the bases into an infinite number of infinitely thin laminae (prisms). There is good evidence, as we have seen, that Democritus was capable of using this method; but there is nothing in any of the Egyptian

APPENDIX 521

mathematical documents so far brought to light to suggest that the Egyptians had attained to this order of ideas in geometry. At present, therefore, I should prefer to think that, the pyramid on a square base being for them a figure of unique importance, they had arrived at an expression for its content by measuring how much corn, or what not, would go into a vessel of that shape. They must have seen that the height and the area of the base would enter into the calculation. But, even assuming that they had obtained the formula $\frac{1}{3}ha^2$ for the pyramid in this way, it was an achievement on their part to prove that the formula $\frac{1}{3}h(a^2+ab+b^2)$ represents the volume of the frustum, of height h, which is the difference between two similar pyramids. Suppose that k, k' are the heights, and (a^2), (b^2) the bases, of the two pyramids; it is necessary to prove that, if $k' : b = k : a$, then

$$\tfrac{1}{3}(ka^2 - k'b^2) = \tfrac{1}{3}(k-k')\,(a^2+ab+b^2)$$

An alternative method of obtaining this result would be the following (cf. Heron, *Metrica*, ii. 8). Draw perpendiculars to the base of the frustum from the corners of the upper square face (b^2); this determines a square in the base. Draw this square, and produce its sides in both directions. The frustum is then divided into (1) a parallelepiped with b^2 as base and height, h, the same as that of the frustum, (2) four prisms each of which has one face coincident with one of the four upright faces of the parallelepiped, and (3) four pyramids at the corners, each of which has a square base of area $\{\tfrac{1}{2}(a-b)\}^2$ and height h. Adding the volumes, we have

$$b^2h + 4\cdot\tfrac{1}{4}(a-b)hb + 4\cdot\tfrac{1}{3}\{\tfrac{1}{2}(a-b)\}^2 h = b^2h + (a-b)hb$$
$$+ \tfrac{1}{3}h(a^2 - 2ab + b^2) = \tfrac{1}{3}h(a^2+ab+b^2).$$

The problem has been discussed at length by Battiscombe Gunn and T. Eric Peet in the *Journal of Egyptian Archaeology* (November 1929, vol. xv, pts. iii–iv, pp. 176–85).

2. Even more baffling is the similar question arising on Problem 10 in the Moscow Papyrus. In this problem the sur-

face of a hemisphere is calculated, step by step, in accordance with the following formula, where d is the diameter,

$$S = \{2d - \tfrac{1}{9} . 2d - \tfrac{1}{9}(2d - \tfrac{1}{9} . 2d)\} . d.$$

The effect of this is that

$$S = \tfrac{8}{9} . \tfrac{8}{9} . 2d . d = \tfrac{1}{2}(\tfrac{256}{81})d^2, \text{ or } 2(\tfrac{256}{81})r^2,$$

where r is the radius.

On the basis of the Egyptian value ($\tfrac{256}{81}$) for π, this is correct. However the result was arrived at, it is highly remarkable, seeing that in Greek mathematics we do not find on record any attempt to determine the surface of a sphere before the time of Archimedes, who was the first to prove scientifically that the surface of a sphere is equal to four times the area of a great circle in it.

It may be that, even when the papyrus has been published, these puzzles will still remain unsolved.

2. *Ancient Babylonian Mathematics* (cf. pp. 96, 398, 472)

Considerable additions to our scanty knowledge of ancient Babylonian mathematics have been made within the last few years, and especially in 1928 and 1929, through fresh studies and interpretations of ancient Babylonian tablets by a number of scholars, among whom may be mentioned E. F. Weidner, A. Ungnad and H. Zimmern, C. Frank, W. Struve, and especially O. Neugebauer of Göttingen.

The Babylonian geometry is mostly of the nature of mensuration. As in Egyptian documents, no general theorems or rules are stated as such; but we are given calculations in concrete cases, from which we have to infer the rules followed.

As we have seen (p. 96), certain calculations of the length of chords of a circle from their *sagittae* and vice versa make it difficult or impossible to doubt that the Babylonians of about 2,000 B.C. knew the property of the square on the hypotenuse of a right-angled triangle, at all events in one case, that of a triangle with sides in the ratios of the numbers 3, 4, 5.

APPENDIX 523

With regard to the circle, the Babylonians used 3 as the ratio of the circumference of a circle to its diameter. This approximation to the value of π is inferior to that implied in the estimate which the Egyptians give of the area of a circle, namely $\frac{64}{81}$ times the square on the diameter, corresponding to a value for π of 3·1605. The Babylonians had, however, a correct idea of the relation of the area of a circle to its circumference; they made the area to be $\frac{1}{12}C^2$, where C is the length of the circumference, which, on the assumption that $\pi = 3$, is right.

But some other cases of mensuration recently explained are remarkable as being more akin to algebra in the respect that, from a certain number of data in the shape of dimensions, other (unknown) dimensions, lengths, or areas, are calculated. Some problems, as we shall see, really involve the equivalent of the solution of quadratic equations.

In the first number of the new series of *Quellen und Studien zur Geschichte der Mathematik* (Abt. B, Studien, i. 1), 1929, O. Neugebauer has explained some problems included in cuneiform texts from the Library of Strassburg University published by C. Frank (1928). One case, the solution of which is given, is interesting. The figure attached to the problem in the tablet is crude, and, to understand the procedure, it is necessary to draw a better diagram in the light of the actual text. We are given a parallel-trapezium, as $ABDC$, in which AB, CD are the parallel sides, while a third side AC is perpendicular to them. The trapezium is divided into two trapezia by the straight line EF parallel to AB. (The letters in the diagram and the dotted lines are my addition; the original diagram has no letters, but merely shows the areas of the two parts as 13 3 and 22 57 respectively in the sexagesimal notation, meaning 783 and 1377,

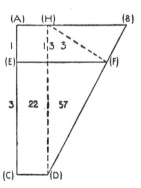

and has the numbers 1 and 3 along the parts of the perpendicular denoting the ratio of AE to EC).

We are given (1) the areas of the two trapezia, (2) the difference (HB) between AB and CD, (3) that $AE = \frac{1}{3}EC$, (4) the fact, geometrically consequent on (3), that

$$AB - EF = \tfrac{1}{3}(EF - CD);$$

and we have to find in numbers the five lengths AB, EF, CD, AE, EC.

If these lengths be denoted by x, y, z, u, v respectively, the data, divorced from geometry, are equivalent to the five algebraical equations

$$u \cdot \tfrac{1}{2}(x+y) = 783,$$
$$v \cdot \tfrac{1}{2}(y+z) = 1377,$$
$$u = \tfrac{1}{3}v,$$
$$x-z = (x-y)+(y-z) = 36,$$
$$x-y = \tfrac{1}{3}(y-z).$$

To say that the writer solves these algebraical equations would no doubt be misleading; what he does is rather arithmetical computation with reference to the facts represented in the figure, with the aid of the correct formulae for the area (in terms of bases and heights) of the rectangle, the triangle, and the parallel-trapezium. The results obtained are $x = 48$, $y = 39$, $z = 12$, $u = 18$, $v = 54$.

It is important to observe that in the original text all the numbers are consistently expressed on the sexagesimal system. For example, 2160 is written as 36 simply (meaning 36.60), 1296 is 21 36; as there is no sign for 0, the fact that, in the first case, 36 means 36.60 has to be gathered from the context; 21 36 is of course $21 \cdot 60 + 36 = 1296$. Similarly with fractions: $\frac{1}{2}$ is 30, meaning $\frac{30}{60}$; $\frac{1}{18}$ is 3 20, meaning $\frac{3}{60}+\frac{20}{60^2}$; and so on. To calculate a certain fraction of a number, the Babylonian turns the fraction into a sum of successive sexagesimal fractions and *multiplies*; and this he does with certainty (notwithstanding the absence of a sign for 0) by keeping the particular

APPENDIX

denominations distinctly in his mind while carrying out the operation. Thus $\frac{1}{72} \cdot 864$ $[=(\frac{0}{60}+\frac{50}{60^2}) \cdot 864]$ is obtained by multiplying 14 24 $[= 864]$ by 50, and the result is expressed as 12 simply.

There follow in Neugebauer's article other problems relating to a right-angled triangle divided by a number of straight lines parallel to one of the perpendicular sides into parallel-trapezia and a triangle (there are five such trapezia in the case next taken); and, while some dimensions are given, the rest have to be found by processes similar to the above.

Problems reducible to quadratic equations

Even more remarkable in one respect is the next type of problem, where the right-angled triangle is divided into one parallel-trapezium and a triangle. In the first case we are given that, in the subjoined figure, $AB = 30$, area of DEC is 4 30 $[= 270]$, and $DC - AD = 10$.

No solution is given in the text; but, in order to see the nature of the problem, let us suppose that $AD = x$.

Then $270 = \frac{1}{2} DE \cdot (x+10)$, and, by similar triangles,

$$30 : DE = (2x+10) : (x+10).$$

Eliminating DE, we easily obtain

$$x^2 - 16x - 80 = 0,$$

which gives $x = 20$.

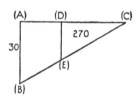

We are further assured by Neugebauer that a certain problem in *Cuneiform Texts in the British Museum* (ix. 12) leads to the equation $x^2 - \dfrac{2F}{d} x + \dfrac{2Fh}{d} = 0$, and that the root is calculated precisely according to the formula

$$x = \frac{F}{d} - \sqrt{\left\{\left(\frac{F}{d}\right)^2 - \frac{2Fh}{d}\right\}}.$$

But more remains behind. Neugebauer is, in another paper

(*Quellen und Studien*, Abt. B. i. 2) continuing the work of Frank upon the Strassburg cuneiform texts, and proving line by line that the texts do give solutions of quadratic equations amounting to the substitution of the specific numbers of the particular question in formulae identical with what we should arrive at by our ordinary algebraic working.

We are introduced to equations in *two* unknowns which reduce to quadratics with one unknown. The first two simple cases are remarkable in that, though dealing with squares, rectangles, and sides thereof, they are purely *algebraic*, since they add numbers representing a square and its side respectively without any regard to *dimensions*. This divorce of equations from geometric connotation does not seem to occur in Greek mathematics before Heron (cf. the case on pp. 430–1 above). The second of two cases given by Neugebauer is this:

'The 7th part of the length, the 7th part of the breadth, and the 7th part of the area gives 2. Length and breadth added give 5 50. Calculate length and breadth. 3 30 (is) the length, 2 20 the breadth.' (The numbers are expressed in the sexagesimal notation and are therefore 2, $5\frac{5}{6}$, $3\frac{1}{2}$, and $2\frac{1}{3}$ respectively.)

If x, y are the length and breadth respectively, the equations are
$$\tfrac{1}{7}x + \tfrac{1}{7}y + \tfrac{1}{7}xy = 2,$$
$$x + y = 5\tfrac{5}{6};$$

They reduce to the quadratic in x,
$$x^2 - \tfrac{35}{6}x + \tfrac{49}{6} = 0.$$

Only the result ($x = 3\tfrac{1}{2}$, $y = 2\tfrac{1}{3}$) is given in the text, not the working.

The working is, however, given in the remaining three cases and appears to prove Neugebauer's thesis up to the hilt.

We have only space for one case.

The problem is: 'An area (consisting of) the sum of two squares (is) 16 40 [= 1000]. (One) square (is) $\tfrac{2}{3}$ of (the other) square, while 10 is subtracted from the lesser square. Calculate the squares.'

APPENDIX 527

(It will be noticed that the problem is of the same type as the Egyptian one, from the Berlin Papyrus 6619, cited on pp. 467–8 above. That case is, however, much simpler, leading to a *pure* quadratic equation.)

The word for 'square' is ambiguous; it may mean the *side* of a square as well as square. Nor does the phrase '10 is subtracted from the lesser square', by itself, make its import clear. But the working shows that it is the *side* of one square which is two-thirds of the *side* of the other; it also shows that, if we call the squares x^2, y^2, the equations solved are actually of the form

$$x^2 + y^2 = A, \qquad \qquad (1)$$

$$y = \frac{\alpha}{\beta} x - d. \qquad \qquad (2)$$

We should solve by substituting in (1) the value of y in terms of x as given in (2). We have then the equation

$$\left(1 + \frac{\alpha^2}{\beta^2}\right) x^2 - \frac{2 d\alpha}{\beta} x + d^2 = A.$$

Remembering that 2/3, the particular value assigned in the question to the ratio α/β, is written in the sexagesimal notation as '40' (with sixtieths understood), and therefore β is, in the particular case, 60, we can simplify the equation by substituting $\xi = x/\beta \, [= x/60]$, and the equation becomes

$$\xi^2 - \frac{2 d\alpha}{\alpha^2 + \beta^2} \xi = \frac{A - d^2}{\alpha^2 + \beta^2},$$

the solution being

$$\xi = \frac{1}{\alpha^2 + \beta^2} [d\alpha \pm \sqrt{\{d^2 \alpha^2 + (\alpha^2 + \beta^2)(A - d^2)\}}].$$

Bearing in mind that, in the particular case, $A = 1000$, $d = 10$, $\alpha = 40$, $\beta = 60$, let us now compare with this the working given in the text by putting the working on the left side of the page and the corresponding algebraical expressions on the right. The text gives all the numbers in the sexagesimal notation. Thus 22 24 26 40 means

$$22.60^3 + 24.60^2 + 26.60 + 40 = 4840000;$$

21 40, where it occurs, means (in the absence of a sign for 0) $21.60^3 + 40.60^2 = 4680000$; and 30 multiplied by 1 in one place means $\frac{30}{60}.60 = 30$; the denominations, integral and fractional, have always to be inferred from the context. To make the solution easier to follow, we will express all numbers in our notation.

' You proceed thus:

Square 10, this gives 100: subtract 100 from 1000, this gives 900. $\qquad A - d^2 = 900$

Square 60, this gives 3600; 40^2 is 1600: $3600 + 1600 = 5200$. $\qquad \alpha^2 + \beta^2 = 5200$

Multiply 5200 by 900: this gives 4680000. $\qquad (\alpha^2 + \beta^2)(A - d^2) = 4680000$

Multiply 40 by 10, this gives 400. $\qquad \alpha d = 400$

Square 400: this gives 160000. $\qquad \alpha^2 d^2 = 160000$

Add 160000 to 4680000, this gives 4840000. $\qquad \alpha^2 d^2 + (\alpha^2 + \beta^2)(A - d^2)$
$$= 4840000$$

The square root of this is 2200. $\qquad \sqrt{\{\alpha^2 d^2 + (\alpha^2 + \beta^2)(A - d^2)\}}$
$$= 2200$$

Add 400 already found, this gives 2600. $\qquad d\alpha + \sqrt{\{\alpha^2 d^2 + (\alpha^2 + \beta^2)(A - d^2)\}}$
$$= 2600$$

What part of 5200 gives 2600? Answer: one-half (30 in text).
$$\frac{d\alpha + \sqrt{\{\alpha^2 d^2 + (\alpha^2 + \beta^2)(A - d^2)\}}}{\alpha^2 + \beta^2} \ [= \xi] = \tfrac{1}{2}$$

$\tfrac{1}{2}$ multiplied by 60 gives 30 as [side of] greater square. $\qquad \tfrac{1}{2}\beta = \xi\beta = x = 30$

Multiply $\tfrac{1}{2}$ by 40, this gives 20. $\qquad \tfrac{1}{2}\alpha = \xi\alpha = 20$

Subtract 10 from 20 and this gives 10 as [side of] lesser square.' $\qquad \xi\alpha - d = y = 10$

The exact correspondence of the working to the algebraical formulae is manifest.

The last two problems (which are alike except that the concrete numbers taken in the two cases are different) are a little more complicated, giving rise to equations of this form:

$$\left.\begin{array}{r}x^2+y^2 = A, \\ x = u+d_1, \\ y = \dfrac{\alpha}{\beta}u+d_2,\end{array}\right\}$$

where u is an auxiliary unknown.

If we put $\xi = \dfrac{u}{\beta}$ ($\beta = 60$), we obtain, by substitution, the quadratic in ξ

$$\xi^2 + \frac{2(d_1\beta+d_2\alpha)}{\alpha^2+\beta^2}\xi = \frac{A-(d_1^2+d_2^2)}{\alpha^2+\beta^2},$$

the solution being

$$\xi = \frac{1}{\alpha^2+\beta^2}[\pm\sqrt{\{(d_1\beta+d_2\alpha)^2+(\alpha^2+\beta^2)(A-\overline{d_1^2+d_2^2})\}}$$
$$-(d_1\beta+d_2\alpha)],$$

after which we find $x\ (=\beta\xi+d_1)$ and $y\ (=\alpha\xi+d_2)$.

Here again the working in both cases corresponds line by line with the algebraical expressions composing the result.

There is no doubt that we have here discoveries of extraordinary interest. That the Babylonians of 2000–1800 B.C., or thereabouts, should have advanced so far in what is real algebra without (so far as the texts at present published show) any algebraical notation or any sign of a general *theory* underlying the examples is wonderful in itself; it is equally extraordinary that these developments in arithmetic and algebra should have remained, for most of 1,800 years at all events, unknown to, or at least without (so far as we can judge) any traceable effect upon, the Greek pioneers in the same subjects. The first extant Greek treatises which give arithmetical solutions of quadratic equations are those of Heron and

APPENDIX

Diophantus. If any Greek mathematician can be supposed to have known, or assimilated, the ancient Babylonian arithmetic and algebra, we can think of no one more likely to have done so than Hipparchus, who evidently knew in all detail the results of Babylonian observations and research in astronomy. The discoveries about the Babylonian algebra in any case add fresh interest to the Arabian attribution to Hipparchus of a work on 'the art of algebra', upon which Abū 'l Wafā wrote a commentary. Did Hipparchus write a book on algebra by way of making known what he found of the nature of algebra in ancient Babylonian tablets and adapting it for use in astronomical and other problems? The passage of Plutarch (*De Stoicorum repugnantiis*, c. 29) in which he classes Hipparchus among the 'arithmeticians' may perhaps be thought to favour such a supposition; but, unless and until fresh documents come to light, the question will presumably remain unanswered.

3. *Hipparchus and Chaldaean Astronomy* (cf. pp. 396–7)

In considering the work of Hipparchus and the other Greek astronomers we must bear in mind the admitted indebtedness of Greek to Chaldaean astronomy. New light has recently been thrown upon this question by systematic study of new material in the shape of Greek texts printed for the first time between 1906 and 1926, and of Babylonian texts published from 1900 to 1926. It is established that a continuous record of dated observations began with the reign of Nabonassar (747–734 B.C.) and came into the possession of the Greeks. In particular, a collection of such observations was sent by Callisthenes to his uncle, Aristotle, after the capture of Babylon by Alexander in 331 B.C.: and this collection must have been accessible to Callippus. The names of two Babylonian astronomers, Naburianos (fl. about 500 B.C.) and Cidenas (Kidinnu), are worthy to be mentioned in the history of astronomy beside those of Hipparchus and Ptolemy. Cidenas, about 383 B.C., determined the length of the synodic month as 29·530594 days as compared with Naburianos' 29·530614, Hipparchus' 29·530585, and the true value of 29·530596 days. The Babylonians made continuous observations and calculated constants; the Greeks apparently used these as a basis for their cycles and for the geometrical hypotheses by which they sought to account for the apparent irregularities of motions (Dr. J. K. Fotheringham in *The Observatory*, vol. li, no. 653, Oct. 1928, and in the *Nautical Almanac* for 1931).

GREEK INDEX

ἄβαξ, ἀβάκιον 25.
ἀγεωμέτρητος, -ον: ἀγεωμέτρητος μηδεὶς εἰσίτω 3, 171.
ἀδύνατος, -ον: ἀπαγωγὴ εἰς ἀδύνατον, *reductio ad impossibile* 215.
αἴτημα, postulate 216.
ἄλογος, -ον, irrational 51: περὶ ἀλόγων γραμμῶν καὶ ναστῶν, work by Democritus 54.
ἀνάλημμα 412.
ἀνάλογον, proportional 51.
ἀναλυόμενος (τόπος), Treasury of Analysis 434.
ἀνάπαλιν: ἀνάπαλιν λύσις, 'solution backwards' 452: (in proportions) *inversely* 225.
ἀναστρέψαντι (ἀναστρέφω) = *convertendo* 226.
ἀναστροφή, conversion, *ib*.
'Αναφορικός, work by Hypsicles 384.
ἀντικείμενος, -η, -ον: ἀντικείμεναι (τομαί) = double-branch hyperbola 359.
ἀόριστος, -ον, undefined: πλῆθος μονάδων ἀόριστον = unknown quantity (x) 476.
ἀπαγωγή, reduction (of a problem or theorem) 215: ἀπ. εἰς ἀδύνατον 215.
ἀπόδειξις, proof 214.
ἀποκαταστατικός, -ή, -όν, recurring 68.
ἄρβηλος, 'shoemaker's knife' 284, 442–3.
ἀριθμητική, theory of numbers, opp. to λογιστική 5.
ἀριθμός, number: in Diophantus = unknown quantity (x) 476–7.
ἀριθμοστόν (Dioph.), reciprocal of ἀριθμός (= $1/x$) 478.
ἀρτιάκις ἄρτιος, even-times even number (Pythagorean) 39.
ἀρτιάκις περιττός 39, or ἀρτιοπέριττος, even-odd 40: = number of form $2(2m+1)$ 40.

'Αρχαί, Principles, lost work of Archimedes 327–8.
'Αστροθεσίαι or 'Αστρονομία, work by Eratosthenes 345.
ἀστρολάβον ὄργανον of Hipparchus 397.
ἀσύμπτωτος, -ον, non-secant, *asymptote* 2.
Αὐτοματοποιητικῆς, Περί, work by Heron 417.

Βαρουλκός, work by Heron 417, 462.
Βελοποιϊκά of Heron 415, 416.
βωμίσκος, 'little altar', properly a wedge-shaped solid = σφηνίσκος 427.

γεωδαισία, mensuration 6.
γνώμη, two meanings of (Democritus) 118–19.
γνώμων, gnomon, q. v.: κατὰ γνώμονα, a perpendicular 45, 115.
γνώριμος, -ον, known: γνωρίμως, 'in the well-known manner' 324.
γραμμικός, -ή, -όν, linear (of prime numbers) 40.

δεικνύναι, to prove: ὡς ἑξῆς δειχθήσεται 216.
δεῖν: δεῖ δή in διορισμός 215.
δεύτερος, -α, -ον: δευτέρα μυριάς, second myriad = 10000^2 18: δεύτερος καὶ σύνθετος, secondary (opp. to prime) and composite (number) 40.
διαιρεῖν: διελόντι (in proportions), *separando* or *dividendo* 226.
διαίρεσις: (λόγου) *separation* of ratio 226: περὶ διαιρέσεων βιβλίον, On divisions (of figures), by Euclid 258–62.
διάστασις, dimension: περὶ διαστάσεως, work by Ptolemy 414.
δίαυλος, 'race-course' = representation of square number as sum of series of terms 72.

GREEK INDEX

δικόλουρος, -ον, twice-truncated: of pyramidal number 67.

δίοπτρα, dioptra, q.v.

διορίζω: διωρισμένη τομή, *Determinate Section*, work by Apollonius 366.

διορισμός, definition, delimitation: (1) a constituent part of a theorem or problem 214: (2) determination of conditions of possibility of solution of problem 215.

διπλοϊσότης, 'double equation' (Diophantus) 487.

διπλοῦς, -ῆ, -οῦν, double: διπλῆ μυριάς $= 10000^2$ (Apollonius) 19: διπλῆ ἰσότης or ἴσωσις, double equation (Dioph.) 487.

δοκίς, beam, a class of solid number 67.

δύναμις, square or square root: with Theaetetus equivalent to 'surd' 133: square of unknown quantity ($= x^2$) (Dioph.) 477–8.

δυναμοδύναμις, 'square-square' = fourth power of unknown, x^4 (Dioph.) 478.

δυναμόκυβος, 'square-cube', equivalent to x^5 (Dioph.) 478.

δυναμοστόν, δυναμοδυναμοστόν, &c., reciprocals of foregoing respectively 478.

εἶδος, figure: 'figure' of a conic (Apollonius) 359: 'species' = particular power of unknown, or *term*, in equation (Dioph.) 480.

ἔκθεσις, *setting-out*: constituent part of proposition 214.

ἔλλειψις, *falling-short* (in application of areas) 100: name given to *ellipse* by Apollonius 359.

ἐλλιπής, -ές, defective: of number contrasted with 'perfect' number 64.

ἐναλλάξ (in proportion), *alternando* 225.

ἔννοια, notion: κοιναὶ ἔννοιαι, common notions = axioms 195, 216.

ἐξελιγμός 392.

ἐξήγησις, elucidation 390.

ἑξηκοστόν, or πρῶτον ἑξ., a 60th = minute: δεύτερον ἑξ., a second, &c. 23.

ἐπάνθημα, 'bloom' (of Thymaridas): solution of system of simple equations 57–8, 72.

ἐπαφή, contact: 'Ἐπαφαί (of Apollonius), *Tangencies* 366.

ἐπί, upon: = multiplied by 30, 31.

ἐπιμερής, -ές, *superpartiens*, = ratio $1+(m/n)$, 65.

ἐπιμόριος, -ον, *superparticularis*, = ratio $1+(1/n)$, 53, 64.

ἐπίτριτος = ratio 4/3, 64, 65.

ἑτερομήκης, -ες, oblong: of numbers of form $m\,(m+1)$, 49.

ζυγόν, lever of balance: περὶ ζυγῶν, lost work of Archimedes 285.

ἡμιόλιος, -α, -ον, *sesquialter*, = ratio 3/2, 64.

θυρεός, shield: old name for ellipse 265, 348.

ἰσόμετρος, -ον, of equal contour (Zenodorus) 382.

ἴσος, -η, -ον, equal; δι' ἴσου, *ex aequali* (in proportions) 226: δι' ἴσου ἐν τεταραγμένῃ ἀναλογίᾳ, 'ex aequali in disturbed proportion' *ib.*

καμπτήρ, turning-point (in racecourse) 72.

καμπύλος, -η, -ον, curved or bent: of curves 157.

κανών, table (astron.): Προχείρων κανόνων διάτασις καὶ ψηφοφορία, work by Ptolemy 412: also = rule or scale, Κατατομὴ κανόνος, musical work attributed to Euclid 268.

κατασκευή, *construction* (constituen part of a proposition) 214.

Κατατομὴ κανόνος, v. κανών.

κατονόμαξις τῶν ἀριθμῶν, system of naming numbers (Archimedes) 284.

κεντροβαρικά, problems on centre of gravity 285.
κέντρον, centre: ἡ ἐκ τοῦ κέντρου, *radius* 221.
κερατοειδής, -ές: κ. γωνία, 'horn-like angle' = 'angle of contact' 223.
κλάειν, to inflect: κεκλάσθαι 196.
κόλουρος, -ον, truncated: of pyramidal number 67.
κόσκινον, *sieve* (of Eratosthenes) 63.
κυβόκυβος, cube-cube, = 6th power of unknown, or x^6 (Dioph.) 478.
κυβοκυβοστόν, reciprocal of κυβόκυβος 478.
κύβος, cube: used for cube of unknown, or x^3 (Dioph.) 477, 478.
κυκλικός, -ή, -όν, circular: used of square numbers ending in 5 or 6, 68: Κυκλικὴ θεωρία, *De motu circulari*, by Cleomedes 510–12.

λείπειν: forms of, used to express *minus*, and sign for (Dioph.) 479.
λεῖψις, a *wanting* = a negative term (Dioph.) 479: λείψει = *minus*, *ib*.
λεπτόν, a fraction (Heron) 21: = a *minute* (Ptol.) 23.
λογισμός, calculation 5.
λογιστική, art of calculation, opp. to ἀριθμητική 5–6.
λόγος, ratio: *Λόγου ἀποτομή*, *Sectio rationis*, by Apollonius 363–5.

μάθημα, subject of instruction: μαθήματα first appropriated to mathematics by Pythagoreans 5: τοὶ περὶ τὰ μαθήματα (Archytas), 'mathematicians' 5.
μαθηματικός, -ή, -όν, mathematical: Μαθηματικὴ σύνταξις of Ptolemy 402.
μετέωρος, -ον: *Περὶ μετεώρων*, *Meteorologica*, by Posidonius 386–7.
Μικρὸς ἀστρονόμος, or ἀστρονομούμενος (τόπος), *Little Astronomy*, dist. from Μεγάλη Σύνταξις of Ptolemy 200, 403, 450.
μοῖρα, part or fraction: 1/360th of circumference or a *degree* 23, 33:

μοῖρα τοπική, χρονική (Hypsicles) 384.
μονάς, unit: def. of 38: μονάδων σύστημα, def. of number *ib*.: μονὰς θέσιν ἔχουσα, 'unit having position', def. of point *ib*.
μόριον, part or fraction: μορίου or ἐν μορίῳ, 'divided by' (Dioph.) 22, 480.
μυριάς, myriad (with or without πρώτη or ἁπλῆ) = 10000: μ. δευτέρα or διπλῆ = 10000^2, &c. 18–19.

νεύειν, to *verge* (towards): a subject of formal definition in Aristotle's day 196.
νεῦσις, *inclinatio* or 'verging': a type of problem 128, 148–9, 150, 151: Νεύσεις, two Books, by Apollonius 372–5.
νύσσα, goal (of race-course) 72.

'Ολυμπιονῖκαι, work by Eratosthenes 346.
ὄρθιος, -α, -ον, right or perpendicular: ὀρθία πλευρά, *latus rectum* of conic 2, 359: with πλαγία πλευρά, 'transverse side' (= corresponding diameter), makes up 'figure' of conic *ib*.
ὁρίζειν, define, determine: πλῆθος ὡρισμένον, def. of number (Eudoxus) 38: ὁρίζων κύκλος, dividing circle: ὁρίζων (without κύκλος), *horizon* (Eucl.) 266.
ὅρος, definition 216.
οὐδεμία or οὐδέν: O stands for, in Ptolemy 23.

πᾶ βῶ καὶ κινῶ τὰν γᾶν, 'give me a place to stand on and I will move the earth' (Archimedes) 279.
παρ' ἣν δύνανται (αἱ καταγόμεναι τεταγμένως), ἡ, expression for *parameter* (of lines drawn *ordinatewise*) (Apollonius) 359.
παραβολή, application: π. τῶν χωρίων, application of areas 100:

τὰ ἐκ τῆς παραβολῆς γινόμενα σημεῖα = foci of central conic 362: παραβολή first applied to *parabola* by Apollonius 103-4, 359.
Παραδοξογράφοι 519.
παράλληλος, -ον, parallel 2.
παρισότης, nearness to equality: παρισότητος ἀγωγή, 'inducement of approximation', a method of Diophantus 493.
πεμπάζειν, to 'five' = count 11.
πένταθλος 343.
περαίνουσα ποσότης, 'limiting quantity', def. of unit 38.
πέρας, limit, extremity 174: πέρας συγκλεῖον, def. of 'figure' (Posidonius) 385.
περισσάρτιος, *odd-even*: with Neo-Pythagoreans = a number of form $2^{n+1}(2m+1)$, 40.
περιττὰ ἀρτιάκις, περιττάκις 39.
πηλικότης, size 224.
πλάγιος, -α, -ον, transverse: πλαγία πλευρά or διάμετρος, *transverse side* or *diameter* 359.
Πλατωνικός, work by Eratosthenes 343.
πλῆθος, multitude: πλῆθος ὡρισμένον, def. of number 38: πλῆθος μονάδων ἀόριστον, def. of unknown quantity ($= x$) (Dioph.) 476.
πλινθίς, *brick*: a certain form of solid number 67.
πολλαπλασιεπιμερής, -ές, *multiplex superpartiens* = ratio of form $p + \{m/(m+n)\}$, 65.
πολλαπλασιεπιμόριος, *multiplex superparticularis*, = ratio of form $m + (1/n)$, 65.
πολλαπλάσιος, multiple 64.
πόλος 270.
πολύσπαστος, a compound pulley 279.
πόρισμα, *porism*: (1) = corollary 215: (2) a special type of proposition 215-16, 263.
προμήκης, *prolate* (= oblong), but distinct from ἑτερομήκης, used of numbers 49.

πρότασις, *enunciation* (constituent part of proposition) 214.
πρῶτος, *prime* 40.
πυραμίς, pyramid 79.
πυρεῖον, πυρίον, burning-mirror: Περὶ πυρείων, work by Diocles 167, 379.

ῥοπή, inclination of lever of balance: Περὶ ῥοπῶν, mechanical work by Ptolemy 414.

σάλινον (Archimedes?), possibly 'salt-cellar' 339.
σήκωμα 28.
σκάφη ('boat'), a sun-dial 270
σκηνογραφική, scene-painting 7.
σπεῖρα, *spire* or *tore* (anchor-ring) 380: varieties of, συνεχής, διεχής, ἐμπεπλεγμένη 381: sections of (Perseus) 381-2.
στηλίς, *column*, a class of solid number 67.
στιγμή, point: στιγμὴ ἄθετος, 'point without position', def. of unit 38.
στοιχειωτής, writer of Elements (στοιχεῖον): ὁ στ. = Euclid 204.
συμπέρασμα, *conclusion* (constituent part of proposition) 214.
σύνθεσις (λόγου), *composition* (of a ratio) 225-6.
σύνταξις, collection: Μαθηματικὴ σύντ. of Ptolemy 402: Μεγάλη σύνταξις 403.
συντιθέναι: συνθέντι (in proportions) = *componendo* 225-6.
σφαιρικός, -ή, -όν, spherical: used of cube numbers ending in 5 or 6 68.
σφηνίσκος, 'little wedge', a certain form of solid number 427.
σχέσις, relation 224.

ταράσσειν, disturb: δι' ἴσου ἐν τεταραγμένῃ ἀναλογίᾳ, *ex aequali* in disturbed proportion 226.
τάσσειν: τεταγμένος, assigned or ordered, αἱ καταγόμεναι τεταγμένως, *ordinates* in conics 2, 359.

GREEK INDEX

τεταρτημόριον, ¼-obol 27.
τετραγωνίζειν, to square: ἡ τετραγωνίζουσα (γραμμή), quadratrix of Hippias 142.
τετρακτύς, tetractys, = triangular number $1+2+3+4$ 42-3.
τμῆμα, segment: (1) of a circle 124: (2) used of 1/360th parts of circumference (degrees) and 1/120th parts of diameter (Ptolemy, &c.) 23.
τομεύς, shoemaker's knife: term for sector of circle 222.
τόπος, locus: τόποι πρὸς ἐπιφανείᾳ, 'loci on a surface' 265: τόπος ἀναλυόμενος (Pappus), 'Treasury of Analysis' 434: μικρὸς ἀστρονομούμενος (τόπος), Little Astronomy 200, 403, 450.
τρικόλουρος, thrice-truncated: of pyramidal number 67.

ὕπαρξις, 'forthcoming' = positive term (Dioph.) 479.
ὑπεπιμερής, subsuperpartiens 65.
ὑπεπιμόριος, subsuperparticularis 64.
ὑπερβολή, exceeding (in application of areas) 100: name given to hyperbola by Apollonius 103-4 359.

ὑπερτελής, 'over-perfect' (number) 64.
Ὑποθέσεις τῶν πλανωμένων, work by Ptolemy 412.
ὑποπολλαπλάσιος, ὑποπολλαπλασιεπιμερής, &c. 65.
ὑποτείνειν, subtend: ἡ ὑποτείνουσα (πλευρά), hypotenuse 2.
ὕσπληξ, starting-point (of racecourse) 72.

Φάσεις ἀπλανῶν ἀστέρων, work by Ptolemy 412.

χαλκοῦς, ⅛th of obol 26, 27.
χείρ, manus, in sense of a number of men 11.
χροιά, colour or skin: Pythagorean name for surface 174.
Χρονογραφίαι, work by Eratosthenes 346.
χρῶμα, colour (in relation to surface) 174.
χωρίον, area: παραβολὴ τῶν χωρίων, application of areas 100: Χωρίου ἀποτομή, Sectio spatii, work by Apollonius 365.

Ὠκυτόκιον, '(means of) quick delivery', work by Apollonius 147, 376.

ENGLISH INDEX

Abacus 24–8.
Abraham Echellensis 353.
Abū 'l Fatḥ al-Iṣfahānī 353.
Abū 'l Wafā al-Būzjānī 530.
Achilles of Zeno 192.
Addition in Greek notation 28.
Adrastus 513–14.
Aëtius 185.
'Aganis' 205.
Ahmes (Papyrus Rhind) 76, 467.
Alexander Aphrodisiensis 122.
al-Fakhrī, by al-Karkhī 68.
Algebra: *ḥau*-calculations 466–7: Pythagorean 55–60: *Epanthema* of Thymaridas 57–9: in Greek Anthology 468–9: indeterminate problems (Greek) 470–2: Diophantus 472 sqq.
Algebra, geometrical 100–4, 110–11: (application of areas) 100–4, 231–5.
al-Ḥajjāj, translator of Euclid 209: of Ptolemy 403.
al-Karkhī 68–9.
Almagest 403.
Alphabet, Greek, derived from Phoenician 15–16.
Alphabetic numerals 15–18.
Amenemhet III 76.
Ameristus 90.
Amyclas (better Amyntas) 158, 184.
Analemma of Ptolemy 412: of Diodorus 434.
Analysis: used by Eudoxus 184, absent from Euclid's *Elements* 215: defined by Pappus 452–3.
Anatolius 61.
Anaxagoras 112, 113, 191: explanation of eclipses 113–14: moon borrows light from sun 113, 515: centrifugal and centripetal forces 114: tried to square circle 114, 140: magnitudes divisible without limit 114: in *Erastae* 115.

Anaximander 90–1: introduced gnomon 91: A.'s astronomy *ib.*: distances of sun and moon *ib.*: map of inhabited earth 90.
Anaximenes 515.
Anchor-ring: see *Tore*.
Angle *of* a segment 197, 222: *of* a semicircle 196, 222: 'angle of contact' ('hornlike') 118, 222–3.
Anharmonic property: (1) of four great circles through a point on sphere 401, (2) of four straight lines through one point, *ib.* 458.
Anthemius of Tralles, 518, 519.
Antiphon the Sophist 122, 140–1.
Āpastamba-Śulba-Sūtra 97.
Apices 25.
Apollodorus ὁ λογιστικός, distich of 86, 95.
Apollonius of Perga 4, 31, 205, 232, 352–76. Arithmetic: 'tetrads' (powers of 10000) 19, 437: Ὠκυτόκιον 147, 376: approximation to π 147: Astronomy 376.

Conics 352–63: text and translations 353: genesis from any oblique cone 355–9: property equivalent to Cartesian equation (oblique axes) 103, 359: new names *parabola, hyperbola, ellipse* 103–4, 352, 359: double-branch hyperbola 359–60: tangent properties 360: asymptotes of hyperbola 361–2: transformation of coordinates 361: rectangles under segments of intersecting chords, harmonic properties of pole and polar 362, focal properties of central conics 362–3: normals as maxima and minima 354, construction and number of *ib.*, derivation of evolute *ib.*

Sectio rationis 363–5, *Sectio spatii* 365–6, *Determinate Section* 366: *On contacts* 366–7 and problem of three circles 367–71: *Plane Loci* 371–2: Νεύσεις 372–5 and rhombus-problem 373–5: Comparison of dodecahedron and icosahedron 375: 'General treatise' 375–6: *On the cochlias* 376: on irrationals, on burning mirrors *ib.*: duplication of cube 166–7.

Application of areas 100–4: equivalent to solution of general quadratic equation 101–4, 231–5.

Approximations: to $\sqrt{2}$ (by 'side'- and 'diameter'-numbers) 55–7, Indian 97: to $\sqrt{3}$, Ptolemy 23–4, Archimedes 309–10: to π 146–7: to surds generally (Heron) 422–4: to cube root 430.

Apuleius of Madaura 62.

Archibald, R. C. 258.

Archimedes 4, 31, 135, 141: traditions 277–80, engines *ib.*, mechanics 279, 284–5, general estimate 4, 280–3.

Works: character of 281–2, list of, 283–6, text and editions 286–7: the *Method* 115, 119, 282, 287–93: *On the Sphere and Cylinder* 293–305: *Measurement of a circle* 305–9: *On Conoids and Spheroids* 310–17: *On Spirals* 144–5, 317–22: *Sand-reckoner* 19–20, 327–30: *Quadrature of Parabola* 330–2: *Plane Equilibriums* 323–7: *On Floating Bodies* 332–6, problem of crown 333–5: *Liber assumptorum* 284, 338–40: cattle-problem 284, 336–7: *Catoptrica* 285.

Arithmetic: 'octads' 19–20: fractions 21: value of π 146: approximations to $\sqrt{3}$, 309–10.

On Aristarchus's heliocentric hypothesis 270–1: conics, propositions in 351–2: cubic equation solved by conics 303–4: on Democritus 115–16, 193, 283: anticipations of integral calculus 4, 299–300, 314–17, 320–2: Lemma or Axiom of Archimedes 193–4, 224, 294: area of triangle in terms of sides 340: on semi-regular solids 337–8: trisection of any angle 151–2: construction of regular heptagon in a circle 340–2.

Archytas 4, 5, 52, 112, 134–7, 187: on μαθήματα 5: on means 51, 52: no mean proportional number between numbers in ratio $n/(n+1)$, 53, 136–7: on music 135–6: mechanics 135: solution of problem of two mean proportionals 155–7.

Aristaeus: *Solid Loci* 203, 265, 349–51: comparison of five regular solids 254.

Aristaeus of Croton 52.

Aristarchus of Samos 270–6: σκάφη (sun-dial) 270: anticipated Copernicus 270–1: other hypotheses 272–3: *On Sizes and Distances of Sun and Moon* 272–6: trigonometrical purpose 272, 273: numbers in 18: fractions in 21.

Aristophanes 26.

Aristotelian treatise on Indivisible Lines 200.

Aristotle 5, 6, 37, 55, 120, 121–2, 175, 186, 190, 194–200, 205: on mathematical subjects 6: on first principles, hypotheses, definitions, postulates, axioms 194–6: on Pythagoreans and mathematics 36, 37: on Pythagorean astronomy 109.

Arithmetic: 2 even and prime 41: on Pythagorean numbers 37: on gnomons 44–5.

On the continuous and infinite 198–9: proof of incommensurability of diagonal 55: on principle of exhaustion 199: on Hippocrates of Chios 8: on Democritus 116.

ENGLISH INDEX

Geometry: on parallels 197: definitions not in Euclid 196: propositions not in Euclid 198: proofs differing from Euclid's 196–8: on quadrature 121–2: on Hippocrates' quadratures of lunes 122, 130–1.
Mechanics 199, 269: parallelogram of velocities 199: 'Aristotle's Wheel' 432.
Aristoxenus 37.
Arithmetic (1) = theory of numbers (opp. to λογιστική) 5–6: early Elements of Arithmetic 53, 137: systematic treatises, Nicomachus 61–70, Theon of Smyrna 70–1, Iamblichus' comm. on Nicomachus 71–2, Domninus 517.
(2) Practical arithmetic 24–35: arithmetic in Greek education 7.
Arithmetic mean 51.
Arithmetica of Diophantus 473–507.
Arithmetical operations, *see* Addition, Subtraction, &c.
Arithmetical progression: Egyptian case 467: in Euclid 240.
Arrow of Zeno 192.
Āryabhaṭṭa, value of π 147.
Asclepius of Tralles 62.
Astronomy: in elementary education 5: as secondary subject 7.
Asymptote 2.
Athelhard of Bath, first translator of Euclid 209–10.
Athenaeus 95.
Athenaeus of Cyzicus 185.
'Attic' or 'Herodianic' numeral system 14–15.
August, E. F. 208.
Autolycus of Pitane 200–1: *On the Moving Sphere ib.*: relation to Euclid *ib.*
Axioms: Aristotle on, 194–5: = Common Notions in Euclid 218: 'Axiom of Archimedes' 193–4, 224, 294.

Babylonians: systems of numerals, decimal and sexagesimal 12–14: sexagesimal fractions 13, 23: consistent use of sexagesimal systems 2000–1800 B.C. 524–5 sq.: 'perfect' proportion 51–2.
Bachet, editor of Diophantus 475.
Bacon, Roger 211–12.
Barrow, I., editor of Euclid 213, 224.
Bathycles 92.
Bāudhāyana 97.
Benedetti, G. B. 269.
Besthorn, R. O. 209.
Billingsley, Sir H. 213.
Björnbo, A. A. 210, 400.
Bobiense, Fragmentum mathematicum 379–80.
Boëtius 53, 62.
Boissonade 517.
Bombelli, Rafael 475.
Borelli, G. A. 353.
Breton (de Champ) 263.
Bretschneider, C. A. 98.
Brougham, Lord 263.
Bryson 141–2.
Burnet, J. 183.

Calculation, practical: the abacus 24–8, addition and subtraction 28; multiplication (i) Egyptian 29, 'Russian peasant' 29 *n*, (ii) Greek 30–1: division 31–2, extraction of square root 32–5.
Callimachus 92.
Callippus' Cycle 396, 530.
Campanus, translator of Euclid 210, 212.
Canonic = theory of musical intervals 7.
Carpus of Antioch 146.
Catoptric = theory of mirrors 7.
Catoptrica attributed to Euclid 268: *C.* by Archimedes 268, 285, by Heron 417, 432–3.
Cattle-problem of Archimedes 284, 336–7.
Cavalieri, B. 280.
Censorinus 396.

ENGLISH INDEX

Chace, Arnold B. vi.
Chalcidius 514.
Chaldaeans: measure of angles by ells 384: order of planets 513: debt of Hipparchus to 530–1.
Chasles, M. 280: on Euclid's *Porisms* 263.
Chords, Tables of 23, 147, 398, 399.
Chrysippus 119.
Cicero 206, 278, 280.
Cidenas (Kidinnu) 530.
Circle: division into degrees 23, 384: squaring of 139–47, Antiphon 140–1, Bryson 141–2, by spiral of Archimedes 144–5, by *quadratrix* of Hippias 142–4, by Apollonius and Carpus 145–6, approximations to π 146–7.
Cissoid of Diocles 168–9.
Clavius 223.
Cleomedes, *De motu circulari* 510–12: 'paradoxical' case of eclipse 511–12.
Cleonides 268.
Cochlias (cylindrical helix) 146, 376.
Cochloids 150: 'sister of cochloid' 145.
Commandinus, F., translator of Euclid 208, 212–13, 258, of Archimedes 287, of Apollonius 353, of works of Ptolemy 412, of Pappus 436.
Common Notions = Axioms 195, 218.
Conchoid of Nicomedes 150–1.
Conclusion (of proposition) 214.
Cone: Democritus on 119, and on volume of 115–16, 193, 282–3: volume of frustum of (Heron) 427.
Conic sections: discovered by Menaechmus 158, 347: Euclid's *Conics* and Aristaeus' *Solid Loci* 265, 349–51: focus-directrix property assumed by Eucl. 153, 266: conics in Archimedes 351–2: modern names due to Apollonius 103, 352, 359: Apollonius' *Conics* 352–63: conics in *Fragmentum mathematicum Bobiense* 379–80, in Anthemius 519.
Conon of Samos 277.
Construction (formal constituent of proposition) 214.
Conversion of ratio (*convertendo*) 226.
Corollary 215.
'Counter-earth' 109.
Croesus 81.
Ctesibius 135: relation to Philon and Heron 415.
Cube, called 'geometrical harmony' (Philolaus) 51.
Cube, duplication of: history of problem 154–5: reduction by Hippocrates to problem of two mean proportionals 121, 131, 155: solutions in that form by Archytas 155–7, Eudoxus 157–8, Menaechmus 158–61, 'Plato' 161–2, Eratosthenes 162–3, Nicomedes 164–6, Apollonius, Philon, Heron 166–7, Diocles 167–9, Sporus and Pappus 169–70: approximation by plane methods 437.
Cube root: approximation to (Heron) 430.
Cubic equation solved by conics 303–4.

Dactylus, 1/24th of ell 384.
Damastes of Sigeum 117.
Darius vase 27–8.
De levi et ponderoso 269.
Decagon inscribed in circle, side of 251: area of (Heron) 425.
Decimal system 11.
Dee, John 213, 258.
Definitions: Pythagorean 38, 92: in Plato 174–5: Aristotle on 195: in Euclid 216–17: *Definitiones* of Heron 418.
Democritus 74, 115–20, 191: date, travels 116, Aristotle on, *ib.*: list of works 117: *On irrational lines and solids* 54, 117: on nature of

ENGLISH INDEX 541

contact of line with circle and plane with sphere 117-19: on circular sections of cone 119-20: discovered volume of cone and pyramid 115-16, 193, 282-3: atoms mathematically divisible *ad inf.* 120: on Great Year 117.
Dercyllides 514-15.
Descartes 42, 454.
Dicaearchus 513.
Dichotomy of Zeno 192.
Diels, H. 93 *n.*
Digit 11.
Dinostratus 142, 158, 184, 185.
Diocles, inventor of cissoid 167: *On burning-mirrors* 167, 379: solution of Archimedes *On the Sphere and Cylinder* ii. 4, 304.
Diodorus (math.), on parallel-postulate 205: *Analemma* of 434.
Diodorus Siculus 75, 92, 93.
Diogenes Laërtius 26, 86, 90, 95, 187.
Dionysius, Plato's teacher 115.
Dionysodorus 304: on volume of *tore* 385, 428.
Diophantus of Alexandria 472-509: date 472, works and editions 473-6. *Arithmetica*: notation and definitions 476-80: signs for unknown and powers 477-8, for sub-multiples 478, for *minus* 479 and ἴσος *ib.*: fractions in 20, 21, 22: methods 482-94: *Porisms* 474, 494-5: propositions in Theory of Numbers 496-8: *On Polygonal Numbers* 507-9.
Dioptra 397: Heron's *Dioptra* 431.
Division, Greek method 31-2.
Divisions (of Figures), On, treatise by Euclid 258-62: similar problems in Heron 428-9.
Dodecagon, area of 425.
Dodecahedron: discovery by Pythagoreans 105 sq.: early occurrence of 106: Plato's construction of 106, 177: inscribed in sphere (Euclid) 252-3: Apollonius on 375.
Domninus of Larissa 517.
Dositheus 310, 317, 330.
Duhem, P. 269.

Earth: measurement of, by Eratosthenes 343-5, by Posidonius 386-7.
Ecliptic: obliquity discovered by Oenopides 114-15, 514: estimate of inclination (Eratosthenes, Ptolemy) 345.
Ecphantus 271.
Edfu, Temple of Horus at 77.
Egypt: supposed origin of geometry in 75.
Egyptian mathematics: numeral system 11-12, multiplication, &c. 29: geometry (mensuration) 76-80, rectangles, triangles, quadrilaterals 76-8: circles and cylinders 78-9, pyramids 79-80, frustum of pyramid 520-1, hemisphere 521: *hau*-calculations 466-7: arithmetical progressions 467, *regula falsi* 467.
'*Elefuga*' 211.
Elements: as known to Pythagoreans 109-11: progress in, to Plato's time 112-13, 131-2: writers of Elements, Hippocrates of Chios 74, 121, Leon, Theudius 184: other contributors to, Leodamas, Archytas, Theaetetus 112, Hermotimus of Colophon, Eudoxus 185: *Elements* of Euclid 204-54, so-called 'Books XIV, XV' 254.
Ell as measure of angles 384.
Empedocles on Pythagoras 36.
Enneagon, regular, in a circle: Heron's measurement of side and area 425-6.
Epanthema of Thymaridas (system of simple equations) 57-8, 72, 469.
Equations, simple: equivalent in Papyrus Rhind 466-7 in *Epan-*

ENGLISH INDEX

thema of Thymaridas and in Iamblichus 57–60: in Greek Anthology 468–9: Indeterminate, *see* Indeterminate Analysis: *see also* Quadratic, Cubic.

Eratosthenes 277, 343–6: *sieve* for finding primes 63: on Duplication of Cube 154–5, 162–4: *On Means* 343, *Platonicus*, *ib.*: measurement of earth 343–5: chronology, geography 345–6: astronomy 345.

Erycinus 439.

Euclid 4, 10, 202–3: traditions 10, 202, 203: Pappus on 203.

Arithmetic: classification and defs. of numbers 39–40: perfect numbers 41–2, 240: formula for right-angled triangles in rational numbers 48, 470.

Conics 203, 265, 349–51: focus-directrix property 153, 266: on ellipse as section of any cone or cylinder 265, 347–8.

Data 255–8.

Divisions (of Figures), On 258–62.

Elements: texts 207–8, ancient commentaries 204–5, Arabic and Latin translations 209–10: first printed editions 212–13: first English translation (Billingsley) 213: Euclid in Middle Ages, at Oxford and Cambridge 211–12: analysis of 216–54: arrangement of Postulates and Common Notions 208, 217–18: so-called 'Books XIV, XV' 254.

Mechanics 269: Music 268, *Sectio Canonis*, *ib.*: *Optics* 266–8: *Phaenomena* 266: *Porisms* 262–5: *Surface-Loci* 265–6: *Pseudaria* 262.

Eudemus: *History of Geometry* 73, 83, 94, 100, 115, 121: on Hippocrates' quadratures of lunes 121–31: *History of Astronomy* 89, 114, 514–15.

Eudoxus 4, 38, 52, 105, 120, 155, 171, 185, 187–94, 283: new Theory of Proportion 4, 190–1, 224–5: Method of Exhaustion 4, 120, 191, 193–4: problem of two mean proportionals 157–8: discovered three new means 52: so-called 'general theorems' 184, 189–90: hypothesis of concentric spheres in astronomy 188–9: *Phaenomena* and *Mirror* 188.

Eugenius Siculus, Admiral 413.

Euler, L. 42, 497, 498.

Euphorbus (= Pythagoras) 92.

Euphranor 52.

Eurytus 38.

Eutocius 30–1, 154, 157, 286, 352.

'Euthymetric' (= prime) numbers 40.

Even numbers, defined 39.

'Even-odd' numbers 39–40.

Exhaustion, method of 4, 120, 191, 193–4, 248, 249–50: development of, by Archimedes 294–5.

False hypothesis: Egyptian use 467: in Diophantus 499–500.

Fermat, P. 42, 280, 475, 476, 497, 498.

Fractions: Egyptian (all submultiples except $\frac{2}{3}$) 12, Greek systems 20–2: sexagesimal fractions (Babylonian) 13, 23, 524–5: in Greek 23–4.

'Friendly' numbers 42.

Galilei 223, 269.

Geëponicus, Liber 417, 419.

Geminus 387–92: on divisions of Optics 7: on fifth Postulate 389–40: on parallels 205, 390: on original production of the three conics 347: encyclopaedic work on mathematics 388: *Introduction to Phaenomena* 391–2: comm. on Posidonius' *Meteorologica* 390–1.

Geodesy (γεωδαισία), originally mensuration (dist. from geometry) 6.

Geometric mean defined (Archytas) 51: one such between two squares (or similar plane numbers), two between cubes (or similar solid numbers) 53, 178, 238: no rational mean between two consecutive integers 53, 136–7.
'Geometrical harmony' (Philolaus' name for cube) 51.
Geometry: supposed origin in Egypt 75: in Greek education 6, 7–9.
Georgius Pachymeres 474–5.
Gerhardt, C. J. 436.
Gherard of Cremona, translator of Euclid and an-Nairīzī 210–11, of Menelaus 400.
Ghetaldi, Marino 373.
Girard, Albert 476.
Gnomon: history of term 45: gnomons of square numbers 44, of polygonal numbers 45–6, of oblong numbers 48–9: in application of areas 103, 233: in Euclid 45, 104: sun-dial with vertical needle 91.
Great Year of Oenopides 115, of Democritus 117.
Gregory, D. 208, 353.
Gruppe 183.
Guldin's theorem, anticipated by Pappus 454, known to Heron 428.
Gunn, Battiscombe 76, 521.

Halicarnassus inscriptions 15, 16.
Halley, E., editions of Apollonius' *Conics* 353 and *Sectio Rationis* 363: of Menelaus 400.
Halma, editor of Ptolemy 403.
Hammer-Jensen, Ingeborg 416.
Hankel, H. 98.
Harmonic mean (originally 'subcontrary') 51.
Hārūn ar-Rashīd 209.
Hau-calculations (Egypt) 466–7.
Hecataeus of Miletus 117.
Heiben, J. L. 146.

Heiberg, J. L. 208, 209, 266, 283, 287, 353, 395, 403, 412, &c.
Helicon of Cyzicus 155, 171. .
Hendecagon in a circle: side and area of (Heron) 426.
Heptagon, regular, in a circle: Archimedes' construction of 340–2; measurement of side and area (Heron) 425.
Heraclides of Pontus 171, 186–7: maintained rotation of earth about axis 186, and revolution of Mercury and Venus round sun 187, 271.
Heraclitus of Ephesus 36.
Hermotimus of Colophon: on Elements and loci 185.
'Herodianic' or 'Attic' numerals 14–15.
Herodotus 26, 36, 75.
Heron of Alexandria 45, 75, 146, 285, 310: date 415–16: relation to Ctesibius and Philon 415: character of works and list 416–18.
 Arithmetic: fractions in 20–2: method of approximation to surds 422–3, to cube root 430: quadratic equations 430–1.
 Geometry: *Definitiones* 418: comm. on Euclid's *Elements* 417–18: proof of formula for area of triangle

$$\varDelta = \sqrt{\{s(s-a)(s-b)(s-c)\}}$$

420–2: solution of problem of two mean proportionals 166–7.
 Metrica 419–30: mensuration of triangles 420–2, regular polygons 424–6, circle and segment 426: of solid figures, cone, pyramid (and frustum) 427, sphere and segment thereof and Archimedes' two special solids 427, anchor-ring 427–8, five regular solids 428: divisions of figures 428–30, of sphere, cone, and frustum thereof 429–30.
 Mechanics 431–2: on Archi-

ENGLISH INDEX

medes' mechanical works 285: centres of gravity 432.
Belopoeica, &c. 416: *On Dioptra* 431: *Catoptrica* 417, 432–3: *Automatopoeëtice* 417: *On Water Clocks* (lost work) 417.
Heronas 62.
Hicetas 109.
Hieronymus 82.
Hilāl b. Abī Hilāl al-Ḥimṣī 353.
Hill, M. J. M. 228.
Hilprecht, H. V. 13.
Hipparchus 271, 278, 384, 395–9: discovery of precession 396: on mean lunar month 396–7: catalogue of stars 397: debt to Chaldaeans 530–1: geography 397: trigonometry 398–9: supposed work on algebra 398, 530.
Hippasus 51, 52, 105: construction of dodecahedron in sphere 105.
Hippias of Elis: taught mathematics 9: varied accomplishments *ib.*: inventor of *quadratrix* 4, 112, 120–1, 142–4.
Hippocrates of Chios 3–4, 112, 121, 235: taught for money 8: first writer of Elements 121: assumes νεῦσις equivalent to solution of quadratic equation 126–8: quadratures of lunes 123–30: proved theorem of Eucl. XII. 2. 123, 132, 194: reduced duplication of cube to problem of finding two mean proportionals 3, 121, 131, 155.
Hippopede of Eudoxus 189.
Hoche, R. 63.
Homer 11.
Horizon: word first used technically by Euclid 266.
Horsley, Samuel 373.
Hultsch, F. 310, 383, 436.
Hunrath, K. 310.
Hypatia 474, 516.
Hypotenuse: meaning 2: theorem of square on 95–100: Proclus on discovery of theorem 95, cases of it known to Babylonians and Indians 96–7.
Hypsicles 383–4: author of 'Book XIV' of Euclid 254: def. of polygonal number 50, 383–4, 507: astronomical work 'Ἀναφορικός (*Ascensiones*) 384; first Greek division of zodiac circle into 360 equal parts *ib.*

Iamblichus 51–2, 62, 71–2: works 71–2: comm. on Nicomachus 72: on ἐπάνθημα of Thymaridas 58–60: on square as δίαυλος (racecourse) 72: property of sum of consecutive numbers $3n-2$, $3n-1$, $3n$, 72.
Ibn al-Haitham, *On burning-mirrors* 379, 474.
Icosahedron 106: discovery attributed to Theaetetus 108, 134: Plato on 106, 177: construction by Euclid for 252: comparison of, with dodecahedron (Aristaeus) 254, (Apollonius) 375: volume of (Heron) 428.
Incommensurable: discovery of, by Pythagoreans 54, 104–5, proof of incommensurability of diagonal of square 55.
Indeterminate analysis: first cases of: right-angled triangles in rational numbers, formulae of Pythagoras 47, 469, Plato 47, 180, 469, Euclid 48, 243, 470: 'side-' and 'diameter-' numbers 55–7: rectangles with area and perimeter numerically equal 60: indeterminate equations, first degree 469: second degree, in Heronian collections 470–2, in Diophantus 484–90, of higher degrees (Dioph.) 491–3.
India: rational right-angled triangles known to 97–8: approximation to $\sqrt{2}$, 97, to π (Āryabhaṭṭa) 147.
Irrational: discovered by Pythagoreans with reference to $\sqrt{2}$,

ENGLISH INDEX

54–55: Democritus on irrationals 54, 117: Theodorus on 54, 132–3: Theaetetus' extensions 133–4: Plato on 181: in Euclid, Book X 241–7: work on, by Apollonius 376.
Isḥāq b. Hunain, translator of Euclid 209 and Ptolemy 403.
Isidorus of Miletus 254–5, 285.
Isocrates, on mathematics in Greek education 8.
Isoperimetric figures: Zenodorus 382–3, Pappus 383, 449.
Isosceles 1–2.

Jacob b. Machir, translator of Menelaus' *Sphaerica* 400.
Jan, C. 268.

Kahun Papyri 78–9.
Kant, I. 1, 114.
Kepler 280, 337.
Koppa (Ϙ, for 90) = Phoenician Qoph 16.

Lagrange 498.
Laplace 114.
Lawson 263.
Leibniz 280.
Lemma 216.
Leodamas of Thasos 112, 183.
Leon 184.
Leon of Constantinople 286.
Leonardo of Pisa 211, 258.
Lepsius, C. R. 77.
Leucippus 36, 115.
Linear (of numbers) = prime 40.
'Linear' loci and problems 139.
Livy 278.
Loci: classification of, as plane, solid, and 'linear' 139 (cf. 437–8): surface-loci or loci on surfaces 265–6: *Solid Loci* (= conics as loci) by Aristaeus 203, 265, 349–51.
Loftus, W. K. 13.
'Logistic' (opp. to 'arithmetic') 5–6.
Loria, G. 264.
Lucian 43, 62, 108, 279.

Mamercus or Mamertius 90.
al-Maʾmūn, Caliph 209.
al-Manṣūr, Caliph *ib.*
Manus for a number 11.
Marinus 268, 375.
Martianus Capella 206.
Mastaba tombs 80.
Mathematics: first made a science by Greeks 1: meaning of term 5: classification of subjects 5–7: branches of applied mathematics 6–7: mathematics in Greek education 7–10.
Means: arithmetic, geometric, and subcontrary (harmonic) known in Pythagoras' time 51, defined by Archytas *ib.*: three more due to Hippasus or Eudoxus 52, four more to Myonides and Euphranor *ib.*: the ten set out by Nicomachus and Pappus 52–3, 69–70: no rational geometric mean between numbers in ratio $n/(n+1)$, 53, 136–7.
Mechanics, divisions of, 7: writers on, Archytas 135, Aristotle 199, Euclid (?) 269, Archimedes 284–5, 323–7, Ptolemy 414, Heron 417, 431–2, Pappus 460–2.
Menaechmus 184: discoverer of conic sections 4, 158–9, 185, 347: solutions of problem of two mean proportionals 159–61.
Menelaus of Alexandria 399: Table of Chords 399, *Sphaerica* 400–2: other works 400: 'Menelaus' theorem' 401–2, anharmonic properties 401: 'paradoxical' curve 400.
Mensa Pythagorea 25.
Mensuration: in primary education 7: in Egypt 76–80, 520–1: Babylonian 522 sq.: in Heron 418–28.
Metrodorus 468.
Minus, sign for, in Diophantus 479.
Muḥammad Bagdadinus 258.
Multiplication: Egyptian method

ENGLISH INDEX

29, 'Russian peasant' 29 n., Greek 30–1.
Musical intervals and numerical ratios 37, 135, 136.
Myriads, first, second, &c., or 'simple', 'double', &c. (powers of 10,000), notation for 18–19.

Naburianos 530–1.
an-Nairīzī: comm. on Euclid 210, 211.
Naṣīraddīn aṭ-Ṭūsī: version of Euclid 209, of Ptolemy 403.
Neoclides 183.
Ner (Babylonian) = 600, 14.
Nesselmann, G. H. F. 473.
Neugebauer, O., viii, 96, 522, 523, 525, 526.
Newton 213–14, 280, 367.
Nicomachus of Gerasa 51, 52, 61: works 61: *Introductio arithmetica* 61–70: classification of numbers 63–8: on 'perfect' numbers 41, 64: on ten means 52–3: on a 'Platonic' theorem 53: sum of series of natural cubes 68–9.
Nipsus, M. Junius 84–5.
Nix, L. 417.
Number: defined by Thales 38, by Pythagoreans, Eudoxus, Euclid *ib.*: classification of numbers 39–41, 63–8: perfect, over-perfect, and defective numbers 41–2, 64: 'friendly' numbers 42: figured numbers 43–6: oblong, prolate 48–9, polygonal numbers and gnomons thereof 45–6, 66–7, 383–4, 507–9: solid numbers, pyramids, cubes, &c. 67–8.
Numerals: systems of, decimal, quinary, vigesimal 11: Egyptian 11–12, Babylonian (1) decimal 12, (2) sexagesimal 13–14, 524–5: Greek (1) 'Attic' or 'Herodianic' 14–15: (2) alphabetical 15–18, date of introduction 16, how written 15–17: notation for large numbers 18,

Apollonius' 'tetrads' 19, Archimedes' system (octads) 19–20.
Nymphodorus 135.

Oblong numbers 48–9: gnomons of 49.
Ocreatus 70.
'Octads' of Archimedes 19–20.
Octagon, regular: area of 425.
Octahedron 106, 177: discovery attributed to Theaetetus 109, 134: construction of, by Euclid 251.
'Odd' number, defined 39: 'odd-even', 'odd-times odd', &c. numbers 39–40.
Oenopides of Chios 112, 114–15: discovered obliquity of ecliptic 114–15, 514: Great Year of 115: called perpendicular 'gnomon-wise' *ib.*: two propositions in elementary geometry *ib.*
One, the principle of number 38.
Optics, divisions of 7: *Optics* of Euclid 266–8, of Ptolemy 413, of Theon of Alexandria 266–7.
Oxyrhynchus Papyri 92.

Pamphile 83, 85–6, 93.
Pappus 62, 203, 204, 285, 434–65: on Apollonius' 'tetrads' 19, 437: on ten means 52–3: on mechanical works of Archimedes 285: on *Conics* of Euclid, Aristaeus, and Apollonius 203, 265, 349–51: proved focus-directrix property 153: comm. on Euclid 204, 241, 434, 435, on *Data*, Ptolemy's *Syntaxis* and *Planispherium*, Diodorus' *Analemma* 434: on Euclid's Post. 4, 435: proof of Euclid I. 5, *ib.*: classification of problems and loci 139, 437–8: on Euclid's *Surface-Loci* 153, 265–6, 459: on Euclid's *Porisms* 262–5, 458–9: def. of Analysis and Synthesis 452: on books forming 'Treasury of Analysis' 452: on cochloids 150, 444, and *quadratrix* 144, 444–5: on spiral of Archimedes

443–4: spiral on a sphere 445–6: on trisection of any angle 148–50, 152–3, 446–7: on problem of two mean proportionals 169–70, 437: on isoperimetry 383, 449: on astronomical works 450–1: on mechanics 460–1: problems connected with ἄρβηλος 442–3: generalization of Eucl. I. 47, 440–1: 'Pappus' Problem' 453–4.
'Paradoxes' of Erycinus 439–40.
Parallelogram, meaning 1.
Parallelogram of velocities 199, 432.
Parapegma of Democritus 117.
Pebbles, used in calculations 24–6.
Peet, T. Eric vi, 77, 78, 97, 521.
Peletarius 223.
Pentagon, regular: construction Pythagorean 107–8: area of (Heron) 424–5.
Pentagram, star-pentagon, Pythagorean 108.
Perfect numbers 41–2: list of first twelve *ib.*: 10 the perfect number for Pythagoreans 42–3.
'Perfect' proportion 51–2.
Pericles 113.
Perseus' *spiric* sections 380–2.
Peyrard 207–8.
Phaenomena = observational astronomy 7, 188: Euclid's *Phaenomena* 200–1, 265, 266: Hipparchus *On Phaenomena of Eudoxus and Aratus* 396.
Phidias, father of Archimedes 277.
Philippus of Medma or Opus 50, 185: works, astronomical and mathematical 185.
Philolaus 37: on odd, even, and even-odd numbers 39: called cube 'geometrical harmony' 51: Pythagorean astronomical system attributed to 109.
Philon of Byzantium 135: duplication of cube 166–7: relation to Ctesibius and Heron 415.
Philoponus, Johannes 62.
Phocus of Samos 90.

Phoenician alphabet: used by Greeks for numerals 15–16.
'Piremus' or 'peremus' in pyramid 79–80.
Pistelli 63.
'Plane' loci and problems 139.
Planudes, Maximus 475.
Plato 5, 6, 7, 70, 74, 112, 134: ἀγεωμέτρητος μηδεὶς εἰσίτω 3, 171: on education in mathematics 7, 171–2: principle of lever 181: on optics 182: astronomy 181–2: classification of numbers, odd, even 39–40: on arithmetical problems 7, on geometry 172: ontology of mathematics 172–3, hypotheses in mathematics 173–4, definitions 174–5: on points and indivisible lines 175: formula for rational right-angled triangles 47, 180, 469: 'rational' and 'irrational diameter of 5' 54, 181: on the irrational 180–1: on regular and semi-regular solids 105–6, 175–8: Plato and duplication of cube 172: on geometrical means between two square and two cube numbers respectively 53, 131, 178: on 'perfect' proportion 52: a proposition in proportions 175: two geometrical passages in *Meno* 178–80: propositions on 'the section' 181, 184.
'Platonic' figures = the five regular solids 105–6, 175–8.
Playfair, John 218, 263.
Pliny 382.
Plutarch 50, 82, 95, 117, 119, 155, 172, 183, 185: on Archimedes 277, 278, 279.
Point: def. as 'unit having position' 38: Plato on points 175.
Polybius 26, 277, 382.
Polygon: sums of exterior and interior angles respectively 94: mensuration of regular polygons (Heron) 423–6.
Polygonal numbers 45–6, 66–7, 383–4, 507–9.

ENGLISH INDEX

Polyhedron, *see* Solids.
Porism (1) = corollary 215: (2) = a special kind of proposition *ib.*: *Porisms* of Euclid 261-5, of Diophantus 473-4, 494-5.
Porphyry 95, 205.
Posidonius 385-7: definitions of figure 385 and parallels (as equi-distants) 205, 385-6: book against Zeno of Sidon 205-6, 386: *Meteorologica* 386: *On the Ocean ib.*: estimate of circumference of earth 386-7: on size of sun 345, 387.
Postulates, Aristotle on 195: in Euclid 217-18: of Archimedes in mechanics 323, in hydrostatics 332-3.
Powers, R. E. 42.
Prestet, Jean 42.
Prime numbers 40-1: numbers prime to one another *ib.*: 2 prime for Aristotle and Euclid, not for Pythagoreans 41.
Problems: classification of, as plane, solid, and 'linear' 139, 437-8: problems and theorems distinguished 263.
Proclus 62, 90, 100, 105, 181, 189-90, 202, 205, 206, 382, 516-17: comm. on Eucl. I. 205: Proclus' 'Summary' 73-5, 92, 112, 133, 183-4, 184-5: on discoveries of Pythagoras 51, 105: on Eucl. I. 47, 95-6: on Porisms 263: comm. on *Republic* 57, 517: *Hypotyposis of astronomical hypotheses* 517.
Prolate, of numbers 49.
Proof (formal constituent of proposition) 214.
Proportion: theory discovered by Pythagoras 51, for commensurable magnitudes only 110: def. of numerical proportion 124, 235: 'perfect' proportion 51-2: Euclid's universally applicable theory due to Eudoxus 190-1, 224: definition of 225.

Proposition, geometrical: formal divisions of 214-15.
Protagoras 9: against geometers 118-19.
Psammites or *Sand-reckoner* of Archimedes 19, 327-30.
Psellus 472.
Pseudaria of Euclid 262.
Pseudo-Boëtius 11, 207, 209.
Pseudo-Eratosthenes: letter on history of duplication of cube 154-5.
Ptolemy I, story of 202.
Ptolemy, Claudius 402-14: sexagesimal fractions in 23-4: approximation to π 147: attempt to prove Parallel-Postulate 414: *Syntaxis* 402 sq.: commentaries and editions 403, contents 404-5: trigonometry in 405-11: Table of Chords and propositions preliminary to it *ib.*: *Analemma* 412, *Planispherium* 412-13, *Optics* 413, other works 413-14.
Pyramids: possible origin of name 79: measurements of, in Rhind Papyrus 79-80: pyramids at Dahshur and Gizeh 80; measurement of height by Thales 82-3: volume of pyramid 115-16, 193, 248-50: volume of frustum (Heron) 427, known to Egyptians 520-1.
Pythagoras 3, 9-10, 36-7, 43: motto 10, 92: Heraclitus, Empedocles, and Herodotus on 36: made mathematics part of liberal education 37, 92: called geometry 'inquiry' 109: began with definitions 92: Proclus on discoveries of 51, 105: arithmetic (theory of numbers) 38-50: introduced 'perfect' proportion 51-2: discovered arithmetical ratios corresponding to musical intervals 37: formula for right-angled triangles in rational numbers 47: theorem of Pythagoras

ENGLISH INDEX

95–100, general proof, how developed 97–9, known, at least in particular cases, to Babylonians 97, and to Indians 98: astronomy 108–9: earth spherical, independent movement of planets *ib.*

Pythagoreans 3, 4, 5: *quadrivium* 5: first to advance mathematics 36: 'all things are numbers' 37: number of an object 38: defs. of unit and number 38: classification of numbers 39–41: figured numbers, triangular 43–4, pentagonal, &c. 45–6, oblong numbers 48–50: discovered the incommensurable with reference to $\sqrt{2}$, 54: 'side'- and 'diameter-' numbers giving approximations to $\sqrt{2}$, 55–7: sum of angles of any triangle 86–8: geometrical theorems attributed to 93–108: application of areas and geometrical algebra 100–4: theory of proportion (numerical) 51, 110: construction of regular pentagon 107–8: astronomical system (non-geocentric) 109, 182: summary 109–11.

Qay en ḥeru (height of pyramid) 80.
Quadratic equations: solved by Pythagorean application of areas 101–4, 231–5: numerical solutions (Heron) 430–1, (Dioph.) 482–4: already solved numerically in accordance with modern formulae by Babylonians of 2000–1800 B.C. 525–9.
Quadratrix of Hippias 142–4.
Quadrivium of Pythagoreans 5, 63.
Quinary system of numerals 11.
Quṣṭā b. Lūqā 474.

Rangabé, A. R. 26.
Ratdolt, Erhard: first printed edition of Euclid 212.
Reductio ad absurdum 215, 453.
Reduction of a problem 215.

Reflection: angles of incidence and reflection equal 267, 285, 433.
Refraction: first attempt to formulate law (Ptolemy) 413.
Regiomontanus 475.
Regula Nicomachi 70.
Rhabdas, Nicolas 19.
Rhind Papyrus, mensuration in 76–80.
Right-angled triangle: 'inscribed by Thales in circle' 83, 85–8: theorem of Pythagoras 95–6, known (at least in certain cases) to Babylonians 97, and Indians 98.
Right-angled triangles in rational numbers: formula of Pythagoras 47, 469, of Plato 47–8, 469, of Euclid 48, 470: Indian examples 98: Diophantus' problems on 505–7.
Robertson, Abram 287.
Ruler and compasses restriction 115.

Salaminian table 26–7.
Salinon in Archimedes 339.
'Sampi' (ϡ = 900) derived from Ssade 15.
Sar (Babylonian for 60^2) 14.
Sar-gal ('Great *Sar*') = 60^4, 14.
Savile, Sir H., lectures on Euclid 213.
Scalene: of triangles 92–3: of certain solid numbers 67: of odd numbers (Plato) 174: of oblique cone (Apollonius) 355.
Schmidt, W. 417.
Schöne, R. 416.
Scholiast to *Charmides* 466.
Schooten, van 42.
Schoy, C. vii.
Schulz, O. 476.
'Secondary' numbers 40.
Sectio Canonis 53, 136, 268.
Seleucus 271.
Semicircle, angle in, is right (Thales) 83, 85–8.
Senkereh Tables 13.

ENGLISH INDEX

Senti, base (of pyramid) 80.
Seqeṭ, 'that which makes the nature' (of pyramid) = cotangent of angle of slope ('batter') 79–80, 82.
Serenus, On Sections of Cylinder and Cone 515–16.
Sexagesimal system of numbers and fractions, Babylonian 12–13, 524–5: sexagesimal fractions in Greece 23–4, 405–11.
'Side'- and 'diameter-' numbers 55–7, 104, 221, 512.
Simplicius: extract from Eudemus on Hippocrates' quadratures of lunes 121–31: on Eudoxus' system of concentric spheres 188–9, 518: comm. on Euclid's *Elements* 205, 518: on Antiphon 517: on mechanical works of Archimedes 285.
Simson, R.: editions of Euclid's *Elements* (and *Data*) 213, 214, 228: on Euclid's *Porisms* 262–3.
'Solid' loci and problems 139: *Solid Loci* of Aristaeus 203, 265, 349–51.
'Solid' numbers classified 67.
Solids, five regular: discovery attributed to Pythagoras or Pythagoreans 105–6, alternatively as regards octahedron and icosahedron to Theaetetus 108, 134: all investigated by Theaetetus 134: Plato on 106, 175–8: Euclid's constructions for 250–3: volumes of 428.
Solon 26.
Sophists: taught mathematics 9.
Sosigenes 188.
Soss (sussu) = 60 (Babylonian) 14.
Speusippus 43, 50, 61: on Pythagorean numbers 43, 186.
Sphaeric (mathematical astronomy) 5: treatises on, by Autolycus and Euclid 200–1, presuppose earlier text-book 201: *Sphaerica* of Theodosius 201, 393–5, of Menelaus 400–2.

Sphere-making, lost treatise on, by Archimedes 278.
Spiric sections (Perseus) 380–2.
Sporus: criticism on *quadratrix* 144: solution of problem of two mean proportionals 169–70.
Square numbers: formation of, by adding successive odd numbers (gnomons) 44: any square is sum of two triangular numbers 50, 66, 508: eight times a triangular number $+1$ = a square 50, 504, 508.
Square root: extraction of 32–3, ex. in sexagesimal fractions 33–5: method of approximation to surds 309–10, 422–3.
Ssade, Phoenician sibilant, became ℷ = 900, 15.
Star-pentagon, pentagram, of Pythagoreans 108.
Stereographic projection (Ptolemy) 412–13.
Stevin, Simon 476.
'Stigma', name for numeral ϛ, originally Ϲ (digamma) 16.
Strabo 75, 386.
Strato 270.
'Subcontrary' (=harmonic) mean defined 51.
Subtraction in Greek notation 28.
Surds: Theodorus on 54, 132–3: Theaetetus' generalization 133, 181: method of approximation to 309–10, 422–3.
Surface-Loci 139, 265–6.
Sussu (Babylonian) = 60, 14.
Synthesis, defined (Pappus) 452.

Tables of Chords 23, 147, by Hipparchus 398, Menelaus 399, Ptolemy 398, 405–11.
Tannery, P. 84, 190, 381, 474, 476, 477, 478, 497.
Teos inscription 15.
'Tetrads' of Apollonius 19.
Tetrahedron: construction of (Euclid) 251.

ENGLISH INDEX

Thābit b. Qurra, translator of Euclid, &c. 209, 340; of Archimedes' *Liber assumptorum* 284, of Apollonius V–VII 353, of Ptolemy 403.

Thales 1, 3, 81–2: introduced geometry into Greece 75: geometrical theorems attributed to 83–8: measured height of pyramid 80, 82–3, and distance of ship from shore 84–5: definition of number 38: astronomy 89–90, predicted solar eclipse 89.

Theaetetus 4, 112, 133–4, 185: on surds 133, 181: investigated five regular solids 108, discovered octahedron and icosahedron 108, 134: on irrationals 134, 241, 243.

Theodorus of Cyrene: on surds $\sqrt{3}, \sqrt{5}\ldots\sqrt{17}$ 54, 132, 181.

Theodosius 201, 393–5: *Sphaerica* 393–5: other works 395.

Theologumena arithmeticēs 60.

Theon of Alexandria: examples of multiplication and division 31–2, extraction of square root 33–5: edition of Euclid's *Elements* 207–8 and *Optics* 266: *Catoptrica* 268: comm. on Ptolemy's *Syntaxis* 383, 403.

Theon of Smyrna, treatise of 70–1, 512–15: on 'side'- and 'diameter'-numbers 55–7, 512: on forms of numbers which cannot be squares 71.

Theophrastus on Plato's views regarding earth (rest or motion) 183.

Theudius 184–5.

Thévenot 416.

Thrasylus 116, 512.

Thucydides 382.

Thymaridas: def. of unit 38: of prime numbers as 'rectilinear' *ib.*: ἐπάνθημα ('bloom'), solution of system of simple equations 57–8.

Timaeus of Locri 52.

Tittel, on Geminus 387.

Tore or anchor-ring: use of, by Archytas 155–6: sections of, or *spiric* sections (Perseus) 380–2: volume of (Dionysodorus) 385, 428.

Torelli, J. 287.

Transversals ('Menelaus' theorem' for spherical and plane triangles) 401–2.

'Treasury of Analysis' 255, 434, 452–9.

Triangle: theorem on sum of angles (Pythagorean) 85–8, 93–4: Geminus and Aristotle on theorem 88–9.

Triangle, spherical: called by Menelaus 'three-side' (τρίπλευρον, dist. τρίγωνον) 400: propositions analogous to Euclid's on plane triangles 400–1: sum of angles > two right angles 401.

Triangular number: formation of 43–4.

Trigonometry: beginnings 273–4: development of, by Hipparchus and Ptolemy 398–9, 407–11.

Trisection of any angle 147–54: Pappus on 148, 446–7: solutions 148–54.

Tycho Brahe 187.

Ukha-thebt (side of base in pyramid) 79–80.

Unit: defs. of, by Pythagoras and others 38.

Venatorius, Thomas Gechauff, *editio princeps* of Archimedes 286.

Vieta 223, 367.

Vigesimal system of numerals 11.

Vincent, A. J. H. 263.

Vitruvius 97.

Wallace, W. 263.

Wallis 213, 282.

Weierstrass 225.

Wescher, C. 416.

Wilamowitz-Möllendorff, U. von 154.
William of Moerbeke 283, 286.
Woepcke 258, 269.

Xenocrates 186: work on Numbers *ib.*: upheld 'indivisible lines' 120, 186.
Xylander (Wilhelm Holzmann) 475.

Zamberti, B., translator of Euclid 212-13.

Zeno of Elea: arguments on motion 186, 192.
Zeno of Sidon, on Eucl. I. 1, 205-6, 386.
Zenodorus 382-3.
Zero, sign O for (Ptolemy) 23.
Zeuthen, H. G. 470.
Zodiac circle: obliquity discovered by Oenopides 114-15, 514: estimate of angle of inclination 345.

A CATALOG OF SELECTED
DOVER BOOKS
IN SCIENCE AND MATHEMATICS

Astronomy

CHARIOTS FOR APOLLO: The NASA History of Manned Lunar Spacecraft to 1969, Courtney G. Brooks, James M. Grimwood, and Loyd S. Swenson, Jr. This illustrated history by a trio of experts is the definitive reference on the Apollo spacecraft and lunar modules. It traces the vehicles' design, development, and operation in space. More than 100 photographs and illustrations. 576pp. 6 3/4 x 9 1/4. 0-486-46756-2

EXPLORING THE MOON THROUGH BINOCULARS AND SMALL TELESCOPES, Ernest H. Cherrington, Jr. Informative, profusely illustrated guide to locating and identifying craters, rills, seas, mountains, other lunar features. Newly revised and updated with special section of new photos. Over 100 photos and diagrams. 240pp. 8 1/4 x 11. 0-486-24491-1

WHERE NO MAN HAS GONE BEFORE: A History of NASA's Apollo Lunar Expeditions, William David Compton. Introduction by Paul Dickson. This official NASA history traces behind-the-scenes conflicts and cooperation between scientists and engineers. The first half concerns preparations for the Moon landings, and the second half documents the flights that followed Apollo 11. 1989 edition. 432pp. 7 x 10.
0-486-47888-2

APOLLO EXPEDITIONS TO THE MOON: The NASA History, Edited by Edgar M. Cortright. Official NASA publication marks the 40th anniversary of the first lunar landing and features essays by project participants recalling engineering and administrative challenges. Accessible, jargon-free accounts, highlighted by numerous illustrations. 336pp. 8 3/8 x 10 7/8. 0-486-47175-6

ON MARS: Exploration of the Red Planet, 1958-1978--The NASA History, Edward Clinton Ezell and Linda Neuman Ezell. NASA's official history chronicles the start of our explorations of our planetary neighbor. It recounts cooperation among government, industry, and academia, and it features dozens of photos from Viking cameras. 560pp. 6 3/4 x 9 1/4. 0-486-46757-0

ARISTARCHUS OF SAMOS: The Ancient Copernicus, Sir Thomas Heath. Heath's history of astronomy ranges from Homer and Hesiod to Aristarchus and includes quotes from numerous thinkers, compilers, and scholasticists from Thales and Anaximander through Pythagoras, Plato, Aristotle, and Heraclides. 34 figures. 448pp. 5 3/8 x 8 1/2.
0-486-43886-4

AN INTRODUCTION TO CELESTIAL MECHANICS, Forest Ray Moulton. Classic text still unsurpassed in presentation of fundamental principles. Covers rectilinear motion, central forces, problems of two and three bodies, much more. Includes over 200 problems, some with.answers. 437pp. 5 3/8 x 8 1/2. 0-486-64687-4

BEYOND THE ATMOSPHERE: Early Years of Space Science, Homer E. Newell. This exciting survey is the work of a top NASA administrator who chronicles technological advances, the relationship of space science to general science, and the space program's social, political, and economic contexts. 528pp. 6 3/4 x 9 1/4.
0-486-47464-X

STAR LORE: Myths, Legends, and Facts, William Tyler Olcott. Captivating retellings of the origins and histories of ancient star groups include Pegasus, Ursa Major, Pleiades, signs of the zodiac, and other constellations. "Classic." – *Sky & Telescope*. 58 illustrations. 544pp. 5 3/8 x 8 1/2. 0-486-43581-4

A COMPLETE MANUAL OF AMATEUR ASTRONOMY: Tools and Techniques for Astronomical Observations, P. Clay Sherrod with Thomas L. Koed. Concise, highly readable book discusses the selection, set-up, and maintenance of a telescope; amateur studies of the sun; lunar topography and occultations; and more. 124 figures. 26 halftones. 37 tables. 335pp. 6 1/2 x 9 1/4. 0-486-42820-6

Browse over 9,000 books at www.doverpublications.com

CATALOG OF DOVER BOOKS

Chemistry

MOLECULAR COLLISION THEORY, M. S. Child. This high-level monograph offers an analytical treatment of classical scattering by a central force, quantum scattering by a central force, elastic scattering phase shifts, and semi-classical elastic scattering. 1974 edition. 310pp. 5 3/8 x 8 1/2. 0-486-69437-2

HANDBOOK OF COMPUTATIONAL QUANTUM CHEMISTRY, David B. Cook. This comprehensive text provides upper-level undergraduates and graduate students with an accessible introduction to the implementation of quantum ideas in molecular modeling, exploring practical applications alongside theoretical explanations. 1998 edition. 832pp. 5 3/8 x 8 1/2. 0-486-44307-8

RADIOACTIVE SUBSTANCES, Marie Curie. The celebrated scientist's thesis, which directly preceded her 1903 Nobel Prize, discusses establishing atomic character of radioactivity; extraction from pitchblende of polonium and radium; isolation of pure radium chloride; more. 96pp. 5 3/8 x 8 1/2. 0-486-42550-9

CHEMICAL MAGIC, Leonard A. Ford. Classic guide provides intriguing entertainment while elucidating sound scientific principles, with more than 100 unusual stunts: cold fire, dust explosions, a nylon rope trick, a disappearing beaker, much more. 128pp. 5 3/8 x 8 1/2. 0-486-67628-5

ALCHEMY, E. J. Holmyard. Classic study by noted authority covers 2,000 years of alchemical history: religious, mystical overtones; apparatus; signs, symbols, and secret terms; advent of scientific method, much more. Illustrated. 320pp. 5 3/8 x 8 1/2. 0-486-26298-7

CHEMICAL KINETICS AND REACTION DYNAMICS, Paul L. Houston. This text teaches the principles underlying modern chemical kinetics in a clear, direct fashion, using several examples to enhance basic understanding. Solutions to selected problems. 2001 edition. 352pp. 8 3/8 x 11. 0-486-45334-0

PROBLEMS AND SOLUTIONS IN QUANTUM CHEMISTRY AND PHYSICS, Charles S. Johnson and Lee G. Pedersen. Unusually varied problems, with detailed solutions, cover of quantum mechanics, wave mechanics, angular momentum, molecular spectroscopy, scattering theory, more. 280 problems, plus 139 supplementary exercises. 430pp. 6 1/2 x 9 1/4. 0-486-65236-X

ELEMENTS OF CHEMISTRY, Antoine Lavoisier. Monumental classic by the founder of modern chemistry features first explicit statement of law of conservation of matter in chemical change, and more. Facsimile reprint of original (1790) Kerr translation. 539pp. 5 3/8 x 8 1/2. 0-486-64624-6

MAGNETISM AND TRANSITION METAL COMPLEXES, F. E. Mabbs and D. J. Machin. A detailed view of the calculation methods involved in the magnetic properties of transition metal complexes, this volume offers sufficient background for original work in the field. 1973 edition. 240pp. 5 3/8 x 8 1/2. 0-486-46284-6

GENERAL CHEMISTRY, Linus Pauling. Revised third edition of classic first-year text by Nobel laureate. Atomic and molecular structure, quantum mechanics, statistical mechanics, thermodynamics correlated with descriptive chemistry. Problems. 992pp. 5 3/8 x 8 1/2. 0-486-65622-5

ELECTROLYTE SOLUTIONS: Second Revised Edition, R. A. Robinson and R. H. Stokes. Classic text deals primarily with measurement, interpretation of conductance, chemical potential, and diffusion in electrolyte solutions. Detailed theoretical interpretations, plus extensive tables of thermodynamic and transport properties. 1970 edition. 590pp. 5 3/8 x 8 1/2. 0-486-42225-9

Browse over 9,000 books at www.doverpublications.com

CATALOG OF DOVER BOOKS

Engineering

FUNDAMENTALS OF ASTRODYNAMICS, Roger R. Bate, Donald D. Mueller, and Jerry E. White. Teaching text developed by U.S. Air Force Academy develops the basic two-body and n-body equations of motion; orbit determination; classical orbital elements, coordinate transformations; differential correction; more. 1971 edition. 455pp. 5 3/8 x 8 1/2. 0-486-60061-0

INTRODUCTION TO CONTINUUM MECHANICS FOR ENGINEERS: Revised Edition, Ray M. Bowen. This self-contained text introduces classical continuum models within a modern framework. Its numerous exercises illustrate the governing principles, linearizations, and other approximations that constitute classical continuum models. 2007 edition. 320pp. 6 1/8 x 9 1/4. 0-486-47460-7

ENGINEERING MECHANICS FOR STRUCTURES, Louis L. Bucciarelli. This text explores the mechanics of solids and statics as well as the strength of materials and elasticity theory. Its many design exercises encourage creative initiative and systems thinking. 2009 edition. 320pp. 6 1/8 x 9 1/4. 0-486-46855-0

FEEDBACK CONTROL THEORY, John C. Doyle, Bruce A. Francis and Allen R. Tannenbaum. This excellent introduction to feedback control system design offers a theoretical approach that captures the essential issues and can be applied to a wide range of practical problems. 1992 edition. 224pp. 6 1/2 x 9 1/4. 0-486-46933-6

THE FORCES OF MATTER, Michael Faraday. These lectures by a famous inventor offer an easy-to-understand introduction to the interactions of the universe's physical forces. Six essays explore gravitation, cohesion, chemical affinity, heat, magnetism, and electricity. 1993 edition. 96pp. 5 3/8 x 8 1/2. 0-486-47482-8

DYNAMICS, Lawrence E. Goodman and William H. Warner. Beginning engineering text introduces calculus of vectors, particle motion, dynamics of particle systems and plane rigid bodies, technical applications in plane motions, and more. Exercises and answers in every chapter. 619pp. 5 3/8 x 8 1/2. 0-486-42006-X

ADAPTIVE FILTERING PREDICTION AND CONTROL, Graham C. Goodwin and Kwai Sang Sin. This unified survey focuses on linear discrete-time systems and explores natural extensions to nonlinear systems. It emphasizes discrete-time systems, summarizing theoretical and practical aspects of a large class of adaptive algorithms. 1984 edition. 560pp. 6 1/2 x 9 1/4. 0-486-46932-8

INDUCTANCE CALCULATIONS, Frederick W. Grover. This authoritative reference enables the design of virtually every type of inductor. It features a single simple formula for each type of inductor, together with tables containing essential numerical factors. 1946 edition. 304pp. 5 3/8 x 8 1/2. 0-486-47440-2

THERMODYNAMICS: Foundations and Applications, Elias P. Gyftopoulos and Gian Paolo Beretta. Designed by two MIT professors, this authoritative text discusses basic concepts and applications in detail, emphasizing generality, definitions, and logical consistency. More than 300 solved problems cover realistic energy systems and processes. 800pp. 6 1/8 x 9 1/4. 0-486-43932-1

THE FINITE ELEMENT METHOD: Linear Static and Dynamic Finite Element Analysis, Thomas J. R. Hughes. Text for students without in-depth mathematical training, this text includes a comprehensive presentation and analysis of algorithms of time-dependent phenomena plus beam, plate, and shell theories. Solution guide available upon request. 672pp. 6 1/2 x 9 1/4. 0-486-41181-8

Browse over 9,000 books at www.doverpublications.com

CATALOG OF DOVER BOOKS

HELICOPTER THEORY, Wayne Johnson. Monumental engineering text covers vertical flight, forward flight, performance, mathematics of rotating systems, rotary wing dynamics and aerodynamics, aeroelasticity, stability and control, stall, noise, and more. 189 illustrations. 1980 edition. 1089pp. 5 5/8 x 8 1/4. 0-486-68230-7

MATHEMATICAL HANDBOOK FOR SCIENTISTS AND ENGINEERS: Definitions, Theorems, and Formulas for Reference and Review, Granino A. Korn and Theresa M. Korn. Convenient access to information from every area of mathematics: Fourier transforms, Z transforms, linear and nonlinear programming, calculus of variations, random-process theory, special functions, combinatorial analysis, game theory, much more. 1152pp. 5 3/8 x 8 1/2. 0-486-41147-8

A HEAT TRANSFER TEXTBOOK: Fourth Edition, John H. Lienhard V and John H. Lienhard IV. This introduction to heat and mass transfer for engineering students features worked examples and end-of-chapter exercises. Worked examples and end-of-chapter exercises appear throughout the book, along with well-drawn, illuminating figures. 768pp. 7 x 9 1/4. 0-486-47931-5

BASIC ELECTRICITY, U.S. Bureau of Naval Personnel. Originally a training course; best nontechnical coverage. Topics include batteries, circuits, conductors, AC and DC, inductance and capacitance, generators, motors, transformers, amplifiers, etc. Many questions with answers. 349 illustrations. 1969 edition. 448pp. 6 1/2 x 9 1/4.
0-486-20973-3

BASIC ELECTRONICS, U.S. Bureau of Naval Personnel. Clear, well-illustrated introduction to electronic equipment covers numerous essential topics: electron tubes, semiconductors, electronic power supplies, tuned circuits, amplifiers, receivers, ranging and navigation systems, computers, antennas, more. 560 illustrations. 567pp. 6 1/2 x 9 1/4. 0-486-21076-6

BASIC WING AND AIRFOIL THEORY, Alan Pope. This self-contained treatment by a pioneer in the study of wind effects covers flow functions, airfoil construction and pressure distribution, finite and monoplane wings, and many other subjects. 1951 edition. 320pp. 5 3/8 x 8 1/2. 0-486-47188-8

SYNTHETIC FUELS, Ronald F. Probstein and R. Edwin Hicks. This unified presentation examines the methods and processes for converting coal, oil, shale, tar sands, and various forms of biomass into liquid, gaseous, and clean solid fuels. 1982 edition. 512pp. 6 1/8 x 9 1/4. 0-486-44977-7

THEORY OF ELASTIC STABILITY, Stephen P. Timoshenko and James M. Gere. Written by world-renowned authorities on mechanics, this classic ranges from theoretical explanations of 2- and 3-D stress and strain to practical applications such as torsion, bending, and thermal stress. 1961 edition. 560pp. 5 3/8 x 8 1/2. 0-486-47207-8

PRINCIPLES OF DIGITAL COMMUNICATION AND CODING, Andrew J. Viterbi and Jim K. Omura. This classic by two digital communications experts is geared toward students of communications theory and to designers of channels, links, terminals, modems, or networks used to transmit and receive digital messages. 1979 edition. 576pp. 6 1/8 x 9 1/4. 0-486-46901-8

LINEAR SYSTEM THEORY: The State Space Approach, Lotfi A. Zadeh and Charles A. Desoer. Written by two pioneers in the field, this exploration of the state space approach focuses on problems of stability and control, plus connections between this approach and classical techniques. 1963 edition. 656pp. 6 1/8 x 9 1/4.
0-486-46663-9

Browse over 9,000 books at www.doverpublications.com